D1067705

Environmental Science

Series editors: R. Allan · U. Förstner · W. Salomons

Springer

Berlin
Heidelberg
New York
Barcelona
Hong Kong
London
Milan
Paris
Singapore
Tokyo

Gerardo M.E. Perillo · María Cintia Piccolo · Mario Pino-Quivira (Eds.)

Estuaries of South America

Their Geomorphology and Dynamics

With 102 Figures and 20 Tables

Springer

Editors

Gerardo M.E. Perillo

Instituto Argentino de Oceanografía, CC 107
8000 Bahía Blanca, Argentina

María Cintia Piccolo

Instituto Argentino de Oceanografía, CC 107
8000 Bahía Blanca, Argentina

Mario Pino-Quivira

Instituto de Geociencias
Universidad Austral de Chile
Casilla 567, Valdivia, Chile

ISSN 1431-6250
ISBN 3-540-65657-X Springer-Verlag Berlin Heidelberg New York

Library of Congress Cataloging-in-Publication Data

Estuaries of South America : their geomorphology and dynamics /
Gerardo M.E. Perillo, María Cintia Piccolo, Mario Pino-Quivira (eds.).
(Environmental science, ISSN 1431-6250)
Includes bibliographical references and indexes.
ISBN 3-540-65657-x (hc : alk. paper)
1. Estuaries--South America.
I. Perillo, G. M. E. (Gerardo M. E.) II. Piccolo, María Cintia.
III. Pino-Quivira, Mario IV. Series: Environmental science (Berlin, Germany) GC97.E75 1999
551.46' 46--dc21

Cover Design: Struve & Partner, Heidelberg
Dataconversion: Büro Stasch, Bayreuth

SPIN: 10708935 32/3020 – 5 4 3 2 1 0 – Printed on acid-free paper

The editors wish to dedicate the present book to the memory of Prof. Mohammed El-Sabh, our friend and colleague of many projects and meeting organizations, with whom we shared so many happy moments during the years we knew him. We think that the words of M. Mario Bélanger, translated by Dr. Phil Hill (both from UQAR and also friends of Mohammed) represent also our feelings.

Gerardo M.E. Perillo, Mariá Cintia Piccolo, Mario Pino-Quivira
February, 1999

Professor Mohammed El-Sabh, 59, died at the Centre hospitalier régional de Rimouski on the evening of February 8, 1999, after a heart attack. Université du Québec à Rimouski (UQAR) and all of Quebec hve lost a great person who made Rimouski known around the world for its marine sciences.

Born in Egypt, Mohammed El-Sabh obtained his Ph.D. from McGill University in Montreal. He arrived at UQAR in 1972 to form part of the first team of researchers given the mission of developing oceanography in Rimouski. Through his teaching and research as well, his numerous publications and involvements, he contributed in a major way to the growing reputation of UQAR both nationally and internationally.

At UQAR, he supervised more than 20 M.Sc. and Ph.D. students in Oceanography. He was often called on by other universities for the evaluation of university work.

From his research, he wrote several major scientific publications including: Natural and Man-Made Hazards (Netherlands, 1986), Oceanography of the St. Lawrence Estuary (Germany, 1990) and Integrated Management and Sustainable Development in Coastal Zones (United States, 1998).

This Rimouski researcher participated greatly in a wide range of colloquia and special meetings around the world, always with the objective of understanding the coastal environment better and contributing to the prevention of natural disasters and to the limitation of their effects. In the course of his work, he developed an impressive worldwide network of professional contacts in marine sciences.

He was also very active in the International Association for the Physical Sciences of the Oceans (IAPSO), where El-Sabh was member of the Executive Committee and Head and participated in several committees such as Natural Hazards and Oceanography for Developing Countries. In this regard, Prof. El-Sabh has dedicated a large part of his active life to promote exchange and collaboration between Canada and developing countries of Africa, Asia and South America.

Mohammed El-Sabh was awarded numerous prizes, notably: the International Tsunami Society Merit Medal (for the organization of the International Symposium on Hazards and Disasters in Rimouski, 1986), the Applied Oceanography prize of the Canadian Society of Meteorology and Oceanography, the Michel-Jurdant prize (Environmental Sciences, ACFAS, 1991), the Alcide C. Horth Distinction (for his contribution to research), the Prix de reconnaissance (UQAR, 1997) and the Science Prize of the International Society for Natural Hazards (1998).

Within the university, he was twice Director of the Department of Oceanography. Last autumn, he took on the administrative and scientific leadership of the PRICAT project, aimed at developing links in the field of Marine Resources between Canada and Tunisia. He was also a member of the Board of Directors of the Foundation of UQAR.

We have lost a major personality. Many people will warmly remember this friendly, enthusiastic man who made a place for himself in the oceanography world. To his friends and colleagues, El-Sabh (The Lion in Arabic) behaved exactly like the meaning of his last name: protector and provider of friendship, advise and defender upon any injustice.

He was the companion of Mme. Pauline Côté, professor of education at UQAR and the father of two children, Youssef and Nadia.

Mario Bélanger
Translation by Philip Hill

Preface

The original idea of this book started when we were making a residual fluxes study of the Paranaguá Coastal Lagoon (Brazil) near the colonial town of Guaraqueçaba. Among the beautiful mangroves of this Brazilian National Park, between profile and profile, we wondered why South American estuaries were little known in the international arena. Besides, most of the papers published in the literature are based on biological research. Practically nothing is known about their geomorphology and dynamics.

That night, while we were walking along the hilly streets of the town, we decided that the only way to have an idea about the degree of advance in the geomorphology and dynamics of our estuaries was to ask the proper South American researchers to write review articles about the estuaries in which they were working or about the general state of the art of the Geomorphology and Physical Oceanography of the estuaries of his/her country. The book grew from then on. Although initially many scientists offer to write a chapter, we ran into the same problem these researchers have to publish in journals, they felt that their English was not good enough and withdrew. However, we are very satisfy about the number and quality of the contributions which also passed a very strong review process.

In this regard, we would like to express our most sincere thanks to those researchers that have devoted their time not only to check the validity of the science but also to willingly improve the English of the papers when it was necessary. They are listed here in alphabetic order: Diana G. Cuadrado, John W. Day, Georges Drapeau, V.N. de Jonge, S. Susana Ginsberg, Eduardo A. Gómez, Iris Grabermman, Federico I. Isla, Guillherme Lessa, Charles Nittrouer, Ned P. Smith, Stephen V. Smith, J. van de Kreeke, John T. Wells. All of them contributed profoundly, providing new insights and criteria that increased the value of each contribution.

We would also like to thank Springer-Verlag for its support, but specially to Mrs. Andrea Weber-Knapp, who worked hard in trying to make the idea of a book a concrete reality, thanks to her support we finally went through the experience. A very particular thanks goes to Dr. Wim Salomons, Environmental Series Editor, who accepted the idea right from the beginning and encouraged us to pursue in our enterprise.

Mario Pino-Quivira
Valdivia, Chile

Gerardo M.E. Perillo · M. Cintia Piccolo
Bahía Blanca, Argentina

June, 1999

Contents

Contributors

Eduardo M. Acha
Instituto de Investigación y Desarrollo Pesquero
Paseo Victoria Ocampo 1
7600 Mar del Plata, Argentina,
and: Departamento de Ciencias Marinas
Universidad Nacional de Mar del Plata, Funes y Peña
7600 Mar del Plata, Argentina
E-mail: macha@lisa.inidep.edu.ar

Maria da Graça Baumgarten
Laboratorio de Hidroquímica
Departamento de Química
Fundação Universidade do Río Grande
Rua Alfredo Huch 475
Río Grande, RS 96.201-900, Brazil
E-mail: dqmhidro@super.furg.br

Otis B. Brown
Rosenstiel School of Marine and Atmospheric Science
University of Miami
4600 Rickenbacker Causeway
Miami, FL, 33149, USA
E-mail: obrown@rsmas.miami.edu

Patrice Castaing
Département de Géologie et Océanographie/URA 197
Université de Bordeaux I
Avenue des Facultés
33405 Talence Cedex, France

María Paula Etala
Servicio Meteorológico de la Armada Argentina
Comodoro Py 2055
1104 Buenos Aires, Argentina
E-mail: etala@rina.ara.mil.ar

Gilberto Fillmann
Laboratorio de Hidroquímica, Departamento de Química
Fundação Universidade do Río Grande
Rua Alfredo Huch 475
Río Grande, RS 96.201-900, Brazil
E-mail: dqmhidro@super.furg.br

Mariana B. Framiñan
Present address: Rosenstiel School
of Marine and Atmospheric Science
University of Miami
4600 Rickenbacker Causeway
Miami, FL, 33149, USA
Permanent address: Servicio de Hidrografía Naval
Montes de Oca 2124
1271 Buenos Aires, Argentina
E-mail: mframinan@rsmas.miami.edu

Raúl A. Guerrero
Instituto de Investigación y Desarrollo Pesquero
Paseo Victoria Ocampo 1
7600 Mar del Plata, Argentina
E-mail: guerrero@lisa.inidep.edu.ar

Björn Kjerfve
Marine Science Program
Department of Geological Sciences
and: Belle W. Baruch Institute
for Marine Biology and Coastal Research
Columbia, SC 29208, USA
and: Departamento de Geoquímica
Universidade Federal Fluminense
CEP 24020-007 Niterói, RJ, Brazil
E-mail: bjorn@sc.edu or bjorn@nitnet.com.br

Bastiaan Knoppers
Departamento de Geoquímica
Universidade Federal Fluminense
CEP 24020-007 Niterói, RJ, Brazil
E-mail: bknoppers@zmt.uni-bremen.de

Jorge López Laborde
Div. Geología Marina
Servicio de Oceanografía e Hidrografía de la Marina
Casilla de Correos 15209
Montevideo, Uruguay
E-mail: jorgeLL@sohma.gov.uy

Carlos A. Lasta
Instituto de Investigación y Desarrollo Pesquero
Paseo Victoria Ocampo 1
7600 Mar del Plata, Argentina
E-mail: clasta@lisa.inidep.edu.ar

Osmar O. Möller Jr.
Departamento de Fisica
Fundaçao Universidade do Río Grande - FURG
Av. Italia Km 8
CP 474, 96201-900 Río Grande - RS, Brazil
E-mail: osmar@calvin.ocfis.furg.br

Gustavo J. Nagy
Departamento de Oceanografía
Universidad de la República
Tristán Narvaja, Montevideo, Uruguay
E-mail: gunab@fcien.edu.uy

Luis Felipe Niencheski
Laboratorio de Hidroquímica, Departamento de Química
Fundação Universidade do Río Grande
Rua Alfredo Huch 475
Río Grande, RS 96.201-900, Brazil
E-mail: dqmhidro@super.furg.br

Gerardo M.E. Perillo
Instituto Argentino de Oceanografía, CC 107
8000 Bahía Blanca, Argentina
and: Departamento de Geología
San Juan 670
8000 Bahía Blanca, Argentina
E-mail: perillo@criba.edu.ar

María Cintia Piccolo
Instituto Argentino de Oceanografía, CC 107
8000 Bahía Blanca, Argentina
and: Departamento de Geografía
12 de Octubre y San Juan
8000 Bahía Blanca, Argentina
E-mail: piccolo@criba.edu.ar

Mario Pino-Quivira
Instituto de Geociencias
Universidad Austral de Chile
Casilla 567, Valdivia, Chile
E-mail: mpino@valdivia.uca.uach.cl

Gilberto Rodríguez
Laboratorio de Biología Marina
Instituto Venezolano de Investigaciones Científicas
Apartado 21827, Caracas, Venezuela
E-mail: grodrigu@oikos.ivic.ve

Herbert L. Windom
Skidaway Institute of Oceanography
University System of Georgia
P.O. Box 13.687, Savannah
Georgia, 31.416, USA
E-mail: herb@skio.peachnet.edu

Chapter 1

What Do We Know About the Geomorphology and Physical Oceanography of South American Estuaries?

Gerardo M.E. Perillo · M. Cintia Piccolo · Mario Pino-Quivira

1.1 Introduction

As long as freshwater is discharged into the sea, there is the potential for the development of an estuary. Although this concept appears to be simple, the marked differences in geomorphologic, oceanographic, atmospheric and biogeochemical conditions that occur along the coasts of the world result in a wide variety of estuarine types. The variety is so large that several ways have been used as mean to define and classify them (Perillo 1995a).

Although work in estuaries can be traced back to the beginning of the century, a formal definition was not available up to the '50s when studies by Ketchum (1951) first, and then by Pritchard (1952), shaped the basic concepts about their geomorphologic and physical characteristics. Since them, the number of researchers and the knowledge about estuaries worldwide have bursted. Fifty years ago the estuarine oceanography was concentrated on Chesapeake Bay and few estuaries in Europe. Nowadays estuarine research has spread so much that we have a very well defined knowledge of their geomorphology and dynamics to the point that we are seriously developing classification schemes that are abstracting this vast knowledge.

Etymologically, estuary derives from the Latin word "*aestus*" which means "of tide". That is to say that the term estuary has to be applied to any coastal feature in which the tide has special significance. Although estuaries may be regarded only by their physiographic parameters: that is, their geomorphology and hydrology, their biological and chemical components should also be considered. Any comprehensive definition must necessarily include all these aspects.

After analyzing over 40 different definitions introduced in encyclopaedia, dictionaries but particularly from the specialized literature, Perillo (1995b) proposed a new definition of estuaries that incorporates features never discussed in previous definitions as:

> "An estuary is a semi-enclosed coastal body of water that extends to the effective limit of tidal influence, within which sea water entering from one or more free connections with the open sea, or any other saline coastal body of water, is significantly diluted with fresh water derived from land drainage, and can sustain euryhaline biological species from either part or the whole of their life cycle."

In considering the estuarine distribution, we are going to employ this definition as the basic criteria in establishing the presence of an estuary along the coastline. The coexistence of tidal action and intrusion of sea water is formally established. In fact, the estuary extends inland up to the effective limit of tidal action, but it is within the segment that stretches from that inland point to the mouth in which seawater dilution can occur. This model permits the differentiation within the estuary of the three sectors proposed by Dionne (1963) (Fig. 1.1) and further described by Dalrymple et al. (1992), and also allows for estuaries that have only one or two of these sectors.

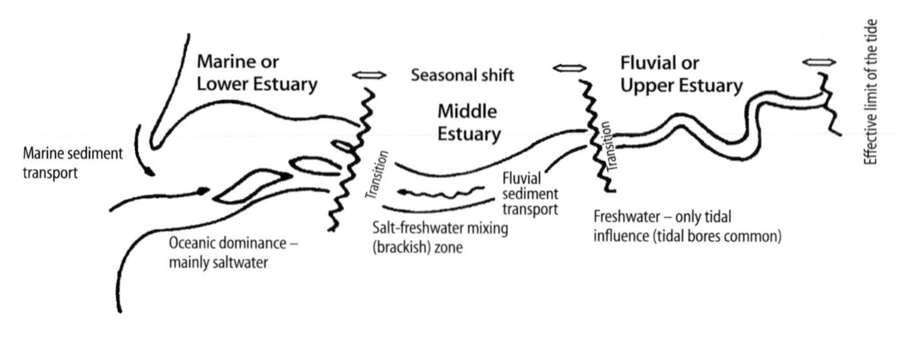

Fig. 1.1. Elements that constitute a standard estuary (after Dionne 1963 and Perillo 1995b)

Based on this definition, Perillo (1995b) proposed a morphogenetic classification to differentiate the various possible settings in which estuaries have developed. This new classification, which is further modified here, opens much more the spectra by covering all possible categories of estuaries that are established by the definition given before. This classification is based on genetic and morphological considerations. The first division is the necessary genetic differentiation of estuaries as either primary and secondary estuaries (Fig. 1.2) following the criteria given by Shepard (1973) in his coastal classification.

Primary estuaries: the basic form has been the result of terrestrial and/or tectonic processes and the sea has not changed significantly the original form. Specifically, these are those estuaries that have essentially preserved their original characteristics up to the present.

Secondary estuaries: the observed form is the product of marine processes and their relative influence over river discharge acting since the sea level has reached nearly its present position.

A brief description of each category is given:

Former Fluvial Valleys formed by sea flooding of Pleistocene-Holocene river valleys during the last postglacial transgression. According to their coastal relief, they have been divided in two subcategories:

- *Coastal Plain Estuaries* normally occupy low relief coasts produced mainly by sedimentary infilling of the river(s). Typical examples are Thames (UK), Gironde (France), Yangh-Tse (China);
- *Rías* are former river valleys developed in high relief (mountainous or cliffy) coasts. Examples of these are the Pontevedra (Spain) and Deseado (Argentina) rias.

Former Glacial Valleys formed by sea flooding of Pleistocene glacial valleys during the last postglacial transgression. Also, based on the coastal relief to which they are associated, they are divided in two subcategories:

- *Fjords* occupy glacially formed troughs located in high relief coasts. Examples are Oslo (Norway), Mercier (Chile);

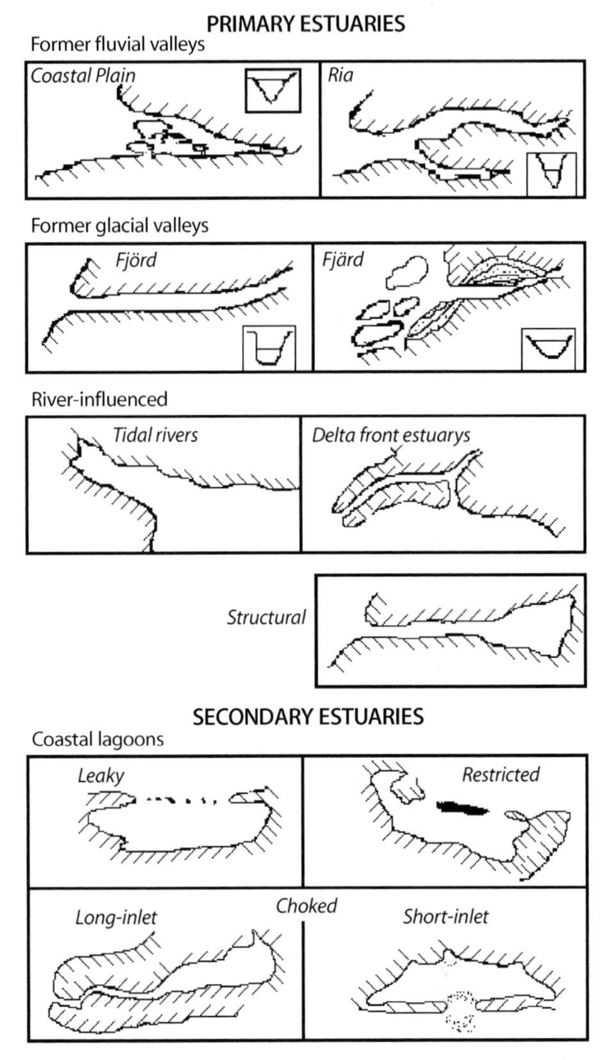

Fig. 1.2. Morphogenetic classification of estuaries (modified from Perillo 1995b)

- *Fjards* occupy glacially formed troughs in low relief coasts. Examples are those formed in the northern coast of Sweden.

River-Influenced. In high discharge rivers like the Amazon (Brazil), Mississippi (USA) and Río de la Plata (Argentina-Uruguay) the valley is not presently drowned by the sea. However, the circulation in the lower portions of the river is highly affected by tidal dynamics, including reversing currents, resulting in characteristic, tidal-dominated morphological patterns. They have been divided in two subcategories:

- *Tidal River Estuaries* include those rivers that are affected by tidal action but salt intrusion may be limited to the mouth or it is totally absent within the valley. Normally these estuaries are associated to large discharge rivers that either by their coastal setting (e.g., de la Plata River) or the relatively strong coastal dynamical processes occurring at their mouth (e.g., Amazon) do not develop a delta. The degree of salt intrusion is seasonally and climatically dependent; however, tidal processes are very important in sediment transport dynamics and morphological evolution within the valley;

- *Delta-front Estuaries.* This category includes the estuaries found in the portions of deltas affected by tidal dynamics and/or salt intrusion. The classic example is the outer Mississippi channels.

Structural Estuaries. Their valleys were formed by neotectonic processes such as faulting, vulcanism, postglacial rebound, isostasy, etc. occurred since the Pleistocene. All the structural processes that give place to the formation of the valley must be active in the present time or being occurring from the Pleistocene. Otherwise, since almost all rivers are controlled by structural (e.g., faults) conditions their corresponding estuaries should all be tectonic. Examples are San Francisco Bay (USA) and Valdivia River (Chile).

Coastal Lagoons (after Kjerfve and Magill 1989). Inland water bodies usually oriented parallel to the coast separated from the sea by a barrier and connected to the ocean by one or more restricted inlets. In the present classification we included the subdivision suggested by Kjerfve and Magill (1989) based on the nature of the entrance:

- *Choked:* only one long and narrow entrance (Dos Patos, Brazil; Mar Chiquita, Argentina);
- *Restricted:* few inlets or a wide mouth (Pamlico Sound, USA; San Sebastián Bay, Argentina, Términos, Mexico);
- *Leaky:* large number of entrances separated by small barrier islands (Belize Lagoon, Mississippi Sound).

However, we introduce here a modification for the choked type in two subtypes depending on the length of the inlet and named *short-inlet* and *long-inlet choked lagoons.* The difference is due to the damping effect that has the inlet on the propagation of the tidal wave. Long inlets are more prone to become occluded and eventually closed (see Knoppers and Kjerfve, this book, for examples of tidal choking and the damping effects of inlets in Brazilian coastal lagoons).

1.2 World Estuary Distribution

Estuaries are one of the most common features in the world coast. The economical activities and biological diversity found in estuaries make over 80% of the corresponding values of the coasts. They are subject to more environmental impact and human influence than beaches or other coastal environments.

A view of the better studied estuaries that there are along the world coasts is given in Fig. 1.3. The figure provides a geographical demonstration of how much work on estuaries has spread out. However, in analyzing the degree of knowledge that exist in every estuary listed in the map, the differences are significant. Whereas, those estuar-

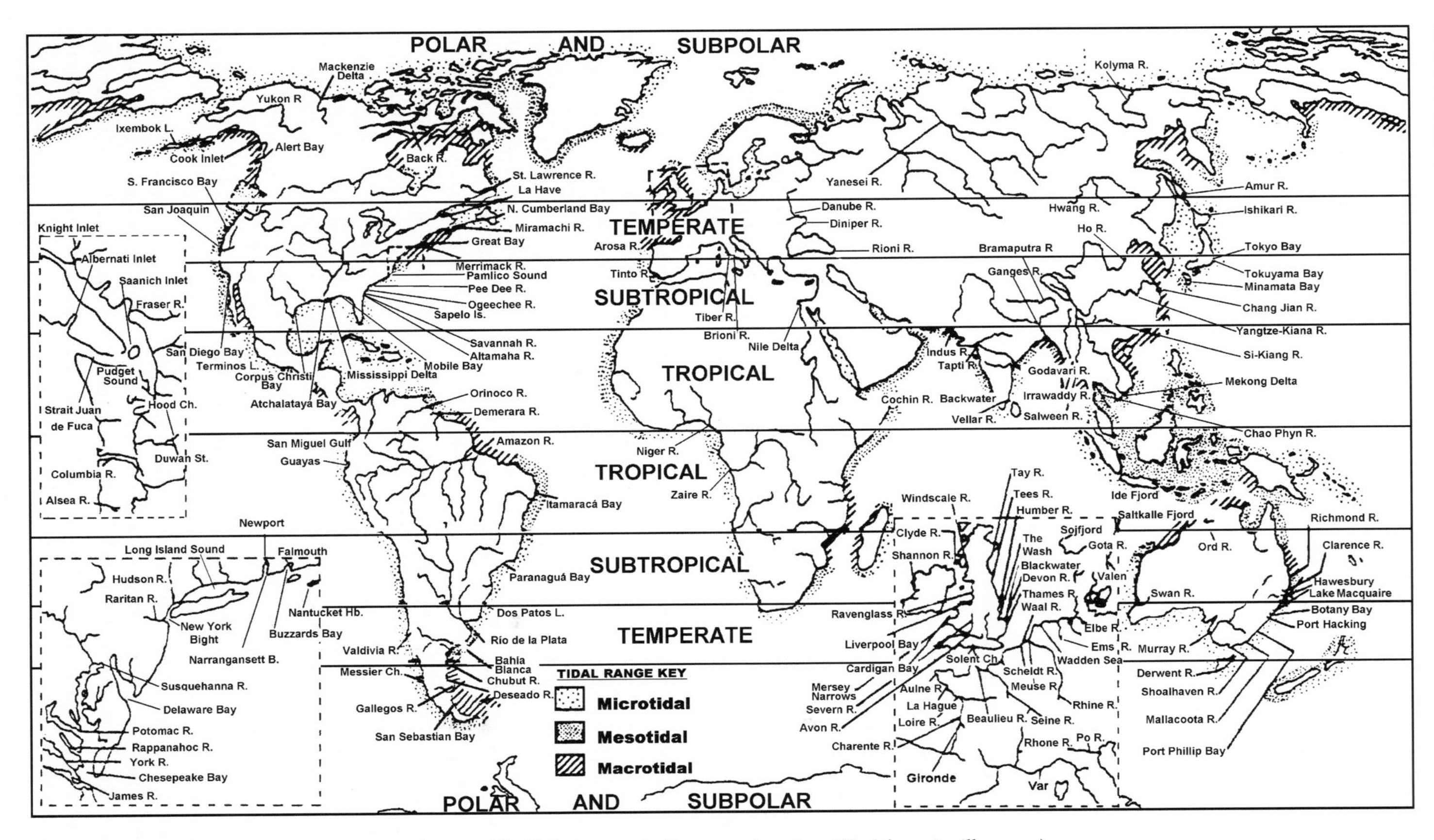

Fig. 1.3. World distribution of estuaries in relation with tidal range and climate regions (modified from Perillo 1995a)

ies depicted along the United States, Canada and Western Europe received a very detailed attention provided by local expertise and advanced technology, most of the ones in the rest of the world have only received elemental attention even though the effort of local researchers and some international interest.

There is still a much larger number of estuaries that has not received even the most basic attention. As we are going to describe in the present chapter, this is the case for most estuaries in South America. As well as it occurs in Central America, Asia and Africa, the number of estuaries about which we have some knowledge is overwhelmed by the number of those from which we know barely their names. Therefore, the objective of the present chapter is actually to show the lack of knowledge and the opportunity for research that there is available in South America. Although this concept is also valid for other estuaries in the world.

1.3 Estuaries of South America

Figure 1.4 provides a general description of the major estuaries present along the South America coast. As estuaries are dependent of fluvial input, the distribution is also closely related to the precipitation conditions on the continent as well as the local topography. The Northern and Eastern regions of the continent are part of the Amazonia with very high levels of precipitation (over 1 200 mm yr $^{-1}$). Therefore, the number of rivers and lagoons that reach the Atlantic Ocean in this area is very large. There are also many small rivers that reach the Pacific Ocean in Colombia and Ecuador, but their length and catchment basins are small because they have their sources in the nearby Andes mountains.

Similar situation occurs as we move down along the Western coast of South America, rivers are relatively small when they are formed in near coast mountains like the Coastal Range in the South of Chile. The larger rivers are those that have sources at the main Andes range (i.e., Bio Bio and Valdivia Rivers in Chile). However, south of the Guayas Estuary, the Pacific coast has mean annual precipitation significatively lower than the Atlantic counterpart down to the Bio Bio River. Thus the number of rivers and their strength is dramatically reduced and associated to the microtidal conditions, practically there are no estuaries in this stretch of the coast.

South of the Valdivia River, precipitation rates becomes high again (2 000 mm yr^{-1}), but most of this rain falls along the Coastal Range of southern Chile. Therefore, there are hundreds of high, torrential type rivers that form small enclosed estuaries, many of them develop choked coastal lagoons (i.e., Queule, Tubul). Further south the Chilean coast up to Tierra del Fuego is cut by deep valleys developed by alpine glaciers that reached its maximum during the last Pleistocene glaciation. Today all those valleys are fjords from which the most remarkable is the Mercier Channel which is estimated as the deepest fjord in the world with maximum depths in excess of 1 200 m.

The Northern coast of South America presents some of the largest rivers in the planet, specifically the Amazon, Demerara and Orinoco Rivers which have extensive basins that cross most of the continent. Although many rivers reach the Atlantic in Northern Brazil down to the Doce River, most of them have their origin at the Brazilian Shield and have steep basins and short lengths.

Fig. 1.4. Distribution of major estuaries in South America. Those names not underlined are described or mentioned in the different chapters of this book

The South of Brazil and North of Uruguay is characterized by a large number of coastal lagoons (for a detailed description see Knoppers and Kjerfve, this book). They included practically all types from choked to leaky and within the choked those with long inlet (typical of the Uruguay coast such as Rocha Lagoon) and short inlet. The best example of the latter is Patos Lagoon (see Moeller and Castaing and Niencheski et al., both in this book) the largest of this kind of water bodies in the world. Most of the Uruguaian lagoons are intermittent, meaning that during the low raining season their mouths are closed by littoral drift. When the freshwater discharge is enough to reopen the mouth, it may change the direction of the inlet depending on the present littoral conditions, developing sharp curved meandering channels.

The Río de la Plata is a very special feature since it is the widest estuary in the world, with dynamical conditions remarkable different of most other estuaries. Although several studies have covered many aspects of this estuary (see Piccolo and Perillo, López Laborde and Nagy, and Framiñan et al., this book), there is still much work to be done to reach a basic understanding of both its geomorphologic and dynamic characteristics. Further to the south, along the Argentina coast, there are over 20 estuaries plus four coastal lagoons basically controlled by the relationship between tidal range and freshwater input (see Piccolo and Perillo, this book).

1.4 International Publications About South American Estuaries

In defining the degree of knowledge that exists about South American estuaries in relation with other countries and/or continents, we made a bibliographic analysis of the papers published in the two journals with main emphasis in estuarine research. They are dedicated to an international audience and open to contributions from all over the world. The selected journals are Estuaries (published by the Estuarine Research Federation, USA) and Estuarine, Coastal and Shelf Science (ECSS) formerly known as Estuarine, Coastal Marine Science (published by the Estuarine, Coastal Science Association, UK).

For the analysis we considered only those articles totally related to one or more estuaries and also to their geographical location. Based on a geographic criteria, the papers were subdivided into seven regions: USA/Canada, Central America/Mexico, Europe, Asia, Africa, South America and Oceania. An eighth subdivision was defined for those papers dealing with theoretical approaches in which no particular estuary is mentioned in the text.

Figure 1.5 provides the percentage distribution for a total of 1120 papers fully devoted to estuaries that were published in the period 1988–1997 in both journals. As a first approach, it is clearly evident from the figure the immense bias existing for papers published on estuaries of USA, Canada and Europe. Almost 77% (861 papers) of the total were related to estuaries found on these three regions. Only a dozen of papers considered comparisons among estuaries in USA and Europe, and only one between Oceania and Europe. In this case, the representatives of each region was given in proportion to the number of estuaries described for each region in the paper (i.e., if two estuaries were mentioned, one for USA and one for Europe, each region received 0.5). Table 1.1 gives the total number of published papers discriminated by region and journal. For this wide scope, we did not separate the papers according to their main subject(s) but during our scrutiny we observed that over 75% of the papers were dealing with biological approaches.

The added percentage of papers on Africa, Asia, Central America/Mexico and South America estuaries (13.29%) is lower by almost 61% those from Europe alone and more

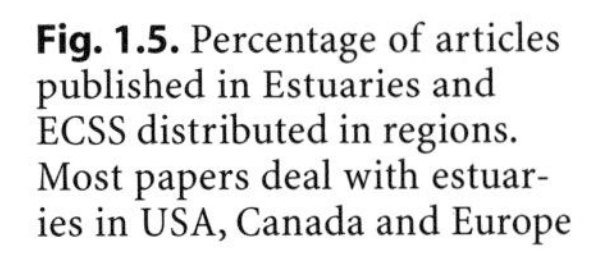

Fig. 1.5. Percentage of articles published in Estuaries and ECSS distributed in regions. Most papers deal with estuaries in USA, Canada and Europe

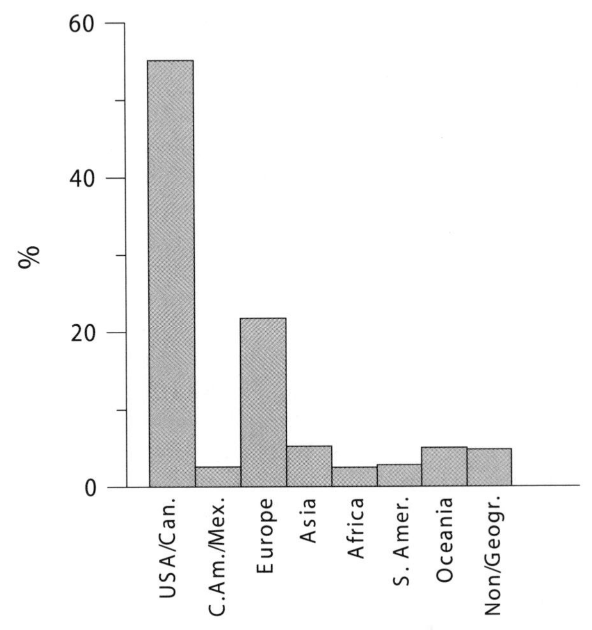

Table 1.1. Number of papers published in the period 1988–1997 in Estuaries and ECSS discriminated by regions. N/G = theoretical paper not applied to any specific estuary

	USA/ Canada	C. Amer./ Mexico	Europe	Asia	Africa	South America	Oceania	N/G	Total
Estuaries	431.0	11	37.5	9.0	5.0	7	10.5	33	544.0
ECSS	186.1	18.0	206.9	50.0	23.0	25.0	46.0	21.0	576.0
Total	617.1	29.0	244.4	59.0	28	33	56.5	54	1120.0
%	55.05	2.59	21.8	5.26	2.5	2.94	5.04	4.82	100

than four times smaller than USA/Canada (Table 1.1) estuaries. In relation with the number of estuaries existent in Australia and New Zealand (100% of the Oceania papers come from there), the number of papers could be considered at the same level of the European output.

Clearly, for regions like South America, there is no relation between the number of estuaries found in their coast and the information published in the international literature. One major reason for this is that researchers tend to publish in regional or national journals favoured by their own native language rather that confronting with the extra task to write his findings in English. Many of them may not be proficient enough or they may need much longer time to produce a good quality English presentation. Even though in many cases editors of international journals are very helpful in trying to cross this barrier, the authors may not be successful enough to adequately express their concepts and papers are rejected discouraging further presentations. However, in other cases, editors reject papers from particular areas because they do not have international appeal.

Although it is not the main concern of this paper, from the analysis of the individual journals (Table 1.1), there is a remarkable bias in the scope of each journal. Almost 80% of the papers published in Estuaries are from USA/Canada and normally European estuaries have a similar treatment (although slightly higher) to the rest of the regions. Whereas, ECSS accepted more papers from Europe and (with a small difference) from USA/Canada. However, authors from the other regions tend to publish more in this journal than in Estuaries.

Considering specifically the papers published in both journals about estuaries from South America, there is a considerable difference between them (7 in Estuaries and 26 in ECSS). In this regard, we did two types of analysis: one dealing with the main subject of the article and another related to the country in which the estuary(ies) is(are) located. Figure 1.6 shows the total percentage distribution based on the eight major subject we classified each paper (Geomorphology, Physical Oceanography, Geomorphology/Physical, Biology, Chemistry, Biochemistry or a combination of both Biology and Chemistry, Biology and Physical, and Biology and Geomorphology). Of the total of 33 papers found in both journals, 23 were fully or partly devoted to Biology as main subject (18 were only on biological matters). Nine papers were fully or partly related to Geomorphology and/or Physical Oceanography, but no paper actually considered both issues together.

Of course, our analysis is not suggesting or indicating that any subject has a lesser importance than the others. What we like to point out is the fact that there is a dramatic bias towards only one of the sciences, which is valid for all over the world probably due to the unbalance that exist on the number of specialist in each discipline. It is necessary to remember that the main approach we tried to follow with the present book was to show the level of knowledge about Geomorphology and Physical Oceanography in South American estuaries. We consider that estuaries should be studied by teams of researchers with a wide variety of disciplines applying an integrated approach.

In Table 1.2 we correlated the main subject of each of the individual articles with the country in which the described estuary is located. Notably of the eleven countries having marine coasts in South America, we found papers dealing with estuaries from only six of them, and two of the countries (Equator and French Guiana) have one paper each. The bulk of the articles comes from Brazil, Chile and Argentina which also have the longer coastlines but also the larger number of estuaries. Again, the bias towards Biology is observed specifically in Brazil and Chile. In the latter all papers but one are purely Biological. Whereas Argentina has more papers dealing with Geomorphology and Physical Oceanography than Biology.

As a further example of this large differentiation, we analysed the references in the following eight chapters of this book. We considered only all articles that strictly refer to South American Estuaries or related coastal features. We did not considered abstracts in congresses or specialized journals. It resulted in a total of 387 references (only 14% were repeated in two or more chapters) resulting that 65.4% were published in Latin American sources (e.g., journals, technical reports, proceedings of congresses, books and chapters of books).

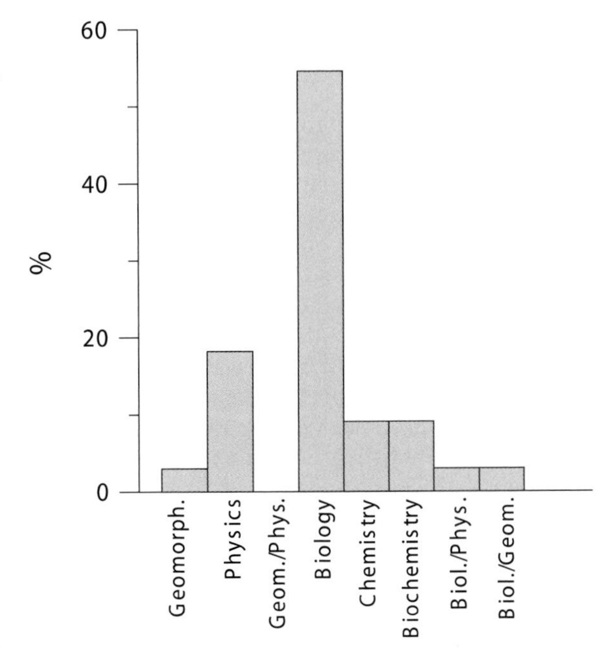

Fig. 1.6. Percentage of articles published in Estuaries and ECSS distributed in disciplines for South American estuaries. Most papers deal with biological issues and rather few of them consider the Geomorphology and/or the Physical Oceanography of these coastal water bodies

Table 1.2. Total number of papers published in the period 1988–1997 in both Estuaries and ECSS discriminated by country and subject

Subject	Argentina	Brasil	Chile	Ecuador	French Guiana	Venezuela	Total
Geomorphology	1	0	0	0	0	0	1
Physics	3	2	1	0	0	0	6
Geomor./Physics	0	0	0	0	0	0	0
Biology	2	6	7	0	1	2	18
Chemistry	0	1	0	1	0	1	3
Biochemistry	0	3	0	0	0	0	3
Biology/Physics	0	1	0	0	0	0	1
Geomor./Biology	0	1	0	0	0	0	1
All	6	14	8	1	1	3	33

Journals and technical reports of relatively low distribution (most of the latter seldom are sent beyond a handful of other institutions) provide 47% out of the Latin American percentage, whereas 23% from the international side corresponds to journals. Interestingly, only 18% out of the 23% corresponds to widest distributed journals, the rest are normally in less common ones published in Germany and France.

1.5 Conclusions

South America has, in comparison with other regions of the world, probably one of the largest number of estuaries per km of coastline. However, very little is known about their characteristics in general and their Geomorphology and Physical Oceanography in particular. There are several reasons why this occur. In the first place, there is a much lower development both in number and specialities of research groups working in South America when compared with North America, Europe and even Australia/ New Zealand. In our continent it has been a tradition, probably inherited from the classical Oceanography of the '50s throughout the '70s that this science was only related to "blue water". Most of our oceanographers were trained with this concept in mind even though the lack of adequate funding and/or ship time and equipment prevented them to do research in that area. Littoral Oceanography ("brown-water oceanography) was always regarded as not a matter of Oceanography, however, this concept is presently changing.

The number of researchers working in estuaries is small in general but by far the largest number is devoted to Biology and very few working in Geomorphology. Only Brazil is now having emerging groups working in the Physical Oceanography of estuaries as well as in basic Geomorphology and sediment transport. On the other hand, Chile has a very small group with interest in the Geomorphology and Dynamics of estuaries. Uruguay is also in the same situation. Argentina has only one group that works full time on the Geomorphology, Dynamics and sediment transport processes in estuaries and at least three other groups that do research in estuaries only part of the time. Two of those groups work mostly on Geomorphology and one in Physical Oceanography.

A second reason that may explain why South American estuaries are little known internationally is the language. It is well accepted that English is the most common international language; however, the number of researchers fluent in writing English articles to the level required for international journals is remarkable low. Then, many investigations on the estuaries seldom pass of the level of local or regional journals or proceedings of congress plus many reports and thesis of restricted circulation and published in Spanish or Portuguese. This situation is also valid for Central America and Mexico as well. Nevertheless, this conditions is also progressing in a positive way as many researchers trained in North America and Europe are coming back to take teaching and research positions within their own country.

We consider that estuaries should be studied from a multi and interdisciplinary approach. However, the present status show a dramatic bias towards biological studies and very little is done regarding their geomorphology and dynamics. However, both aspects are essential to establish the basic conditions in which all other process must be analysed.

Acknowledgements

Research dealing to the present article has been funded by the Inter-American Institute for Climate Change (IAI), National Science Foundation, the European Economic Community, Consejo Nacional de Investigaciones Científicas y Técnicas (CONICET) de la República Argentina and Universidad Nacional del Sur. Lic Walter D. Melo was very helpful in doing some of the figures.

References

Dalrymple RW, Zaitlin BA, Boyd R (1992) A conceptual model of estuarine sedimentation. J Sedim Petrol 62:1130–1146
Dionne JC (1963) Towards a more adequate definition of the St. Lawrence Estuary. Z Geomorph 7:36–44
Ketchum BH (1951) The flushing of tidal estuaries. Sewage Ind Wastes 23:198–209
Kjerfve B, Magill K (1989) Geographic and hydrographic characteristics of shallow coastal lagoons. Mar Geol 88:187-199
Perillo GME (1995a) Geomorphology and sedimentology of estuaries: an introduction. In: Perillo GME (ed) Geomorphology and sedimentology of estuaries, Elsevier Science BV Amsterdam, pp 1–16
Perillo GME (1995b) Definition and geomorphologic classifications of estuaries. In: Perillo GME (eds) Geomorphology and sedimentology of estuaries, Elsevier Science BV Amsterdam, pp 17–47
Pritchard DW (1952) Salinity distribution and circulation in the Chesapeake Bay estuarine system. J Mar Res 11:106–123
Shepard FP (1973) Submarine Geology. Harper & Row, New York, pp 517

Chapter 2

The Maracaibo System: A Physical Profile

Gilberto Rodríguez

2.1 Introduction

Intermontane basins are areas of potential oil accumulation (Hyne and Dickey 1977). The Maracaibo Basin, located on the northern front of South America, is no exception: It is the most important oil accumulation in South America and one of the largest in the world. From 1914 to 1995 the total amount of oil extracted from the basin was 33 000 millions barrels ($5\,238 \times 10^6$ m^3). Most of this production comes from the bottom of Lake Maracaibo itself, the largest lake in South America and the 17th in the world.

The oil interests, together with the large yield of the fisheries in the estuary, more than 40 000 t yr^{-1}, and the accelerated demographic growth, has fostered, since the middle 1950s, a large body of research addressed mostly to two practical problems of management, viz. the possibilities of a process of salinization of the lake caused by the dredging of its bar, and the potential eutrophication of the whole body of water from sewage, pollution and other nutrient-rich sources. Several large scale engineering projects have been considered to alleviate the salinity intrusion through the navigation channels, but as all these imply drastic changes in port facilities and ship routes used at present (Rodríguez 1984), they have to be framed within a program of integrated coastal zone management (Conde and Rodríguez in press).

Most of the research dealing with these problems has been published in private reports to the government and oil companies, and only a small fraction went to main stream journals. The present contribution summarizes the physical framework of the estuary, as discerned through these studies.

2.2 Geomorphology

The large orographic system of the Andes on its northern-most portion is divided into two distinct ranges: the Venezuelan Andes to the north-east and the Sierra de Perijá to the north, forming between them a large pocket of lowlands. This extensive platform has an approximate oval shape, as does the central Lake Maracaibo which acts as reservoir for all the drainage of the surrounding sierras (Fig. 2.1). Lake Maracaibo is a body of water 150 km long and 90 km wide, connected to the sea by the Strait of Maracaibo, a narrow channel about 60 km long and 15 m deep. To the north the Strait open into Tablazo Bay, which is about 27 km wide and 20 km long. Recent marine sediments deposited in the lake have reduced its depth to less than 40 m (Gonzalez de Juana et al. 1980).

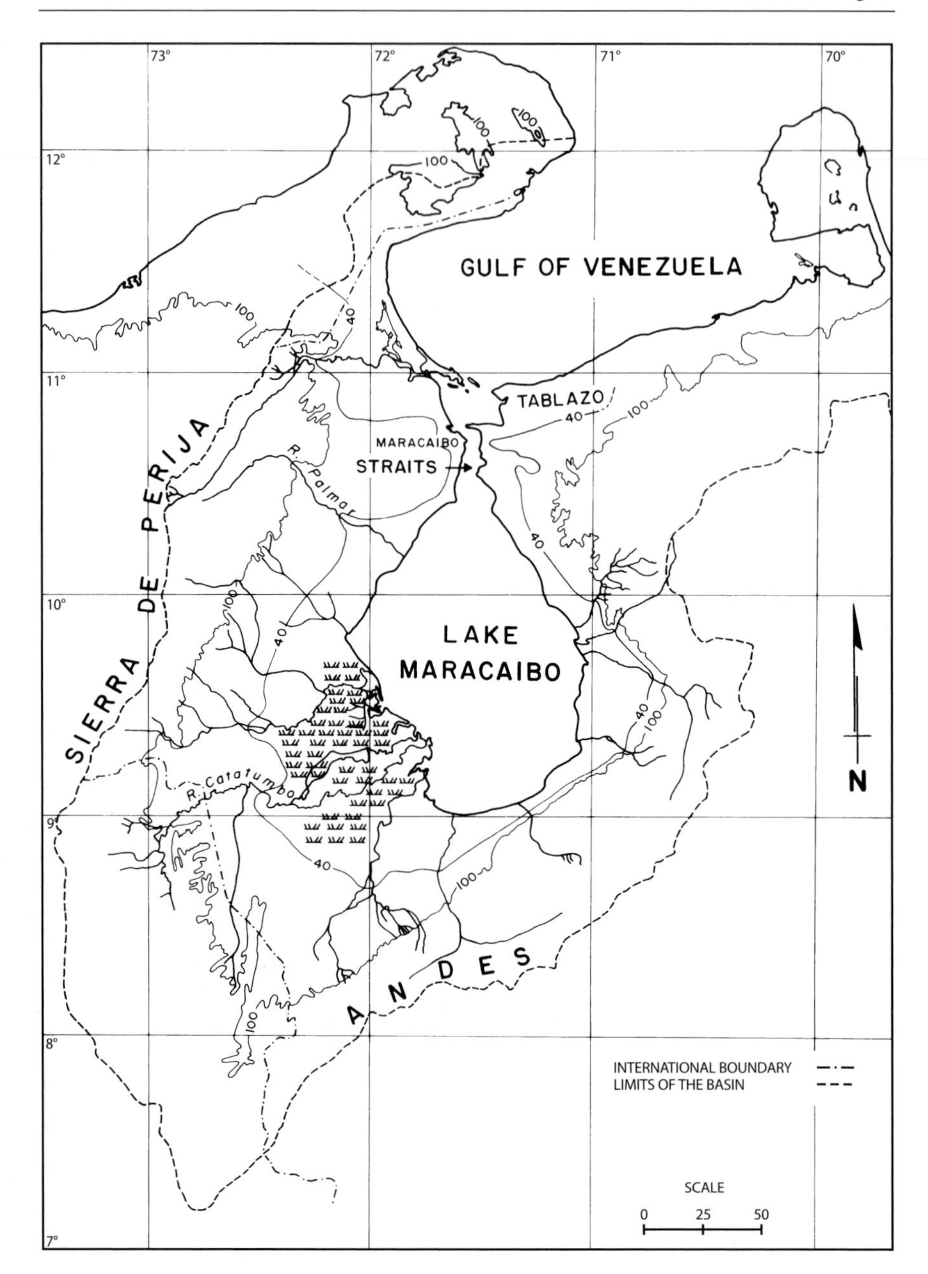

Fig. 2.1. Intermontane basin of the Lake Maracaibo system

The major streams originate in the southern and more extensive part of the basin where the heavily forested slopes of the cordilleras and the large rainfall (more than 3500 mm yr^{-1}) give rise to the rivers which run across swamp land and jungle areas and flow into the western side of the lake. A predominant feature of the south-western corner of the lake is the Catatumbo River, whose birdfoot delta is similar in shape and sedimentology to large oceanic deltas, like the Mississippi's (Hyne and Dickey 1977). The only stream of importance on the western side of the lake is the Palmar River. In the south-east the area between the lake and the bordering mountains is rather narrow and streams are short and fast flowing (Gonzalez de Juana et al. 1980).

The lake bottom is pan-shaped, and nearly flat over two-thirds of its surface. The flat bottom extends to the central and eastern portions and corresponds to water depths of about 30 m. A narrow, relatively steep, shoreward rising of the lake bottom characterizes the eastern and southern borders. In the north-western portion the bottom forms a gentle and broad incline. A sill, about 12 m deep (Redfield et al. 1955), separate the lake bottom from the Strait of Maracaibo.

Even though the lake bottom is nearly flat, in detail it has irregularities. These bottom features consists of fairly large elongated depressions, smaller nearly circular holes, and small mounds. Some of these depressions are 4 km in length with a maximum relief of 10 m. This feature may be due to differential compaction of the Recent sediments. The series of small depressions or holes found in some areas may be due also to compaction phenomena or related to springs or gas seeps (Sarmiento and Kirby 1962).

2.3 Geological History of the Estuary

The Maracaibo Basin occupies an old platform located perypherically to the Precambrian Guiana craton which was covered during most of the Cretaceous transgression by epicontinental seas. It reached a maximum submergence during the Cenomaniense-Santoniense (100 my) This platform has been sinking since the end of the Eocene at a rate of 0.1 mm yr^{-1} (Sutton 1946), and is covered at present by a thickness of sediments of approximately 5 000 m (Zuloaga 1950).

The structural Maracaibo Basin was formed during Mio-Pliocene time, when the Venezuelan Andes and the Sierra de Perijá attained their maximum development. The formation of the basin changed the sedimentary environments from restricted marine and transitional to predominantly transitional and continental. This type of sedimentation has continued into present times resulting in a gradual filling of the slowly subsiding basin (Graf 1969). The Gulf of Venezuela has retained its present form from Pliocene times.

At the beginning of the first eustatic regression of the Quaternary (Nebraska glaciation), the Calabozo Ridge, which partially separate Calabozo Bay from the outer Gulf of Venezuela, was partially emerged, forming a coastal lagoon, or endorheic basin, completely separated from the open sea (Fig. 2.2). During the first interglacial (Aftonian) the sea transgreded again and the interior basin (Lake Maracaibo) was transformed into a coastal lagoon. This cycle repeated itself with the following eustatic variations of sea level, up to the last postglacial transgression which began with the Holocene, approximately 20 000 years B.P. (Graf 1969, 1972).

The Holocene sediments of Lake Maracaibo attain a thickness of as much as 20 m and consists of four stratigraphic units: A clay substratum, probably of Pleistocene origin and belonging to the upper part of the so-called Milagro Formation; a thin unit consisting of peat, clay and sand; soft unconsolidated muds which made the bulk of

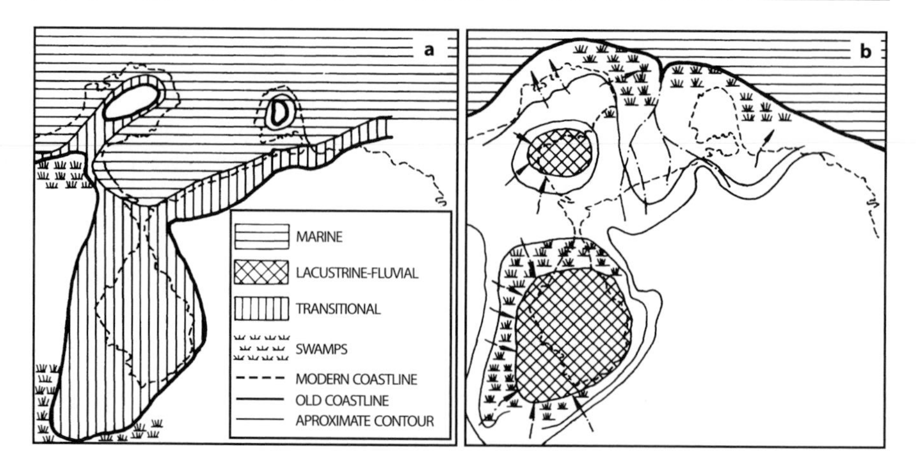

Fig. 2.2. Quaternary history of the Maracaibo basin. **a** second interglacial (Yarmouth); **b** second glaciation (Wisconsin) (data from Graf 1969, 1972)

the Recent sediments; and a horizon of diatomite. The Recent sediments rest unconformable on the clays of the Milagro Formation. ^{14}C determinations on the peats indicate ages of 7 000–10 000 years (Sarmiento and Kirby 1962).

Paleontological evidence shows that the lake throughout its Recent history has fluctuated from fresh to a brackish body of water. The maximum salinity was reached during the rise of the sea level that accompanied the postglacial climate optimum (Sarmiento and Kirby 1962). The date of the last opening of the estuary is given by Redfield and Doe (1964) as 8 000 years B.P. and by Sarmiento and Kirby (1962) between 4 000 and 5 000 years B.P. For the sea to transgrede into the lake it is necessary a rise in sea level of 20 m, which is the value given by Sheppard (1961) for the elevation of the sea level in stable coasts during the last 9 000 yr. After the connection with the sea was established numerous islands and bars were formed at the mouth of the estuary (Redfield 1955).

2.4 Climatology

2.4.1 Temperature

The temperature of the air in the basin is very stable. Mean annual temperature in the north and central part of the basin (mean of 10 yr) is 28.3° C. Mean daily differences is 7° C and mean differences between the hottest and coldest month is approximately 4° C. Surface water temperatures are very uniform throughout the inner shallow area (12–20 m total depth) of the Gulf of Venezuela, with small seasonal and diurnal differences. The column of water is only slightly stratified or completely homogeneous; according to Rodríguez (1973) data for March 1969, the maximum vertical temperature difference observed in 20 m was 1.4° C.

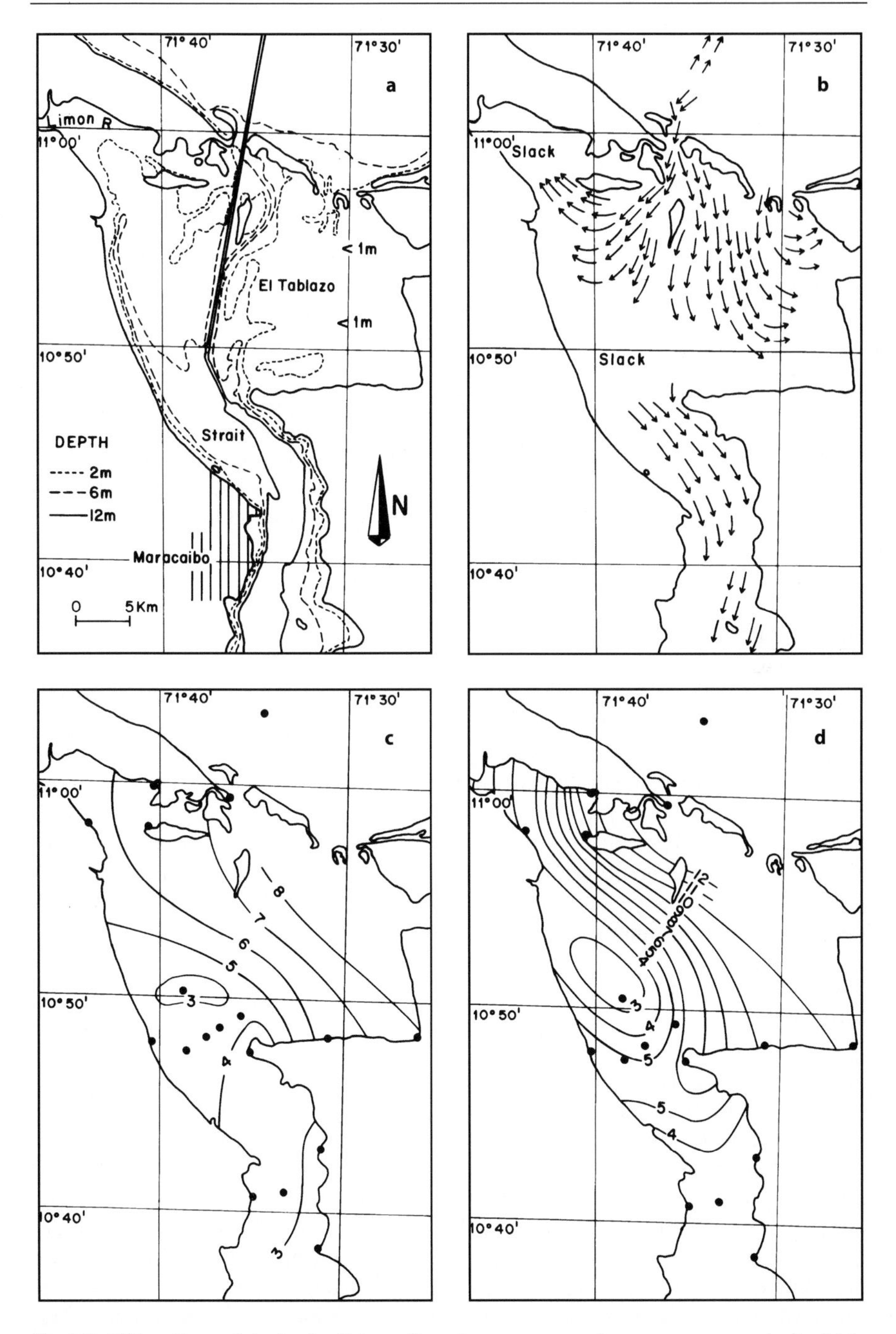

Fig. 2.3. Tablazo Bay and the Strait of Maracaibo: **a** bottom contour; **b** currents pattern 1 h after high-water time; **c** isohalines for January; **d** Isohalines for March (*a, c, d:* data from Rodríguez 1973; *b:* data from Brezina 1975)

The highest temperatures of the system are found in the Strait, reaching sometimes during the day 2° C above the mean temperature in the Gulf and 1–2° C above the temperature of the air (Rodríguez 1973). Along the navigation channel that traverses Tablazo Bay and the Strait (Fig. 2.3a) the temperature of the column of water decreases regularly with depth, and follows, although does not precisely parallel, the increment in salinity.

In the lake itself the extreme minimum mean temperature at 1 m depth is 29.0° C and the maximum 32.5° C, according to data by Escam (1991), which represents the mean of 6 stations covering all sectors of the lake for 2 years. Redfield and Doe (1964) found in the northern portion of the lake a mean temperature at 7 m depth of 30 ±1° C. The yearly temperatures follow the same regular seasonal pattern of other regions of the estuary, with a minimum in February and a maximum in September. The day-time temperature of the water is 1.5° C above the mean temperature of the air in Maracaibo. There is a diurnal increase in temperature up to 1° C at sunset which can reach 3 m depth, but it is not observable at 6 m (Redfield and Doe 1964). The structure and evolution of the thermal stratification in the lake were first described by Redfield et al. (1955) and Redfield and Doe (1964). The seasonal changes in the mixing of the hypolimnium are related to the general hydrography revealed by the chloride distributions and will be discussed under this topic.

2.4.2 Wind Regime

Wind regime within the basin changes throughout the year. From November to April trade winds from the north-east predominate and are replaced during the hottest period of the year – May to October – by local winds. These local winds are the result of the differential heating of the Lake and surrounding lands masses: During the day a greater amount of solar radiation is absorbed by the land; the adjacent air increases its temperature and ascends to the upper atmosphere while the colder and denser air above the lake is stationary resulting in a south-westerly wind. During the night a reverse process takes place and the wind changes direction. The cumulus frequently observed at sunset condense moisture at high altitudes.

2.4.3 Precipitation

The alternation in the predominance of trade and local winds originate a sharp division between rainy and dry seasons, which is typical of the arid, semiarid and central zones to the north where 90% of the rainfall occurs during May–October when a decrease in the trade winds take place (Cockroft 1961). The Gulf of Venezuela has annual means of 100 mm in the dry season and 200 mm during the rainy season. As we proceed southward the amount of rainfall increases and the distinction between dry and rainy season almost disappear. Large annual means of rainfall are found in two areas in the south: The Escalante River basin with 1 400 mm for both seasons, and the Santa Ana River with means of 1 000 and 1 800 mm for the dry and rainy seasons, respectively. The large precipitation over this last area is attributed to a quasi permanent low depression over the mouth of the Catatumbo River. The largest rainfall is found in a locality in the upper Catatumbo, with a peak annual mean of 3 500 mm.

The geomorphology of the Maracaibo basin determines a distinctive synoptic climatology for the whole basin. The trade winds which predominate during the dry season drive the air masses through the north-facing aperture of the basin and force them first in a south-westerly direction over the Sierra de Perijá and then in a south-easterly direction. At the same time the local winds over the lake cause a low pressure area with its center over the mouth of the Catatumbo River, as already mentioned, and creates a closed circulation pattern up to 2 000 m altitude. The air forced over the Sierra de Perijá originate, by convection, clouds of great vertical development which produce frequent rains between 500 and 1 500 m altitude. The low pressure over the lake produces subsidence of the upper air and a scarcity of rain over the lake but at the same time it forces down-valley currents of cold dry air from the Andes, creating a convergence over the southern area of the lake (Gol 1963). This convergence originates heavy rains through the whole year and is responsible for the large output of the Catatumbo and other rivers of the south-western coast of the lake.

2.4.4 Water Balance

The large input of fresh water from the rivers, together with more modest volumes of water from the direct precipitation over the lake, varies on a daily basis according to the seasons. The deficiencies are compensated by the influx of sea water with the tides, density currents and other phenomena. The first approximate water balance of the basin was performed by Carter (1955). He calculated gains of $54.1 \times 10^9\ m^3\ yr^{-1}$ and losses of $54.4 \times 10^9\ m^3\ yr^{-1}$. On a monthly basis he obtained negative values for February and March, compensated by an influx of sea water from the Gulf of Venezuela to the lake.

A more detailed analysis was performed by Corona (1964) using the equation

$$S = V_e + V_1 - E_1 \ ,$$

where V_e is river runoff; V_1 is precipitation over the lake and E_1 is evaporation from the lake. He obtained the the annual balance presented in Table 2.1 , including Tablazo Bay.

In this analysis all the monthly values are positive. The discharge from the mouth of the estuary according to these results varies between 492 and 2 695 $m^3\ s^{-1}$. Based on Carter (1955) analysis and on a balance of the Deuterium content in the waters, Friedman et al. (1956) deduced a residence time of 6 years and 10 months for the water in the lake. Recently the residence time has been more accurately calculated in 5.99 years based on precipitation data from 1958–1977 (Escam 1991).

Table 2.1. Annual balance ($\times 10^9\ m^3\ yr^{-1}$) of the basin

	Gains		Losses
Runoff (V_e)	51.1	To the sea (S)	49.1
Precipitation (V_1)	15.1	Evaporation (E_1)	17.1
Total	66.2		66.2

2.5 Hydrography

2.5.1 Tides

The tides in the outer Gulf of Venezuela, like in the rest of the Caribbean, are of the mixed type in which the diurnal component predominate. The diurnal range of the tide is 0.25 m. In the southern part of the Gulf the semidiurnal components are largely amplified by near-resonance, resulting in semidiurnal mixed tides with a range of 0.80 m (Pelegrí and Avila 1986). As the tidal wave enters Tablazo Bay it is considerable distorted by the dimensions of the bay, but retains its fundamental characteristics. Along the Strait the inequalities between successive tides diminishes and in the northern part of the lake the tidal wave regains its diurnal mixed type.

The analysis of records from tidal gauges located on a line between Tablazo Bay and the Strait (Rodríguez 1970) shows that tidal range diminishes progressively throughout the estuary, from 1.10 m at the entrance of Tablazo Bay, to 0.40 m at the city of Maracaibo. The curve of best fitting is an exponential of the form

$$A = Ne^{-kD} ,$$

where A is the tidal range in cm at a given point, N is the value in cm of A where $D = 0$, k is the damping coefficient in cm^{-1}, and D is the distance from the first tide gauge in cm. The values calculated for this equation are

$$A = 110\, e^{-2.95\times10^{-7} D} .$$

This equation is valid also in the lake as well. Differences in tidal range along the year are negligible. High water takes place in Maracaibo city approximately 1 h after high tide at the mouth of the estuary, 35 km distant, the wave travelling very slowly at the beginning (10 km in 40 min) (Febres 1966).

The tides in the Lake Maracaibo can be analysed according to the theory of tides in narrow embayments (Redfield 1961), as the addition of a primary wave entering from the sea and a secondary wave formed by the reflection of the first wave on a barrier in the southern-most end of the lake. These waves are present along the system in different positions and out of phase. The elevation ($Ç_1$) of the primary wave is

$$Ç_1 = A \cos(st - kx)\, e^{-mx} ,$$

and that of the reflected wave ($Ç_2$)

$$Ç_2 = A \cos(st + kx)\, e^{-mx} .$$

The elevation of the water surface is

$$Ç = Ç_1 + Ç_2 ,$$

where A is the amplitude of each of the waves at the barrier when $t = 0$ and where $x = 0$; s is the change in phase per unit of time; t is time measured from high-water at the barrier; k is the change in phase per unit of distance; x is distance measured from the barrier; m is the damping coefficient which expressed the attenuation of the wave due to frictional and other losses.

Nodes are formed where the waves have opposite phases at distances ¼, ¾, etc., wave lengths from the barrier, and antinodes where the two waves are in equal phases at ½, 1, etc., wave lengths from the barrier. A node for the semidiurnal constituents occurs in the northern part of the lake. An antinode for the semidiurnal constituents occurs in Tablazo Bay at latitude 10°54'N, and may be presumed to coincide with the node for the diurnal constituents.

At the barrier high-water and slack water are synchronous, the amplitude of elevation is maximal and that of current is zero. At the node amplitude of elevation is zero, that of current is maximal and that of high-water changes abruptly by ½ period, with the result that the elevation is rising beyond the node while it is falling between the node and the barrier. At the antinode the amplitude is again maximal and that of the current is zero. The time of slack changes abruptly by ½ period, with the result that high-water and slack water becomes synchronous beyond the antinode and occur ½ period before high-water at the barrier.

2.5.2 Water Levels

The monthly mean tidal levels change as much as 0.24 m in the outer Gulf of Venezuela (Redfield 1955), and by as much as 0.23 m in some of the tide gauges in Tablazo Bay and the Strait (Rodríguez 1973). The pattern is similar to that observed in other localities of the Venezuelan coast and the Caribbean in general (Patullo et al. 1955; Grivel 1979). The mean annual differences across the Gulf of Venezuela is 0.276 m in a distance of 45.6 km. The value for this difference between the Strait and the mouth of the estuary is 0.327 m in 35 km. The difference in level varies with the seasons in the Gulf, being highest (0.33 m) in January, and lowest (0.18 m) in October. Tablazo Bay and the Strait show similar variations with maximum and minimum differences in March (0.38 m) and September (0.248 m), respectively. These seasonal differences are clearly related to the prevalence of trade winds (Redfield 1955).

2.5.3 Currents

According to the water balance given above, there is a surplus volume of $49.1 \times 10^9 \, m^3 \, yr^{-1}$ of fresh water that must be discharged to the sea across the estuary. Up to 56.9% of the fresh water enters the lake on its south-western corner, through the Catatumbo River (Escam 1991). Due to the position of the delta of this river its output is discharged perpendicularly to the main axis of the lake, thus creating a south-east current (Fig. 2.4b). This current is forced by the coastline in a northward direction up to the middle of the lake where a portion of the surface waters moves forward through the Strait and the remnant is deflected to the west. Emery and Csanady (1973) observed similar counterclockwise circulations in 39 out of 40 lakes and marginal seas in the northern hemi-

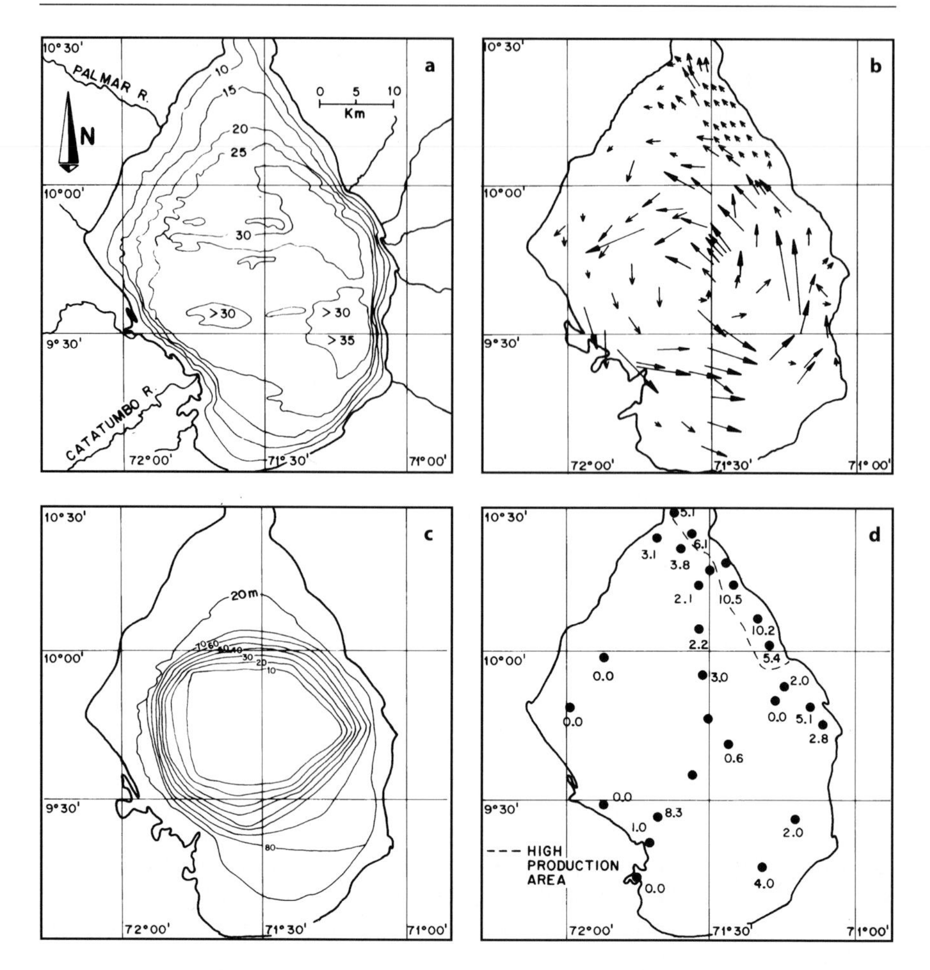

Fig. 2.4. Lake Maracaibo: **a** bottom contour; **b** surface currents; **c** oxygen isopleths at 20 m depth; **d** primary productivity (g C m^2 day^{-1}) (*a*, *b*, *c*: data from Parra Pardi 1979; *d*: data from Battelle Memorial Institute 1973)

sphere, including Lake Maracaibo. They attributed this pattern to the drag of the wind blowing across the bodies of water. In Lake Maracaibo this circulation is initiated by the position of the Catatumbo Delta and facilitated by the quasi-circular shape of the lake.

The velocity of the surface current can reach up to 0.50 m s^{-1}. Its direction does not change with depth, but its velocity is reduced. Towards the center of the lake the velocity decreases rapidly in the first 3 m and from 20 m up to the bottom. In the southern part of the lake this vertical decrement is very regular (Redfield et al. 1955).

There is a fluctuation of the force of the surface current in the northern part of the lake, which follows a semidiurnal cycle with a short inversion of the current direction during a part of the tidal cycle. Along the southern coast the velocity is not reduced but there is a short inversion of the direction which does not exactly match the tidal cycle. The behaviour of these currents, in phase with those in the Strait follows

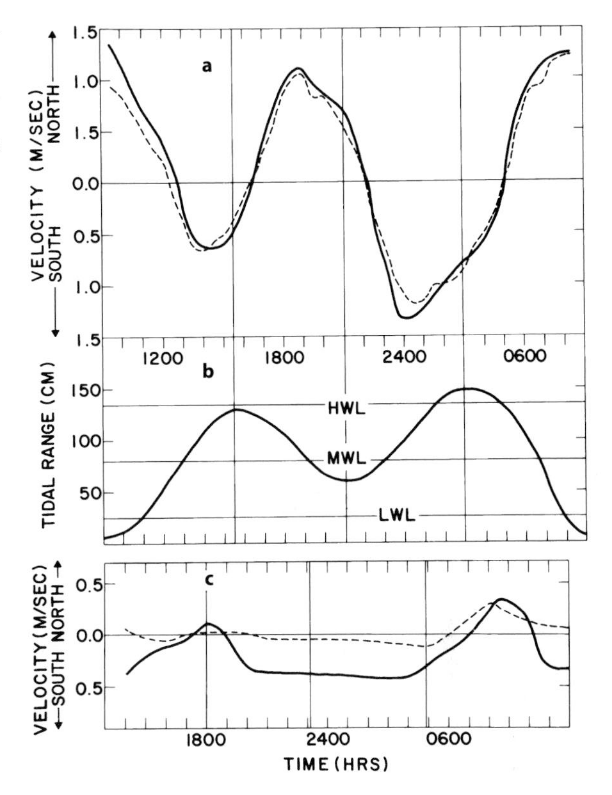

Fig. 2.5. a Direction and velocity of the current at the center of main outlet of the estuary (*solid line:* surface, *broken line:* bottom); **b** contemporary tidal record; **c** velocity and direction of the current near the shore (after Rodríguez 1973)

Redfield's (1961) theory, according to which slack water time must be simultaneous in all points south of Tablazo Bay.

In the Strait the water which escapes from the epilimnium moves northward, forming a discharge of 500 $m^3 s^{-1}$ in the dry season and up to 2 600 $m^3 s^{-1}$ during the rainy season (Corona 1964). The movement of masses of water in the Strait and Tablazo Bay are the result of the following agents: hydraulic forces resulting from differences in the seasonal mean sea level and mean lake level; tidal forces which alternate in direction; gravity forces arising from differences in the distribution of mass within the estuary, for instance those produced by differences in salinity which result in density currents; wind stress acting on the water surface; and frictional resistance on the bottom and on adjacent layers of water.

The interaction of these forces results in a periodical change of direction in the water escaping from the lake. An example is given in Fig. 2.5 from measurements taken in San Carlos Island, at the main outlet of the estuary (Febres 1968). There is a time lag of 5.5 h between high-water time and the time of maximal velocity of current. Measurements taken near the shore in this same area (Fig. 2.5b) show that the water flow southward for almost 24 h, with velocities far smaller that at the center of the channel. This phenomenon have important biological consequences for the immigration of small organisms (e.g. invertebrate larvae) into the estuary, which can penetrate against a net seaward flow by crawling or swimming along the shore.

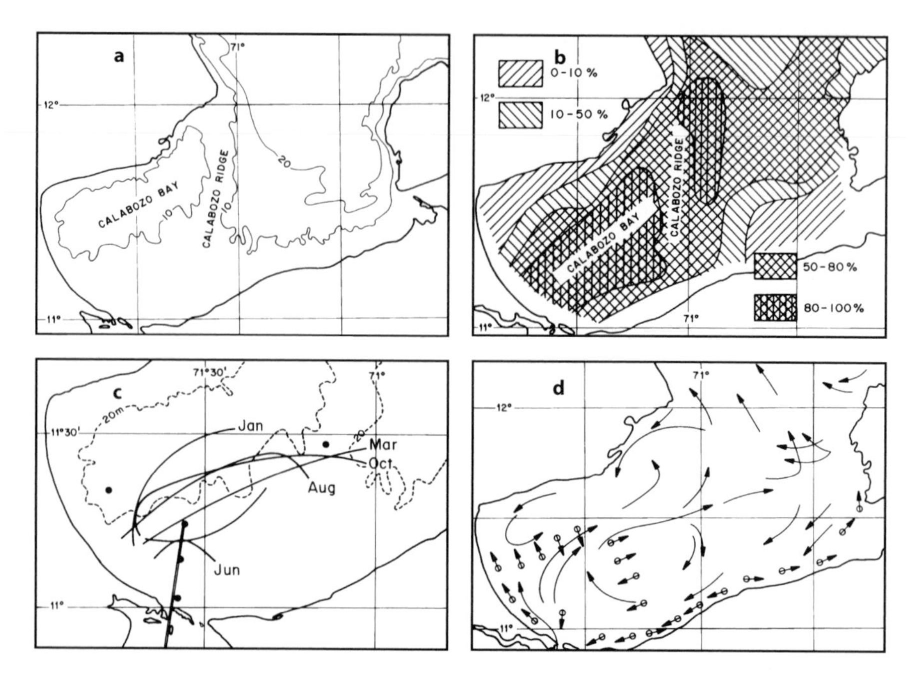

Fig. 2.6. Gulf of Venezuela: **a** bottom contour; **b** silt and clay fraction in the sediments; **c** 30‰ isohalines at 6–7 m depth, inner gulf; **d** surface currents (*a, c:* data from Rodríguez 1973; *b:* data from Ziegler 1964 and Ziegler and Perez Mena 1960; *d:* data from Ziegler 1964 (simple arrows) and Paz Castillo, unpublished (circles))

The density currents caused by the distribution of salinities in the Strait were calculated by Febres (1968), assuming a condition of equilibrium for both, pressure and frictional forces, by means of the following equation:

$$u^2 = \frac{1}{8}\frac{gh^2}{kL}\frac{p_1 - p_2}{p} \quad ,$$

where u^2 is velocity of density currents in a longitudinal direction in the Strait, g is gravity, h is height of water surface, k is the Taylor's damping coefficient, L is the length of the Strait, p_1 and p_2 are the mean densities of two vertical water columns at both ends of the Strait when $p_1 > p_2$, and p is the mean density. The value that this author found for the density current, $u^2 = 0.06$ m s^{-1}, is small as compared with the velocities normally found in the estuary and would tend to increase the velocities of the incoming flow.

The interplay of forces in Tablazo Bay and the Strait creates a rather complicated pattern of circulation in this area (Fig. 2.3b). This has been rationalized by means of a fixed-bed physical model of the lake, the Strait, Tablazo Bay and the southern part of the Gulf of Venezuela (Brezina 1975; Parra Pardi 1977). The model was built at a horizontal scale of 1 : 6 000 and a vertical scale of 1 : 100 (distortion 60 times). It is possibly to discriminate 15 different areas from the Strait to the southern part of the Gulf of Venezuela, with typical patterns of currents, although altered in some occasions by the changes in the tidal wave. These currents are not produced by the ongoing tidal cycle, but are affected by at least the six preceding tidal cycles.

The pattern of circulation in the Gulf of Venezuela (Fig. 2.6d) is due to the winds, the presence of the Calabozo Ridge and the output of water from Tablazo Bay. Coriolis forces seem to have little effect. Redfield (1955) postulated the existence of two estuarine cells, separated from each other by the Calabozo Ridge: The low salinity water from the lake occupies a band on the southern portion of the gulf, and flows north-westerly to the center of the gulf where it merges with the Caribbean water, with a cyclonic circulation on the west and an anticyclonic circulation on the south-east. The effect of this pattern of circulation on the discharge from Tablazo Bay is partly reflected in the distribution of fine-grained sediments in the Gulf of Venezuela (Fig. 2.6b). An upwelling of deep water has been reported on the north-western section of the gulf, where high salinity and low temperature waters flow to the north and surfaces near the Paraguaná Peninsula (Ziegler 1964; Ginés 1982).

2.6 Salinity

2.6.1 Salinity Patterns

The distribution of surface salinities in the Strait, Tablazo Bay and southern part of the gulf throughout the year follows a marked seasonal pattern (Fig. 2.3c,d) with maximal values from February to May and minimal from June to August. The low salinity waters within Tablazo Bay are markedly deflected to the left as a result of the gulf waters that, driven by the wind through the smaller mouth of the estuary, create a zone of high salinities, usually in the range of 20–25 and rarely below 6. At the same time, the output of the Limón River on the north-western corner of Tablazo Bay creates a smaller estuary of its own within the larger estuary; in localities 16 km up-river the salinity can reach up to 8 in the dry season and decrease at the river mouth as low as 0.1 in the wet season. In the Strait the surface salinities are more constant, with fluctuations from 2–6 for most of the year. Throughout the navigation channel that crosses Tablazo Bay and the Strait there is a typical estuarine stratification related to the saline wedge that enters from the gulf (Rodríguez 1973).

The effect of the discharge of the estuary can be observed up to the center of the gulf, 12°N and 71°W, where salinities of 33–34 can be found (Ziegler 1964), although it is in the southern part where this effect is most strongly felt. Figure 2.6c shows the position of the isohaline of 30 at 6–7 m depth in the southern portion the gulf for January, March, June, August and October (Rodríguez 1973). A marked stratification can be observed in Calabozo Bay, with vertical differences of 2 in 12 m, but on the outer gulf the differences are very small. Ziegler (1964) found at 12°N 36.550 and 36.735 for the surface and 60 m depth, respectively.

Tides produce cyclical alterations on the local distributions of salinities in the center of Tablazo Bay where there is a semidiurnal fluctuation at all levels up to 1.5 corresponding to velocity fluctuations in the current.

The access of sea water to the lake through the Strait balances the deficit of freshwater from the large rivers to the south. As already mentioned, the Catatumbo River is responsible for 56.9% of the freshwater that enters the lake (Escam 1991). The salinity, which at the mouth of this river is 0.02 increases rapidly with the distance (Gessner 1956):

Distance from the mouth (km)	0.0	2.4	4.8	7.2
Salinity (*S*)	0.33	0.62	1.03	1.33

However, the low salinity water is limited to a thin layer (<1 m) of the effluent plume which overrides lake water for a distance of 1 km. Below this layer the salinity increases sharply to uniform values in the range of 2–3 (Hyne et al. 1979).

Towards the end of the dry season, brackish water penetrate through the Strait to the deeper portions of the lake. This water contribute to the sharp vertical stratification of the lake which forms a cone-shaped hypolimnium located towards the center of the lake, surrounded by lower density water. The water of the epilimnium moves counter-clockwise around this limnetic cone like a large-scale vortex, with velocities up to 0.55 m s^{-1}. This cyclonic circulation moves water by erosion from the hypolimnium to the epilimnium, where it becomes completely mixed. The volume of the cone, which can reach ¼ of the total volume of the lake is variable, expanding and contracting in a seasonal pattern, according to the variable inputs of brackish water from the Strait or freshwater from the rivers. The saline cone is a permanent feature of the lake and rarely it has been observed to disappear (Redfield and Doe 1964; Parra Pardi 1983)

The history of salinities in the lake is particularly interesting in order to determine whether the navigation channel which was dredged to 14 m in 1953, has produced a salinization of the lake (Rodríguez 1984). The extant records, available since 1937, show that since this year up to 1959 the mean salinity was 1.2 with extreme values of 0.7 and 2.0. Since 1959 and up to 1990 the mean salinity experienced a modest but consistent increase to a mean of 3.1 with extreme values of 2.3 and 4.2 (Parra Pardi 1986; Escam 1991).

Superimposed to the long range increment in salinities, there are also fluctuations in the yearly values, with peak salinities in the years 1939, 1964, 1974, 1979, and 1983. Rainfall records for 35 years (1911–1946) in the city of Maracaibo (Rodríguez 1973), evaluated through periodogram analysis, a method devised to detect rhythmicities in temporal series, show that maxima of precipitation occur with a periodicity of 6 years. A similar result was found in the analysis of the discharge of the Limón River, on the west coast of the lake. It is possible that the fluctuations of salinities in the lake are the result of the periodicities of rainfall over the basin, distorted by the inertia in the large body of water of the lake.

2.6.2 Salt Intrusion

In partially mixed estuaries, like the Maracaibo Estuary, the tidal movements produce a cyclic oscillation and a periodic mixing which result in vertical and longitudinal gradients of salinity. This intrusion, together with the tides, seems to be one of the main factors in the silting of the navigation channel and for this reason it was studied by several authors (Partheniades 1966; Partheniades et al. 1966; Harleman et al. 1966,

Harleman et al. 1967; Cadena 1968). The equation of mass conservation for the salt in unidimensional analysis is

$$\frac{dS}{dt} + u\frac{dS}{dx} = \frac{1}{A}\frac{d}{dS}\left(AD_x\frac{dS}{dx}\right) \quad ,$$

where S and u are the mean salinity and velocity in the cross section of the channel, A is the area of the cross section, and D_x the apparent coefficient of longitudinal diffusion; the velocity of the fluid is the addition of the velocity of the tide (u) and the velocity of the freshwater (U). Introducing certain border conditions they obtained

$$\ln\frac{S_{lws} - S_L}{S_0 - S_L} = \frac{U}{2D_0 B}(x + B)^2 \quad ,$$

where S_L, S_o and S_{lws} are the salinities in the lake, ocean and channel during low-water slack (lws); D_o and B are two characteristic parameters of the longitudinal salinity distribution, the apparent diffusion coefficient in the ocean and the distance to a point in the negative direction of x where the salinity is equal to the oceanic. Both parameters may be determined if the salinity of the low-water slack is known at least at two points.

From these equations it is possible to determine the maximum and minimum intrusion. The salinity distribution parameter is related to a non-dimensional parameter (Estuary Number) which is a measure of the degree of mixing and depends on the tidal prism, Froude Number, freshwater discharge and tidal period. The values for these parameters were determined by Cadena (1968) from field measurements.

2.6.3 Salt Balance of the Lake

The relationship between rainfall and salinity in the lake has been studied by Redfield and Doe (1964) by means of a simplified model of a lake of uniform salinity connected to a constant salinity ocean through a rectangular channel. Beginning with an equation which relates the net rate $F(t)$ at which salt is carried out of the lake through unit area of any section of the channel by the combined effect of advection and diffusion, they established the boundary conditions for distance and salinity, and obtained the following equation for the change in the chlorinity of the lake, at a given moment, during a finite period:

$$\Delta C \cong \left(\frac{R}{V}\right)\frac{C^0 - \overline{C}e^{\frac{RL}{KWH}}}{e^{\frac{RL}{KWH}} - 1} \quad ,$$

where R is the outflow from the lake during a finite period; V is the volume of the lake; L is the length of the channel; K is the coefficient of tidal diffusivity; W and H are the depth and width of the channel respectively; C^0 is the salinity of the sea; C the mean salinity of the lake during the period; is the mean value of $C(t)$ during the period; and $C(t)$ is the salinity of the lake at time t.

The salinity of the lake at any time may be obtained by numerical integration according to the relation

$$C(t) \cong C(t{=}0) + \sum \Delta C \quad .$$

This model was tested using the values of Carter's (1955) hydraulic balance and other. The salinities obtained closely correspond with the measured salinities, notwithstanding the quality of the data and the simplifications of the model.

2.7 Chemistry and Fertility of the Water

2.7.1 Dissolved Oxygen

The dissolved oxygen was carefully sampled by Parra Pardi (1977) in five sections throughout the Strait, during a study of the self-purification of the estuary. Very small variations were observed in the tridimensional pattern thus obtained, with differences rarely above 1.5 ml l^{-1}. Vertically the water is frequently super-saturated on the surface due to local algal blooms, but the oxygen content drops regularly as low as 40% saturation at 10 m, and 30% at 14 m, although on no occasion an anoxic condition was observed. These values fluctuate in 24 hour cycles by as much as 1 ml l^{-1}, but these fluctuations are not clearly correlated with the tidal records. Along the channel, the oxygen content decreases as a consequence of the discharge of sewage and industrial wastes from populated areas, but more normal values were found 18 km north of these areas.

The oxygen content in Tablazo Bay varies widely with the local conditions, but no anoxic areas has been observed. In mangrove areas of this zone Rodríguez (1973) recorded 98.2% saturation.

Anoxic conditions were reported by Gessner (1953), Redfield et al. (1955), Redfield and Doe (1964), and Parra Pardi (1979) at the bottom of the lake, corresponding with the more saline water of the cone-shaped hypolimnium, when the water is completely stratified (Fig. 2.3c). The epilimnium display only moderate variations in the oxygen content. The mean of 6 different locations throughout the lake at 1 m depth, for the years 1982–1984, was 5.14 ml l^{-1} (96% saturation), with extreme values of 3.5 and 6.3 ml l^{-1} (Escam 1991). The lowest values over the cone were observed in December, when the access of saline water is restricted, the base of the cone is reduced and the apex approaches the surface. When the process is completed, the volume of the hypolimnium is considerable smaller, and in some years the water is completely mixed, as was observed by Redfield and Doe (1964) for November 1955 and December 1956.

2.7.2 Nutrients

From the time that Gessner (1953) found in his analysis of the waters of the lake up to 1.3 mg-atoms l^{-1} (40 mg m^{-3}) of phosphate in the surface and 7.1 mg-atoms l^{-1} (220 mg m^{-3})

at 25 m, all authors that have studied the nutrients content in the Maracaibo system have reported similar high concentrations of phosphorus and similar increments with depth. Redfield and Doe (1964) recorded values of about 1–4 mg-atoms l^{-1} of inorganic phosphorous in the epilimnium, and more than 7 mg-atoms l^{-1} in the hypolimnium. Variations throughout the year are considerable. In a study of phosphate in the saline cone, Parra Pardi (1979) found the lowest values, 1.0 mg-atom l^{-1} for the surface and 4.0 mg-atoms l^{-1} for the bottom, in September, and the highest, 1.5 mg-atom l^{-1} for the surface and 15.0 mg-atoms l^{-1} for the bottom, in October. Battelle Memorial Institute (1974) reported a peak value for the bottom at 31 m of 151 mg-atoms l^{-1} (4.9 mg l^{-1}) in January 1973, when the surface water had 2.4 mg-atoms l^{-1} (0.05 mg l^{-1}). The maximum value registered during 1982–1984, at 1 m below the surface, from six different locations throughout the lake, was 10.32 mg-atoms l^{-1} (Escam 1991).

Since the content of phosphorus in the Gulf of Venezuela at the surface is only 0.2–0.4 mg-atoms l^{-1}, and 1.0 mg-atom l^{-1} at the bottom (Redfield 1955), the phosphate in the lake must come necessarily from terrestrial sources. In this regard, Hobson (1979) found that in the rivers of the southwestern part of the Lake, phosphate concentrations in the "white water" of the Santa Ana River ranged from 0.93–2.17 mg-atoms l^{-1}, and averaged 1.5 mg-atom l^{-1} and the "black water" of the Concepción River (with a high concentration of lignin), ranged from 0.93–2.48 mg-atoms l^{-1}, and averaged 1.5 mg-atom l^{-1}; in the Lake itself the concentration ranged from 1.86–6.20 mg-atoms l^{-1}, and averaged 5.27 mg-atoms l^{-1}. One of the sources of phosphate are the mines located near the headwaters of the Catatumbo River and the recently discovered phosphate rocks which cover a large extension in the cordillera south and south-west of the lake (Parra Pardi 1986). On the other hand the increment in phosphorous with depth is attributed to the sinking of organisms from the surface (Redfield et al. 1955) and the slow outward circulation of the cone-shaped hypolimnium. López-Hernández et al. (1980) have suggested that the retention of phosphorus by adsorption in the sediments of the lake contribute to the control of phosphorus in the water.

The main sources of nitrogen compounds in the lake are the sewage and industrial discharges in the vicinity of the city of Maracaibo. This nutrient-rich water is incorporated into the saline wedge that penetrates into the lake through the Strait. The ammonia in the lake follows the same pattern of distribution of the phosphorous once it is incorporated to the cone-shaped epilimnium (Battelle Memorial Institute 1974). In a typical distribution the ammonia does not exceed 6 mg-atoms l^{-1} up to a depth of 10 m throughout the lake, but increases rapidly to reach more than 180 mg-atoms l^{-1} at the base of the cone (Parra Pardi 1979).

Parra Pardi (1979) found the following values for the N total/P relationship: tributary rivers 1–5; sewage 4.5–7.5; industrial discharges 2–10. The mean value obtained for the epilimnium was less than 5. The mean value in the hypolimnium has been estimated in 5, with occasional mean values of over 15. When $N_{total} / P < 5$, the nitrogen is considered as the limiting factor: this can be considered the condition for most of the lake waters. The nitrogen balance in the lake has been established by Delft Hydraulics and Haskoning (1991) as described in Table 2.2.

The primary production of the system is high (Battelle Memorial Institute 1974; Rodríguez and Conde 1989), but its distribution is asymmetrical in the lake, with the eastern side considerable more productive. This is particularly true for a segment of coastal waters between 9°40'N and 10°30'N where the largest concentration of oil pro-

Table 2.2. Nitrogen balance in the lake

	Amount of nitrogen ($\times 10^6$ kg d^{-1})
Import from rivers	0.21
Import through the Strait	0.1
Outflow to the sea	0.07 – 0.4
Storage	0.09 – 0.16

duction occurs. In this sector Battelle Memorial Institute (1974) found at one station a mean of 4.38 g C m^{-2} d^{-1}, but occasionally values over 10 g C m^{-2} d^{-1} were found in two localities within this area. The high productivity is the result of the large algal blooms, with peak values of 18 mg l^{-1} in chlorophyll content (Escam 1991). These algal blooms have also been document through the analysis of remote sensing data using Landsat satellite imagery (Fundación Instituto de Ingeniería 1991).

From the data on productivity and chlorophyll content, Delft Hydraulics and Haskoning (1991) has calculated the total nitrogen requirement for primary production to be 5×10^6 kg N d^{-1}, and the mineralization 1.1×10^6 kg N d^{-1}. According to these values the import of nitrogen given above is largely insufficient for the nitrogen metabolism in the lake, and a self supporting mechanism is postulated as the origin of a nitrogen pool in the lake.

2.8 Conclusions

The Maracaibo system consists of three interactive bodies of water: an oligohaline lake, a deep strait and a shallow semienclosed bay. The final discharge to the adjacent sea occurs through a narrow mouth, at present dredged to 14 m. Since this discharge influences the pattern of currents in the Gulf, the brackish water in the whole system extends for more than 220 km. The hydrography of this system is constrained by its geomorphology: An extensive intermontane north-facing valley, open to the trade winds. This setting causes the moisture to be displaced southward over the slopes of the sierras, producing a heavy rainfall which drains on one corner of the lake. The force and position of the streams set in movement a counter-clockwise current, allowing only a fraction of the circulating water to escape as a thin upper layer through the sharply stratified Strait, while the salinity in the hypolimnium is increased by the saline wedge that enters the lake through the action of the semidiurnal tides and the seasonal changes in the hydraulic balance. The oligohaline water from the lake spreads into the shallow semi-enclosed Tablazo Bay where its salinity is increased by the mixing processes caused by the tides. The final discharge into the water through a narrow mouth establish a two-cell estuarine circulation in the Gulf of Venezuela.

The circular current in the lake creates a cone-shaped hypolimnium that acts as a trap for the nutrients. This nutrient pool account for the large productivity of the lake which is considered as one of the most productive areas in the world. Management and pollution control has to take into account the peculiar dynamics of this large water reservoir.

References

Battelle Memorial Institute (1974) Study of effects of oil discharges and domestic and industrial wastewaters on the fisheries of Lake Maracaibo, Venezuela, V. 1, Battelle Memorial Institute, Pacific Nothwest Laboratory, Richland, Washington

Brezina J (1975) Experiences with a small scale, highly distorted fixed bed model of the Lago de Maracaibo Estuary. Proc. Symposium on Modeling Techniques, Am. Soc. Electric Eng., San Francisco, California

Cadena RE (1968) One dimensional analysis of salinity and sediments of the Maracaibo channel. M.Sc. thesis, Massachusetts Institute of Technology, Cambridge

Carter DB (1955) The water balance of Lake Maracaibo basin. Publications in Climatology Drexel Institute of Technology 8:209–227

Cockroft JS (1961) Weather Manual. Creole Petroleum Corporation, Maracaibo

Conde JE, Rodríguez G (1999) Integrated coastal zone management in Venezuela: The Maracaibo System. In: Salomons W, Turner K, Lacerda LD de, Ramachandran R (eds) Perspectives on integrated coastal zone management. Springer-Verlag, Berlin

Corona LF (1964) Balance hidrológico del Lago de Maracaibo. Mimeo Report, 2 vols, Instituto Nacional de Canalizaciones, Caracas

Delft Hydraulics and Haskoning (1991) Los recursos hídricos del Lago de Maracaibo. Anexos, Caracas

Emery KO, Csanady GT (1973) Surface circulation of lakes and nearly land-locked seas. Proc Nat Acad Sci 70:93–97

Escam (1991) Plan maestro para el control y manejo de la cuenca del Lago de Maracaibo. vol 1 Mimeo Report, Caracas

Febres G (1968) Relaciones entre corrientes y mareas en las bocas San Carlos y Cañonera del Estuario de Maracaibo. Instituto Nacional de Canalizaciones, Publicación Técnica DI 2 5:21–78

Febres G (1996) Hidrografía. Instituto Nacional de Canalizaciones, Publicación Técnica DI-2 5:1–20

Friedman ID, Norton R, Carter DB, Redfield AC (1956) The deuterium balance of Lake Maracaibo. Limnol Oceanogr 1:239–246

Fundación Instituto de Ingeniería (1991) Estudio de contaminantes en el Lago de Maracaibo a través de la técnica de sensores remotos. Caracas

Gessner F (1953) Investigaciones hidrográficas en el Lago de Maracaibo. Acta Científica Venezolana 4:173–177

Gessner F (1956) Das Plankton des Lago Maracaibo. In: Gessner F, Vareschi V (eds) Ergebnisse der Deutschen Limnologischen Venezuela-Expedition 1952. 1:66–92

Ginés H (1982) Carta Pesquera de Venezuela. vol 2 Areas Central y Occidental. Fundación La Salle de Ciencias Naturales, Caracas

Gol AW (1963) Las causas metereológicas de las lluvias de extraordinaria magnitud en Venezuela. Servicio de Metereología y Comunicaciones Ministerio de Defensa Pub Especial N°2 Caracas

Gonzalez de Juana CJ, Iturralde M, Picard C (1980) Geología de Venezuela y de sus cuencas petrolíferas. Ediciones FONINVES, Caracas

Graf CH (1969) Estratigrafía cuaternaria del noroeste de Venezuela. Bol Inf Asoc Venezolana Geol Min Petrol 12:393–416

Graf K (1972) Relaciones entre tectonismo y sedimentación en el Holoceno del Noroeste de Venezuela. Memoria del IV Congreso Geológico Venezolano pp 1125–1144

Grivel F (1979) Variaciones del nivel medio del mar en los puertos del Golfo de Mexico y Mar Caribe. Universidad Nacional Autonoma de Mexico Datos Geofísicos A(5):1–312

Harleman DRF, Corona LF, Partheniades E (1966) Análisis de la distribución de la salinidad en el Canal de Maracaibo. Instituto Nacional de Canalizaciones Publicación Técnica DI-2 2:265–314

Harleman DRF, Corona LF, Partheniades E (1967) Analysis of salinity distribution in the straits of Maracaibo. XII Congress of the International Association for Hydraulic Research, Instituto Nacional de Canalizaciones Publicación Técnica DI-2 3A:149–160

Hobson MM (1979) A preliminary geochemical study of Lake Maracaibo, Venezuela. Ph D thesis University of Tulsa, Tulsa, Oklahoma

Hyne NJ, Dickey PA (1977) El delta contemporáneo del Río Catatumbo, Lago de Maracaibo, un modelo para explicar antiguos deltas intermontañas. Memorias del V Congreso Geológico Venezolano 1:327–337

Hyne NJ, Laidig LW, Cooper WA (1979) Prodelta sedimentation of a lacustrine delta by clay mineral flocculation. J Sediment Petrol 49:1209–1216

López-Hernández D, Herrera T, Rotondo F (1980) Phosphate adsorption and desorption in a tropical estuary (Maracaibo system). Mar Environm Res 4:153–163

Parra Pardi G (1977) Estudio integral sobre la contaminación del Lago de Maracaibo y sus afluentes: Parte I: Estrecho de Maracaibo y Bahia El Tablazo. Ministerio del Ambiente y de los Recursos Naturales Renovables Caracas
Parra Pardi G (1979) Estudio integral sobre la contaminación del Lago de Maracaibo y sus afluentes: Parte II: Evaluación del proceso de eutroficación. Ministerio del Ambiente y de los Recursos Naturales Renovables Caracas
Parra Pardi G (1983) Cone shaped hypolimnium and local reactor as outstanding features in eutrophication of Lake Maracaibo. J Great Lakes Res 9:439–451
Parra Pardi G (1986) La conservación del Lago de Maracaibo. Lagoven, Caracas
Partheniades E (1966) Field investigations to determine sediment sources and salinity intrusion into the Maracaibo Estuary, Venezuela. MIT Report N° 94 RCC-24
Partheniades E, Kennedy JF, Corona LF (1966) The interaction of tides, salinity and sediment in the Lake Maracaibo Estuary in Venezuela. Conference on Structural Engineering American Society of Civil Engineers, Miami, Florida
Patullo J, Munk W, Revelle R, Armstrong E (1955) The seasonal oscillation in the sea level. J Mar Res 14:88–115
Pelegrí JL, Avila RG (1986) Las mareas como sistemas cooscilantes en los Golfos de Venezuela y Paria. Revista Técnica INTEVEP 6:3–5
Redfield AC (1955) The hydrography of the Gulf Venezuela. Deep-Sea Res Sup 3:115–113
Redfield AC (1961) The tidal system of Lake Maracaibo, Venezuela. Limnol Oceanogr 6:1–12
Redfield AC, Doe LAE (1964) Lake Maracaibo. Verh Internat Verein Limnol 15:100–111
Redfield AC, Ketchum BH, Bumpus DF (1955) Report to Creole Petroleum Corporation on the hydrography of Lake Maracaibo, Venezuela. Unpublished manuscript Reference N° 55-9 Woods Hole Oceanographic Institution, Woods Hole Mass, 152 pp
Rodríguez G (1970) Exposure to air of the intertidal zone in the Maracaibo Estuary. Proceedings Second International Congress on Marine Corrosion and Fouling, Athens, pp 245–250
Rodríguez G (1973) El sistema de Maracaibo, biología y ambiente. Instituto Venezolano de Investigaciones Científicas, Caracas
Rodríguez G (1984) Ecological control of engineering works in the Maracaibo system. Water Sci Technol 16:417–424
Rodríguez G, Conde JE (1989) Producción primaria en dos estuarios tropicales de la costa caribeña de Venezuela. Rev Biol Trop 37:213–216
Sarmiento R, Kirby RA (1962) Recent sediments of Lake Maracaibo. J Sediment Petrol 32:75–90
Sheppard FP (1961) Sea level rise during the past 20.000 years. In: Russell RJ (ed) Pacific Island Terraces: Eustatic. Zeit f Geomorph Suppl 3:30–35
Sutton FA (1946) Geology of Maracaibo Basin, Venezuela. Bull Amer Assoc Petrol Geol 30:1621–1741
Ziegler JM (1964) The hydrography and sediments of the Gulf of Venezuela. Limnol Oceanogr 9:397–411
Ziegler JM, Perez Mena R (1960) Distribución de Sedimentos en el Golfo de Venezuela. Bol Geología Publicación Especial 3:895–904
Zuloaga G (1950) World geography of petroleum: Venezuela. Am Geog Soc Spec Pub 31:49–79

Chapter 3

Coastal Lagoons of Southeastern Brazil: Physical and Biogeochemical Characteristics

Bastiaan Knoppers · Björn Kjerfve

3.1 Introduction

Coastal lagoons are one of six categories of inland coastal ocean-connected systems (Kjerfve 1994). Coastal lagoons are shallow water bodies, with depths usually less than 5 m, oriented parallel to the shoreline, separated from the ocean by a barrier, and connected to the ocean by one or more inlets (Phleger 1969). According to their geomorphology and water exchange with the coastal ocean, they are conveniently subdivided into choked, restricted, and leaky lagoon systems (Kjerfve 1986) (Fig. 3.1). Choked lagoons have the longest and leaky lagoons the shortest water residence times (Kjerfve and Magill 1989). Tidal mixing may be negligible or intense, and salinity may vary from that of a fresh water coastal lake to a hypersaline lagoon.

Coastal lagoons are most common along microtidal, high wave energy coasts (Kjerfve and Magill 1989), where the coastal plain has been subjected to submergence during Holocene sea-level rise, and alternatively flooded and exposed by sea-level fluctuations during the late-Quaternary period (Nichols and Allen 1981). The age of present-day lagoons is 5 000–6 000 years, but varies in accordance to the local amplitudes of marine transgression and regression phases, tidal range, shoreface dynamics, availability of coastal sandy sediments, fluvial transport, and climate (Bird 1994; Martin and Domínguez 1994).

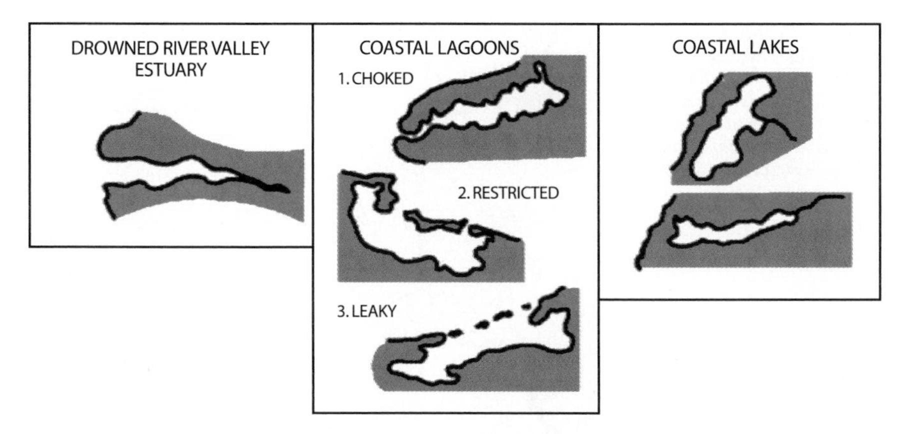

Fig. 3.1. Schematic diagram of drowned river valley estuaries, three types of coastal lagoons (after Kjerfve 1986), and coastal lakes

Coastal lagoons border more than 13%, or 32 000 km, of the world's continental coasts, with lagoon coasts in South America occupying 12.2% (Cromwell 1971). Lagoons play a substantial role in the transport, modification, and accumulation of matter at the land-sea interface (Kjerfve and Magill 1989; Knoppers 1994). Allochthonous matter introduced by the atmosphere, rivers, ground water, the sea, and bordering vegetation, is efficiently retained and recycled as a result of the lagoon enclosure, restricted tidal flushing, and close coupling between pelagic and benthic compartments. Primary production rates usually range from 200–500 g C $m^{-2} yr^{-1}$ (Nixon 1982; Knoppers 1994), and the global contribution of coastal lagoon primary production reaches 10^{11} kg C yr^{-1}, which is approximately the same as the production rate of global coastal upwelling (Knoppers 1994). Fishery yields are commonly 50–120 t $ha^{-1} yr^{-1}$, and a variety of commercially valuable species breed in coastal lagoons (Nixon 1982; Pauly and Yáñez-Arancibia 1994). At present, the symptoms of demographic expansion, such as material waste disposal, cultural eutrophication, engineering works, and uncontrolled management of watersheds, are adversely affecting coastal lagoons (Bruun 1994), including those of Brazil. It is our objective to characterize some of the salient physical and biogeochemical features of the major and most intensely studied coastal lagoons of southeastern and southern Brazil.

3.2
Coastal Lagoons in Brazil

The Brazilian coast stretches for 7 700 km from 4°N to 34°S latitude. It can be divided into four geological sections (Guerra 1962):

1. the northern Quaternary coast (4°N to 3°S),
2. the eastern Tertiary coast (3°S to 20°S),
3. the southern Granitic coast (20°S to 29°S), and
4. the southern Quaternary coast (29°S to 34°S) (Fig. 3.2).

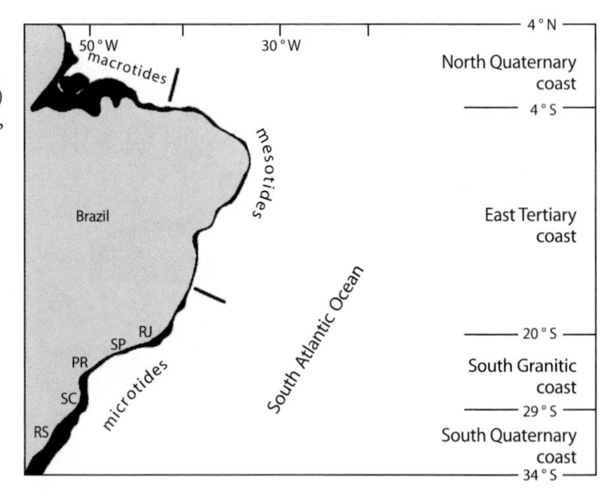

Fig. 3.2. The four main divisions of the Brazil coast (north Quaternary, east Tertiary, south granitic, and south Quaternary) and distribution of macro, meso, and micro tides; *dark shading* indicate Quaternary deposits

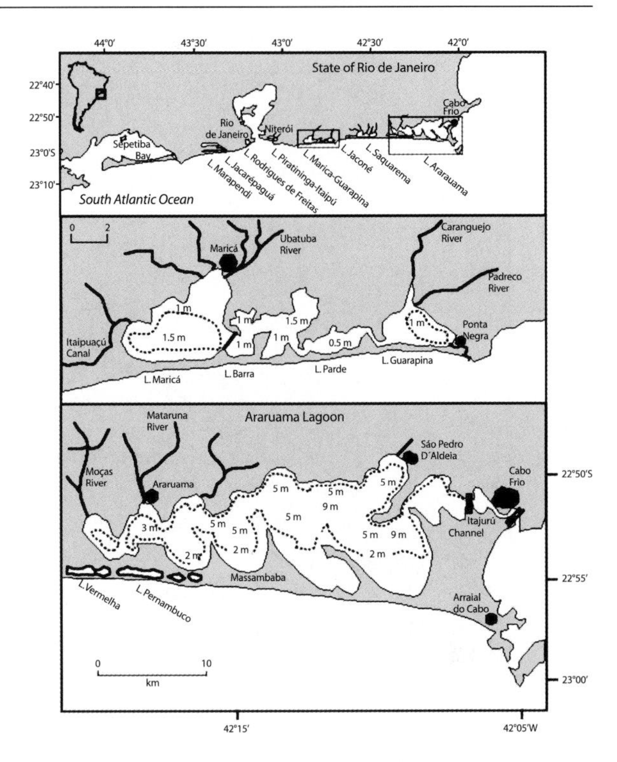

Fig. 3.3. The Fluminense Lagoon coast of the state of Río de Janeiro

Apart from the northern Quaternary coast, 79% of the coast is characterized by a narrow littoral fringe, frequently only a few kilometres wide, occupied by dry coastal ecosystems with a tropical and sub-tropical climate (Lacerda et al. 1993). Wider Quaternary coastal plains are embedded along the northern section of the granitic coast (20°S to 23°S) and the southern Quaternary coast. Whereas coastal lagoons in other parts of the tropics commonly are bordered by extensive mangrove systems (Yáñez-Arancibia 1987), the coastal lagoons of southeastern Brazil are different because they generally lack mangroves. The exception is the Cananéia-Iguape lagoon system in São Paulo, which is bordered by 52 km^2 of mangroves (Herz 1991).

Most lagoons in Brazil are located along the 3 000 km long, largely microtidal (tidal range < 1 m) coast from the delta of Paraíba do Sul River to the Uruguay border, corresponding to the southern granitic and Quaternary coasts (Diegues et al. 1992) (Fig. 3.2). The principal lagoons are shown for the coasts of Río de Janeiro (Fig. 3.3), São Paulo (Fig. 3.4), Santa Catarina (Fig. 3.5), and Río Grande do Sul (Fig. 3.6). Lagoons are largely absent along the northern macrotidal (range >4 m) Quaternary coast and the Tertiary mesotidal (range 1–4 m) coast.

Several dozens of coastal lakes and segmented choked lagoon systems are found in the state of Río de Janeiro alone (Knoppers et al. 1991) (Fig. 3.3), including the large hypersaline Araruama L. (Kjerfve et al. 1996), as well as many smaller lagoon systems with salinities varying anywhere from estuarine to freshwater, including the Saquarema L. system (Urussanga-Fora), Maricá-Guarapina L., Itaipu-Piratininga L., and

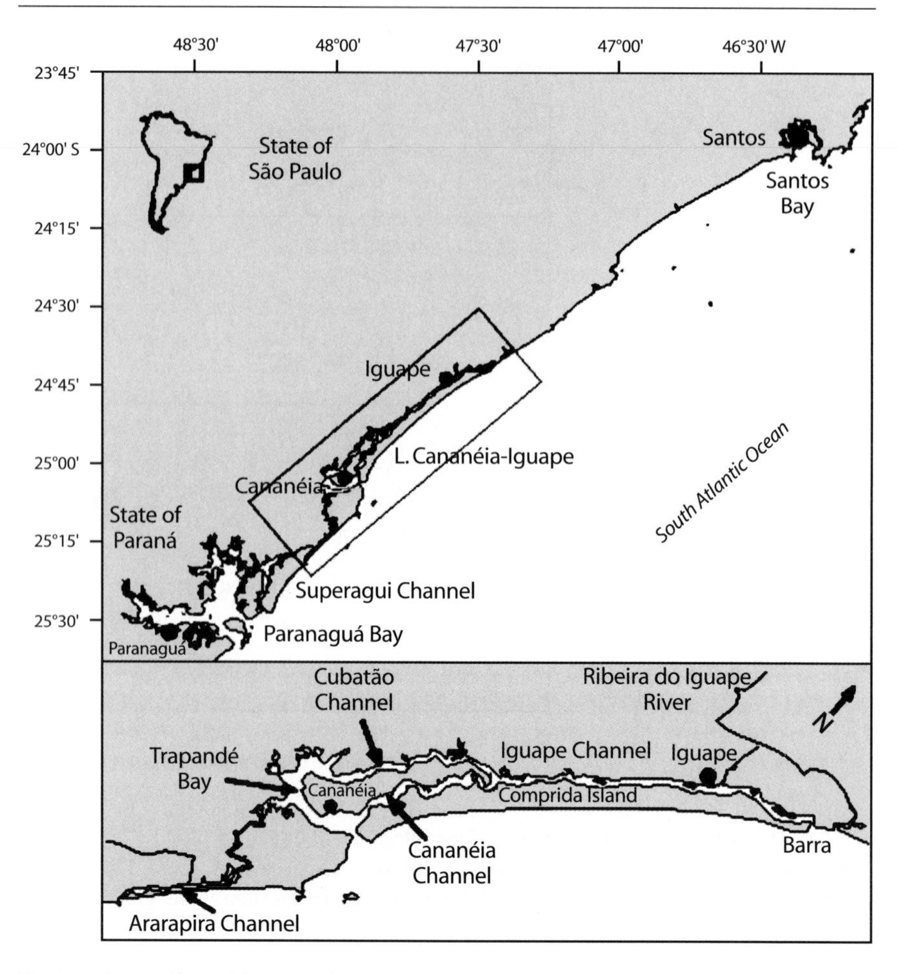

Fig. 3.4. The coastline of the state of São Paulo and its coastal lagoons

Jacarépagua-Marapendi L. The largest coastal lake in Río de Janeiro is Feia L., located north of Cabo Frio but is not included in this comparison because of lack of available data.

The southern section of the granitic coast, including São Paulo (SP) and Paraná (PR), have few coastal lagoons, but it harbours the unique, restricted lagoon system of Cananéia-Iguape L. (SP, Fig. 3.4) linked to Bahía dos Pinheiros (PR). More than 60 coastal lagoons and lakes occupy the coast of Santa Catarina and Río Grande do Sul (Schäffer 1988). The lagoons range from small systems, including Conceição L. (Fig. 3.5), Santo Antônio-Mirim L., Tramandai-Imé L., and Peixe L., as well as the world's largest choked coastal lagoon, Patos L. (Fig. 3.6), into which the similarly large Mirim L. discharges (Kjerfve 1986; Niencheski et al. 1988). The Patos L.-Mirim system is located just north of the Uruguay border. The geographic locations of the coastal lagoons discussed in this paper are summarized in Table 3.1.

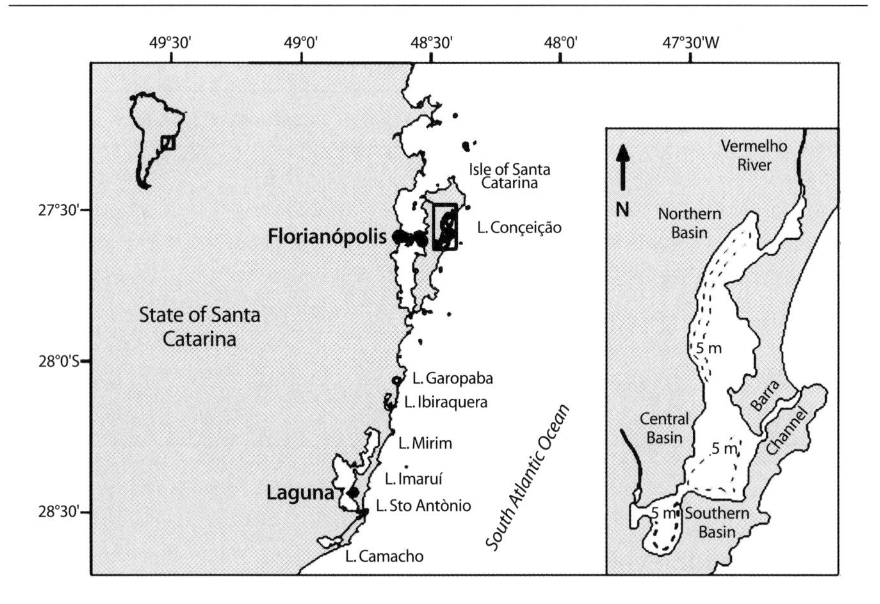

Fig. 3.5. The coastline of the state of Santa Catarina and its coastal lagoons

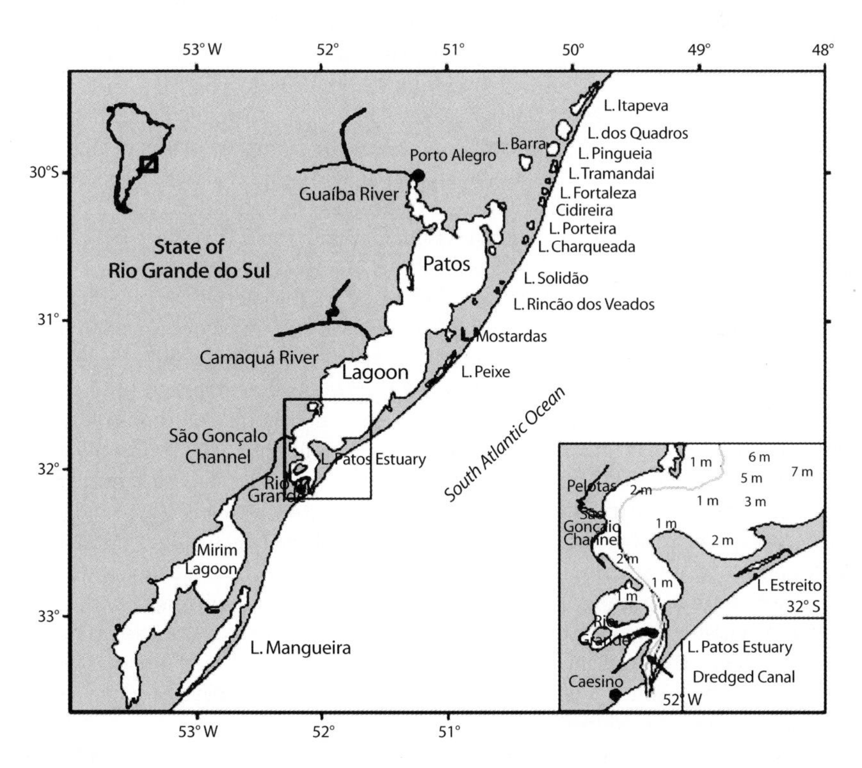

Fig. 3.6. The coastline of the state of Río Grade do Sul and its coastal lagoons and coastal lakes

Table 3.1. Geographical location of eight Brazilian coastal lagoon systems discussed in this study

Lagoon	State	Latitude (S)	Longitude (W)	Type	Cells
Araruama	Rio de Janeiro	22°53'S	42°15'W	choked	3
Jaconé-Saquarema	Rio de Janeiro	22°56'S	42°32'W	choked	4
Maricá-Guarapina	Rio de Janeiro	22°56'S	42°46'W	choked	4
Piratininga-Itaipu	Rio de Janeiro	22°56'S	42°58'W	choked	2
Cananéia-Iguape	São Paulo	25°20'S	47°45'W	restricted	2
Conceição	Santa Catarina	27°34'S	48°27'W	choked	3
Tramandaí-Imbé	Rio Grande do Sul	29°57'S	50°10'W	choked	6
Patos-Mirim	Rio Grande do Sul	30°55' – 32°30'S	50°55' – 52°20'W	choked	4

3.3 Coastal Lagoon Evolution

The evolution of the coastal plains of Brazil, including the formation and maintenance of sand barriers and coastal lagoons, is well documented (Suguio et al. 1980; 1984; Martin et al. 1988; Martin and Domínguez 1994; Muehe 1984). The formation of coastal lagoons attained peak expansion during two major sea-level high stands in the Quaternary. The first, occurring approximately 123 000 years B.P., corresponded to the last Pleistocene interglacial stage with a relative sea level at 8 ±2 m above the present level. The second, occurring 5 100 years B.P., corresponded to the Holocene sea-level high stand, with sea level 5 m above the present (Martin et al. 1980; Suguio et al. 1980). Subsequently, relative sea level fell but may also have experienced 2–3 m oscillations with a period of 200–300 years. These short term sea-level fluctuations would then have modified the sand barriers and the coastal lagoons (Martin et al. 1980; Suguio et al. 1980).

Most expansive of the low-relief coastal plains of southeastern Brazil are the 100 km wide coastal plains of Río Grande do Sul, where the largest lagoons are found. The Paraíba do Sul delta also represents Quaternary coastal deposits, where many coastal lakes and lagoons are embedded within regressive beach ridge and sand terraces formed by coastline progradation during sea-level fall and the continuous fluvial input of sediments (Dias and Silva 1984; Martin and Domínguez 1994). The narrow littoral fringe of the granitic coasts of Río de Janeiro and São Paulo are bordered by numerous lagoons. Most of these were formed during the Late Pleistocene and exhibit a sequence of narrowly spaced Pleistocene and Holocene beach ridges (Muehe 1984; Perrin 1984; Martin and Suguio 1989), including Araruama L., the Saquarema L. system, Maricá-Guarapina L., Piratininga-Itaipu L., Jacarepaguá-Marapendi L., and Cananéia-Iguape L. A string of small land-locked coastal lagoons is enclosed between the Pleistocene and Holocene ridges between Araruama L. and the shoreline (Fig. 3.3). Examples of the evolution of Pleistocene and Holocene sand barrier sequences and lagoons are schematized in Fig. 3.7.

The plains along the coast of Río Grande do Sul are marked by four generations of Quaternary marine sandy deposits, known as Barriers I, II, III, and IV. Barrier I is lo-

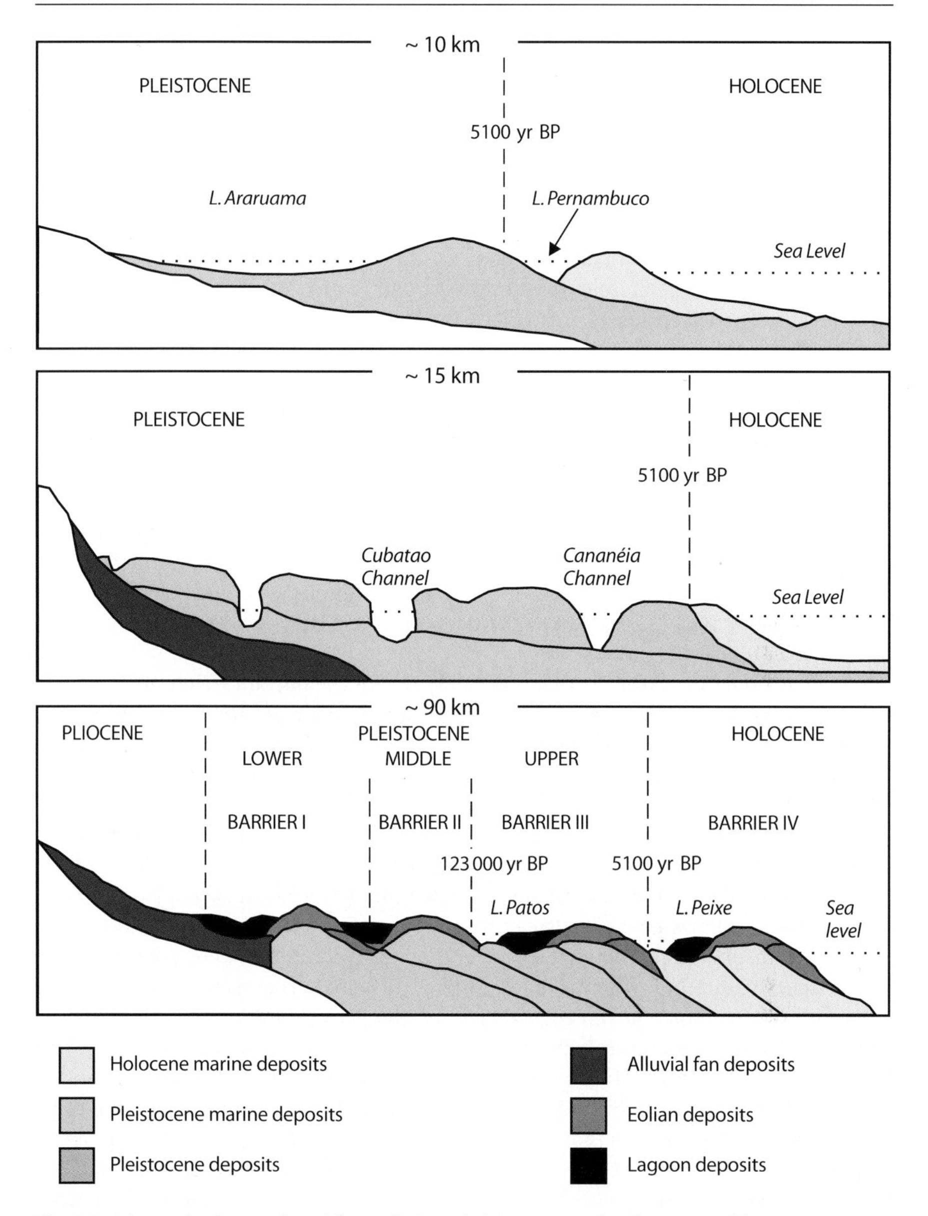

Fig. 3.7. Schematized examples of the evolution of Pleistocene and Holocene sand barrier sequences and lagoons in the state of Río de Janeiro

cated the furthest inland. Barrier IV is along the coastline and corresponds to the last Holocene transgressive-regressive phase. Lagoons and lakes enclosed by this barrier include Peixe L., Mangueira L., and Tramandai-Imbé L. (Schäffer 1988). The Patos-Mirim Lagoon system, located farther inland, is embedded in Barrier III, which represents the penultimate transgressive-regressive period during the Late Pleistocene (Martin et al. 1988). Only the inlet of L. Patos breaches Barrier IV.

3.4 Tidal Choking

Whatever tidal variability occurs in a coastal lagoon is a result of ocean tides (Kjerfve et al. 1990). However, the Brazilian lagoons are largely choked; they extend parallel to the coast, with a single connection to the sea, and consist of two or more elliptical lagoon cells. The exception to this rule is the restricted Cananéia-Iguápe L. system (Fig. 3.4), with two ocean connections. In choked coastal lagoons, the entrance channel serves as a dynamic filter that effectively reduces or eliminates tidal water fluctuations and tidal currents (Keulegan 1967; Kjerfve 1986). In Guarapina L., which is characterized by a 1.5 km long and 40 m wide tidal channel, water level oscillations are usually reduced to 1% or less as compared to the adjacent coastal tide (Kjerfve and Knoppers 1991). Similar conditions of tidal choking are encountered in Fora L. and Urussanga L. (of the Saquarema L. system), and Piratininga L. (Knoppers et al. 1991), Conceição L., (Knoppers et al. 1984; Odebrecht and Caruso 1987), and Tramandaí-Imbé L. (Schwarzbold 1982; Schwarzbold and Schäffer 1984). The lagoons proper of Patos-Mirim L. and Araruama L. lack tides altogether (Herz 1977; Kjerfve et al. 1990; Kjerfve et al. 1996).

Residual tidal circulation is dominant only where the entrance channel does not sufficiently reduce tidal forcing, as in Itaipu L., which exhibits tidal water level oscillations of more than 30% as compared to the coastal tide (Knoppers et al. 1991). In Cananéia-Iguape L., tidal and wind forcing may attain equal importance in driving circulation and mixing (Miyao and Harari 1989). The lower estuarine portion of Patos L. is also affected by tides, but it is governed mainly by wind forcing because of the small range of the tides (Herz 1977; Costa et al. 1988).

A useful indicator of the dominant transport mode in coastal systems is the non-dimensional Peclet number. The Peclet Number is defined by the ratio of diffusive to advective transport times. When it is greater than 1, advective transport takes less time than dispersive mixing; when it is less than 1, diffusive processes dominate (Zimmerman 1981). In most choked coastal lagoons, the Peclet number is on the order of 0.01–0.10, and mixing is more rapid than advective transport (Zimmerman 1981). The Peclet Number, however, often measures 1–100 within entrance channels to choked lagoons, indicating that advective processes dominate in the channel (Kjerfve et al. 1990; Kjerfve and Knoppers 1991), as in the case of Araruama L. (Kjerfve et al. 1996) and most other systems. Ebbing currents in the entrance channel to Guarapina L., for example, are largely uniform in salinity, but flooding currents often exhibit sharp temporal and horizontal salt gradients as stratified water masses from the adjacent ocean invade the lagoon before becoming mixed (Kjerfve et al. 1990).

3.5 Hydrology and Water Balance

Fresh water enters lagoons via river discharge, groundwater seepage, and rainfall, and is lost via evaporation and exchange with the coastal ocean. Seasonal pulses of freshwater input have a profound impact on the ecological viability of coastal lagoons by controlling the salinity, increasing the lagoon water level, and maintaining the entrance channel open. Some smaller coastal lagoons in Brazil, in particular along the coasts of Río de Janeiro near Macaé and Río Grande do Sul, intermittently close, isolating the lagoon from the sea as a result of active wave-infilling of the channel during periods of low runoff.

Although gravitational currents due to longitudinal and vertical density differences are very small in coastal lagoons, usually less than 0.01 m s^{-1}, they can become important in the lagoon entrance channel. Where the channel is relatively deep, which is the case in the canal leading into Lagoa dos Patos, gravitational circulation is important in maintaining the water balance. On the average, lagoon outflow as a result of river runoff is balanced by landward transport in the bottom layer of lagoon inlets. This gravitational circulation is often an order of magnitude greater than the river discharge and is an effective mechanism in ensuring lagoon water exchange.

Rainfall and freshwater runoff have profound impacts on coastal lagoons. During freshets, the water level can increase rapidly and cause extensive flooding of adjacent lands. This effect is most pronounced in choked lagoons, where the water level can rise one or more metres seasonally or in response to storms (Kjerfve et al. 1990). Flooding also causes large shifts in salinity in choked lagoon systems. As the runoff from the drainage basin crests in the lagoon, the resident water mass is rapidly advected out of the system, but when the runoff crest has receded, the salinity in the lagoon returns gradually to the former salinity as a result of lagoonward dispersion of coastal waters (Sikora and Kjerfve 1985).

In arid coastal regions, choked lagoons sometimes become hypersaline, at least seasonally, because of greater evaporation than runoff. A good example of a hypersaline coastal lagoon is Lagoa de Araruama, which remains permanently hypersaline with an average salinity of 52 and also the smaller sized Vermelha L. with a salinity of 180 (Kjerfve et al. 1996).

3.6 Wind Effects

Wind forcing affects water level, seiches, currents, and circulation in choked lagoons. The wind effect consists of both local wind stress on the lagoon and far-field wind forcing on the adjacent coastal ocean. Both contribute to the filling and emptying of the lagoon waters, depending on strength and direction of winds. The local winds generate currents, setup and setdown, and seiches and short-period wind waves. Far-field wind effects are manifested by oscillations in coastal sea level with a period of 2–20 days and result in either landward or seaward pressure gradients to aid in filling and emptying of the lagoons.

Wind stress causes current flow, but because of fluctuations in both wind speed and direction, the resultant currents are usually variable and intermittent. If the wind persists for long periods, however, a surface current will develop and cause a setup at the downwind end of the lagoon. Locally, the setup can cause extensive flooding of adjacent low-lying lands. The surface slope due to wind setup can be anywhere from 10^{-6}–10^{-4} (Kjerfve 1975). Rapidly moving meteorological fronts often cause seiches along the axis of coastal lagoons (Copeland et al. 1968), although the seiche motion in shallow lagoons is quickly overcome by friction.

In choked coastal lagoons, persistent winds can regulate the longitudinal salt distribution by driving oceanic waters into or out of the systems, as in the case of Patos L., which becomes totally fresh within a few kilometres of the ocean entrance during times of persistent winds from the north (Kjerfve 1986). On the other hand, when southerly winds blow, brackish waters push northward and the 5 isohaline can sometimes extend to the innermost reach of the lagoon, more than 250 km from the ocean entrance (Delaney 1963).

3.7 Salinity and Stratification

Salinity in the lagoons varies according to fresh water runoff, local climate, the geomorphology of lagoon cells and ocean connection, and degree of tidal choking. Apart from hypersaline Araruama L., the freshwater Patos L., and the polyhaline Itaipu L., all coastal lagoons in southeastern Brazil are oligohaline to mesohaline (Table 3.2). Most of the exterior lagoons of these systems are mesohaline, and the interior lagoons are either oligohaline or slightly mesohaline. The individual cells of these lagoon systems are usually homogeneously mixed land-sea breezes, and temperature convection as a result of 3–5° C diurnal temperature changes. Wind-induced mixing becomes more important during autumn and winter due to the frequent passage of weather fronts from the south, which commonly cause drastic temperature drops of up to 9° C within 2–3 days (Kjerfve et al. 1990).

However, vertical density (salinity) stratification may still occur in some lagoons. The Patos L. Estuary and the lower section of Cananéia-Iguápe L. are partially stratified (Abreu et al. 1994a,b; Miyao et al. 1986). The central basin of Conceição L. is highly stratified because marine waters enter the lagoon, descend after some mixing to the bottom, and are trapped within the basin by bottom sills (Knoppers et al. 1984; Odebrecht and Caruso 1987). Similar trapping occurs occasionally in Guarapina L. during events of maximum tidal inflow (Knoppers and Moreira 1988). In both cases, bottom waters stagnate and turn anoxic during periods of constant surface water outflow. Renewed tidal intrusion as well as wind-induced erosion of the halocline can result in nutrient-rich bottom waters mixing towards the surface. This feature of coastal lagoons, with relatively deep external basins, has been documented by Mee (1978). Although Araruama L. harbours a basin 17 m deep, the water column remains homogeneously mixed by salt convection from evaporation and intense wind mixing (Kjerfve et al. 1996).

The coastal lagoons exhibit pronounced annual cycles of salinity (Fig. 3.8). Short-term salinity variability is most pronounced in the Patos L. Estuary and is lowest in the hypersaline Araruama L. Salinity changes are always smallest in the internal cells of any lagoon system. Intense rain events may, however, induce drastic salinity changes, as in the case of Guarapina L. and Piratininga L., resulting in marked biogeochemical and ecological responses (Knoppers and Moreira 1988; Carneiro et al. 1993; 1994). Extreme rain events occur sporadically along the microtidal Brazilian lagoon coast.

Table 3.2. Physiographic and hydrological characteristics of lagoons in southeastern Brazil, including runoff ratio, discharges due to river runoff Q_R, direct rainfall Q_P, tidal flushing $|Q_T|$, inflow from landward lagoons Q_X, and flushing half life. Thus the large Q_X value for the Patos L. estuary includes the drainage into both Patos L. and L. Mirim

Lagoon system/lagoon	Lag. area (km^2)	Drain. area (km^2)	Depth (m)	Tide range (m)	Rain ($m\ yr^{-1}$)	Temp (°C)	Typical salinity (‰)	$\Delta f/r$ (–)	Q_R ($m^3\ s^{-1}$)	Q_P ($m^3\ s^{-1}$)	Q_T ($m^3\ s^{-1}$)	Q_X ($m^3\ s^{-1}$)	$T_{50\%}$ (d)
Araruama L.	210	285	3	0.01	0.93	24	52	0.11	1	1	47	3	84
Saquarema													
Flora L.	7	47	1	0.04	1.30	24	15	0.20	0	0	6	2	7
Urussanga L.	13	214	1	0.01	1.30	24	6	0.20	2	1	3	0	22
Maricá-Guarapina													
Guarapina L.	7	70	1	0.03	1.30	24	17	0.20	1	0	5	2	7
Maricá L.	17	280	1	0.01	1.30	24	5	0.20	2	1	4	0	28
Piratininga-Itaipu													
Itaipu L.	2	23	1	0.30	1.40	24	27	0.22	0	0	13	0	1
Piratininga L.	3	22	1	0.01	1.40	24	17	0.22	0	0	1	0	16
Jacarepaguá-Marapendi													
L. Jacarepaguá	8	287	2	0.10	1.50	15	20	0.26	4	0	18	0	5
Marapendi L.	4	8	1	0.02	1.50	20	15	0.26	0	0	2	0	15
L. Cananéia-Iguape	115	600	4	0.70	2.27	21	20	0.45	19	8	1 800	0	2
Conceição L.	19	45	2	0.05	2.16	21	19	0.43	1	1	21	0	11
Patos-Mirim													
Patos estuary	900	100	3	0.10	1.20	20	18	0.24	1	34	2013	3712	3
Patos L.	9330	136 700	5	0.00	1.75	20	0	0.37	3012	518	0	700	82

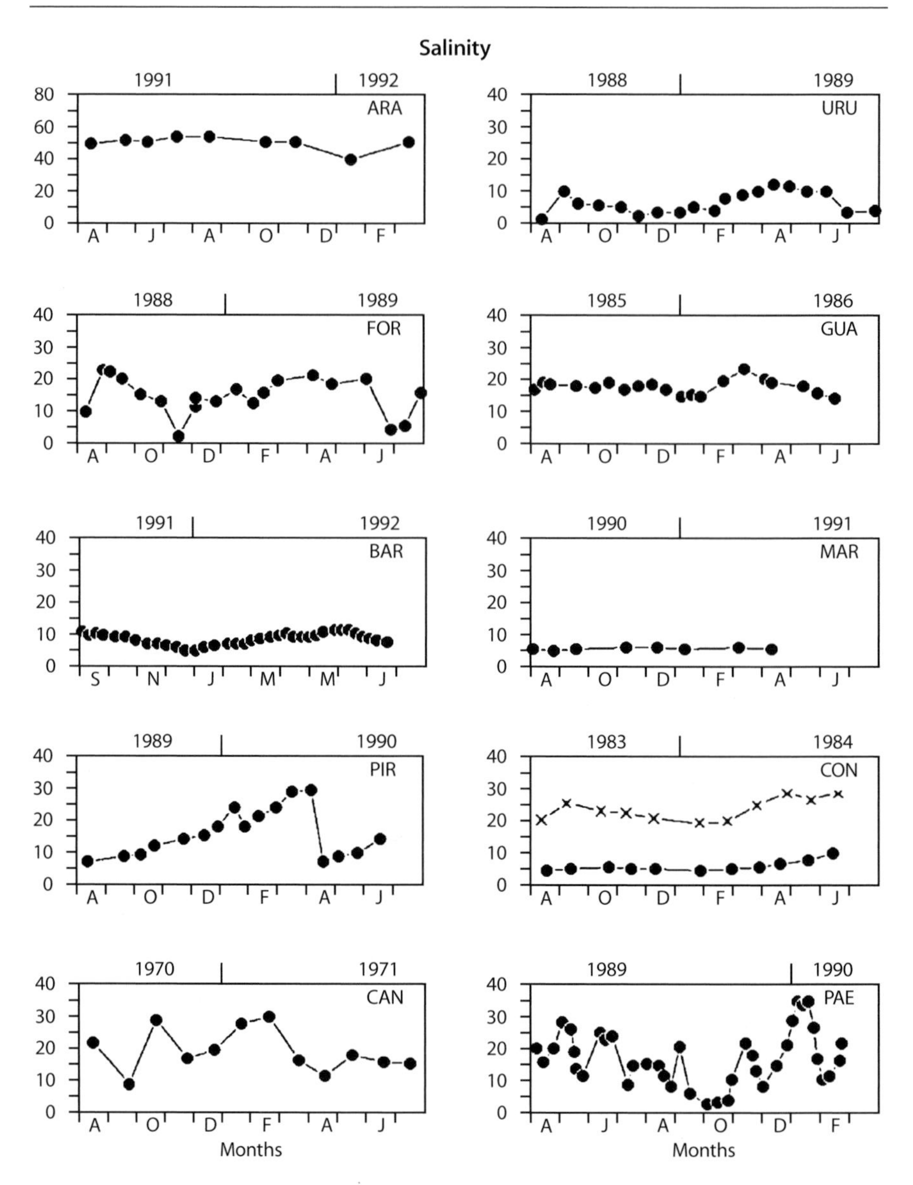

Fig. 3.8. Seasonal distribution of salinity in Araruama L. (*ARA*), Urussanga L. (*URU*), Flora L. (*FOR*), Guarapina L. (*GUA*), Barra L. (*BAR*), Maricá L. (*MAR*), Piratininga L. (*PIR*), Conceição L. (*CON*), Cananéia L. (*CAN*), and the estuary of Patos L. (*PAE*)

3.8 Flushing Time

A useful measure of the state of lagoons is a parameterized water renewal time (Zimmerman 1981; Dronkers and Zimmerman 1982; Merino et al. 1990; Knoppers et al. 1991;

Kjerfve et al. 1996). The flushing half-life ($T_{50\%}$), or the time that it takes to replace half of the lagoon water volume, is one such measure (Pritchard 1961; Knoppers et al. 1991; Kjerfve et al. 1996). Assuming steady state and complete mixing occurring rapidly compared to the flushing half-life, it is possible to write

$$\frac{dV}{dt} = -\kappa V \quad ,$$

(Pritchard 1961), where V denotes the volume of water in the lagoon, t time, and κ a rate constant representing the fraction of lagoon water volume replaced per unit time. Integration from $t = 0$ when the lagoon volume was V_o to a new time, $T_{50\%}$, when the total water volume is the same but only 50% of the original water molecules remain inside the lagoon yields

$$T_{50\%} = \frac{0.69}{\kappa} \quad .$$

Flushing in coastal lagoons depends equally on the sum of water inputs or on the sum of water losses. Selecting to calculate the 50% renewal time in choked coastal lagoons based on water inputs to the lagoon, it is possible to write

$$\kappa = \frac{[Q_R + Q_P + Q_O + |Q_T| + Q_x]}{V}$$

(Kjerfve et al. 1996), where Q_R is runoff from the drainage basin, Q_P is direct precipitation on the lagoon surface, Q_O is the net canal ocean exchange, $|Q_T|$ is the tidal exchange, Q_X is the additional drainage from landward lagoons, and V is the lagoon water volume. Whereas the Q_R, Q_P, and Q_O terms represent net long-term water fluxes, $|Q_T|$ is a tidal oscillating water flux, thus requiring the absolute value sign. The same tidal prism volume enters and leaves the lagoon during one half tidal cycle and represents "new" water entering the lagoon every flooding tide. At least in choked lagoons, it is unlikely that water leaving the lagoon during an ebb tide re-enters the lagoon on the next flood tide because of strong littoral currents. The tidal exchange, in reality, occurs during only half a tidal cycle,

$$Q_T = \pm \frac{A_L \, \Delta h}{[44\,714]} \quad ,$$

where Δh (m) is the mean lagoon tidal range, A_L is the lagoon surface area, and the constant is the duration of a semidiurnal tidal cycle. In the case of a predominantly diurnal tide, the constant should be 89 428 s.

The hydrological balance and water renewal times have been calculated for several Brazilian coastal lagoons, e.g., the Mundaú-Mangauba system in Alagoas (Oliveira and Kjerfve 1993), Araruama L. (Kjerfve et al. 1996), Guarapina L. (Kjerfve et al. 1990; Kjerfve and Knoppers 1991), the Cananéia-Iguape system (Miyao et al. 1986), and Patos L. (Herz 1977; Niencheski and Windom 1994). To ensure a fair comparison, however, we have recalculated the flushing half life ($T_{50\%}$) for the coastal lagoons in southeastern Brazil included in this review, using the method of Kjerfve et al. (1996). The concept of the flushing half-life or the time that it takes to replace half of the lagoon water volume is a robust measure of water renewal, and serves as a good comparative indicator of the hydrodynamic flushing lagoons for trophic state analysis (Knoppers et al. 1991). The hydrological data and results are summarized in Table 3.2.

All the Brazilian lagoons exhibit a positive annual hydrological balance (Knoppers et al. 1991; Miyao et al. 1986) with the exception of the permanently hypersaline Araruama L. Araruama L. is hypersaline as a result of a semi-arid climate with high evaporation and low rainfall rates and an extremely small fresh water input in relation to the lagoon water volume (Kjerfve et al. 1996). The mean flushing half-lives of the lagoons vary from 1–84 days. The shortest flushing half-life occurred in the tidally dominated Itaipu L. (1 day) and the Patos L. Estuary (3 days). The external cells of the choked systems with long tidal channels exhibited flushing half-lives of 5–7 days, while the interior lagoon cells located furthest from the sea without significant tidal exchange had flushing half lives of 15–28 days. The longest flushing half-lives were computed for Araruama L.(84 days) and the Patos L. proper (82 days).

3.9 Nutrient Standing Stock and Particulate Organic Matter

The majority of choked and restricted coastal lagoons in Brazil and other tropical and sub-tropical regions with humid climates are rich in organic materials. Some are entirely detritus-based during some stage of the annual cycle (Nixon 1982; Nichols 1989). The majority are also marked by seasonal changes in their standing stock of biogenic matter and autotrophic biomass (Nixon 1982; Knoppers 1994). However, extreme seasonal variability, as encountered in the coastal lagoons of the west coast of Mexico (Flores-Verdugo 1985; Flores-Verdugo et al. 1988), western Australia, and the Mediterranean, is not encountered in the coastal lagoons of Brazil (Mee 1978; Nixon 1982; Vaulot and Frisoni 1986; Yáñez-Arancibia 1987; Knoppers 1994).

Compatible information on annual cycles of standing stock of biogenic matter for the coastal lagoons of Brazil is limited to distributions of dissolved inorganic nitrogen (*DIN*) and phosphorus (*DIP*), a phytoplankton biomass indicator (chlorophyll a), particulate organic carbon (*POC*), and total suspended solids. Some information has also been published on annual cycles of dissolved organic carbon (*DOC*), dissolved organic nitrogen (*DON*), and dissolved organic phosphorous (*DOP*) for Urussanga L., Fora L., Barra L., Cananéia-Iguape L., and the estuary of Patos L. (Mesquita and Peres 1985; Carmouze et al. 1991; 1993; Abreu et al. 1994a,b). The mean annual concentrations and ranges of values for some nutrient parameters for the eight lagoon systems included in this comparison are summarized in Table 3.3.

The standing stock of particulates generally attains peak concentrations in the late austral summer and early fall, reflecting a trend towards a unimodal seasonal pattern in the development of autotrophic biomass, i.e., chlorophyll a (Fig. 3.9). This is true for all the lagoons of this study with the exception of the Patos L. Estuary. In most of the lagoons, the highest fraction of suspended detrital organic matter is encountered during the less productive period during late fall and winter. Most of the suspended detritus originates from autotrophic production, as indicated by the relatively low particulate organic carbon to nitrogen ratios, with C : N by weight less than 9 : 1, and

Table 3.3. Mean annual values and ranges (max–min) of water temperature, salinity, dissolved inorganic nitrogen (*DIN*), dissolved inorganic phosphate (*DIP*), particulate organic carbon (*POC*), N : P ratio, and chlorophyll a for coastal lagoon systems in SE Brazil

Lagoon	*T* (°C)	*S* (‰)	*DIN* (µM)	*DIP* (µM)	N : P	*POC* ($g\ m^{-3}$)	Chl. a ($g\ m^{-3}$)	References
Araruama	27 21–29	52 50 – 55	11.0 4.0 – 18.0	0.2 0.0 – 1.7	22:01	0.8 0.2 – 2.5	1.5 1 – 4	Landim de Souza (1993)
Urussanga	25 18 – 33	6 2 – 12	3.5 2.0 – 5.0	0.68 0.3 – 1.0	05:01	10.0 2.0 – 14	55 5 – 180	Carmouze et al. (1991)
Fora	25 18 – 33	15 2 – 22	3.5 2.0 – 5.0	0.65 0.3 – 1.0	05:01	7.0 2.0 – 12.0	28 5 – 48	Carmouze et al. (1991)
Guarapina	27 21 – 32	17 11 – 25	2.5 0.0 – 5.0	0.4 0.1 – 1.3	06:01	5.0 2.0 – 13.0	43 15 – 178	Moreira and Knoppers (1990)
Barra	25 21 – 39	9 6 – 11	4.0 1.0 – 10.0	1.0 0.2 – 2.4	04:01	8.0 1.0 – 16.0	89 30 – 375	Carmouze et al. (1993)
Maricá	– –	5 3 – 7	2.8 1.0 – 5.0	– –	–	– –	87 30 – 375	Fernandes et al. (1989)
Itaipu	– –	27 17 – 34	6.0 1.0 – 16.0	0.8 0.2 – 1.3	09:01	5.0 1.0 – 10.0	6 1 – 120	Carneiro (1992)
Piratininga	30 22 – 33	17 8 – 33	8.0 0.0 – 23.0	1.2 0.0 – 3.0	06:01	43.0 10.0 – 54.0	160 40 – 210	Carneiro et al. (1993)
Marapendi	27 –	15 –	29	2	15:01	–	69 15 – 240	Zee et al. (1993)
Cananéia	25 20 – 32	20 10 – 35	1.5 0.4 – 5.0	0.4 0.2 – 1.8	04:01	1.2 0.4 – 2.0	6 2 – 20	Tundisi et al. (1978)
Conceição	23 18 – 30	19 11 – 27	– 4.0	– 0.4	–	– 0.1 – 1.6	52 1 – 1600	Odebrecht (1988)
Patos Estuary	20 12 – 29	18 3 – 34	4.0 0.0 – 7.0	0.6 0.0 – 0.2	07:01	–	4 1 – 11	Abreu et al. (1994)

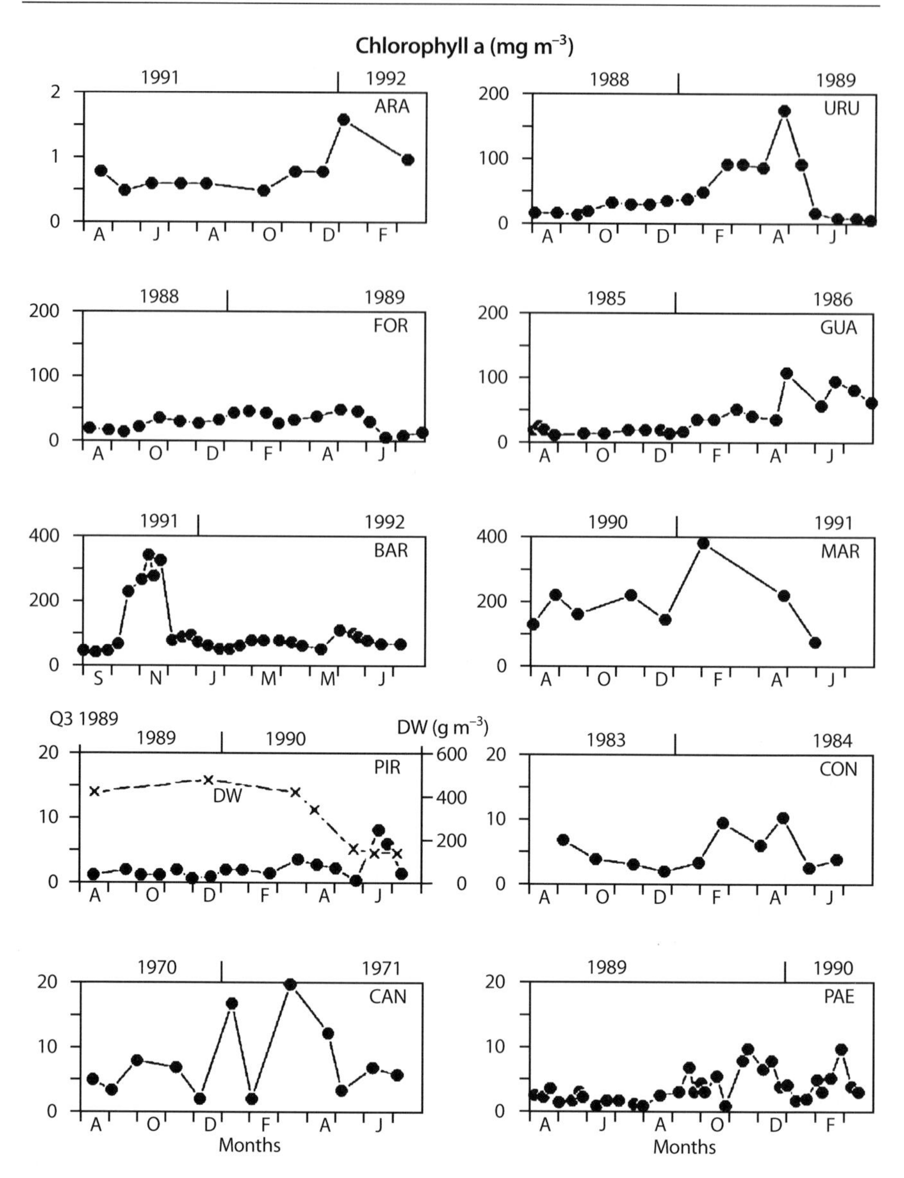

Fig. 3.9. Seasonal distribution of chlorophyll a (mg m^{-3}) in Araruama L. (*ARA*), Urussanga L. (*URU*), Flora L. (*FOR*), Guarapina L. (*GUA*), Barra L. (*BAR*), Maricá L. (*MAR*), Piratininga L. (*PIR*), Conceição L. (*CON*), Cananéia L. (*CAN*), and the estuary of Patos L. (*PAE*)

seems to represent an important pool for the regeneration of nutrients for the sustenance of primary production in the spring (Moreira and Knoppers 1990; Knoppers and Moreira 1990; Carmouze et al. 1991, 1993; Abreu et al. 1994a,b). The presence of large suspended detrital pools has been corroborated for Barra L., Guarapina L., Urussanga L., Flora L., and Conceição L. by direct comparison between the standing stock of particulate organic carbon (*POC*) and phytoplankton carbon (*PPC*). The *POC* : *PPC* ratio varies between 10 : 1 and 4 : 1, depending on the lagoon and season, and may inverse to 0.5 : 1 during exponential growth phases close to the peak of extreme phytoplankton blooms in summer. The *POC* stock varies between 4 and 15 g C m^{-2} (Knoppers et al. 1984; Odebrecht and Caruso 1987; Moreira 1988; Moreira and Knoppers 1990; Knoppers and Moreira 1990; Carmouze et al. 1993).

In contrast, dissolved inorganic nutrient concentrations are generally low and lack clear seasonal trends, mainly as a result of uptake by primary producers throughout the year. With the exception of the hypersaline Araruama L. (Landim de Souza 1993), the polyhaline anoxic central Conceição L. basin (Knoppers et al. 1984), and the limnic sections of the Patos L. Estuary (Proença et al. 1988), the inter- and intraspecific *DIN* varies from 2–10 μM and *DIP* varies from 0.3–1.5 μM in the Brazilian lagoons (Table 3.3). This corresponds to the range established for the majority of tropical and subtropical coastal lagoons with low cultural eutrophication (Nixon 1982). In Flora L. and Piratininga L., the nutrient concentration reaches the upper limit of this range, with ammonia dominating *DIN* as a consequence of domestic effluent discharge and low tidal flushing. In Araruama L., *DIN* may attain 20–30 μM, and in Patos L., 20–50 μM close to the discharge of the Guaiba River. In Araruama L., *DIP* concentrations are frequently at the detection level (Landim de Souza 1993), because phosphorous dynamics are additionally controlled by calcium carbonate reactions (Atkinson 1987; Knoppers et al. 1996). Except for Araruama L. and Patos L. proper, which are limited by phosphorous, the *DIN* : *DIP* atomic ratios in the other lagoons remain well below the Redfield ratio of 16 : 1 (Table 3.3). Thus, irrespective of their natural or cultural nutrient loading, the lagoons are nitrogen limited.

Short-term variability in nutrient concentrations occurs in most of the choked lagoons due to changes in nutrient loading from wash-out from the drainage basins in response to climatic events (Knoppers et al. 1991; Niencheski and Windom 1994), nutrient release from the sediments by wind-induced mixing, and density displacement of pore water during tidal intrusion (Knoppers and Moreira 1988; Knoppers and Moreira 1990; Machado 1989). In Barra L. and Piratininga L., sporadic dystrophic crises and fish kills induce nutrient pulses (Carmouze et al. 1993). Of particular interest is the short temporal variability from intrusion of nutrient-rich coastal water into the Patos L. Estuary. Ocean salinity waters at the surface near the mouth of the lagoon yielded up to 3 μM NH_4-N, 13 μM NO_3-N, and 3 μM PO_4-P in spring and markedly enhanced phytoplankton primary production (Abreu et al. 1995). Similar nutrient enrichment by marine waters has been documented for San Quentin Bay during upwelling events (Alvarez-Borrego and Alvarez-Borrego 1982).

3.10 Primary Producers

Most of the Brazilian coastal lagoons are phytoplankton-based systems. The exceptions are Araruama L., which is dominated by microphytobenthic cyanobacterial mats (Knoppers et al. 1996), and Piratininga L. and Tramandaí-Imbé L., which are dominated by macrophytobenthos (Carneiro et al. 1993; 1994). Macroalgae and submerged macrophytes also grow at shallow depths in some parts of the Patos L. Estuary (Coutinho and Seeliger 1986). The majority of primary producers in the phytoplankton-based coastal lagoons of the state of Río de Janeiro which form the most extensive blooms are cyanobacteria, which proliferate during the warm summer and early fall. In winter and spring, dinoflagellates and diatoms, respectively, prevail (Moreira 1988; Moreira and Knoppers 1990; Carmouze et al. 1993). Similar phytoplankton populations and successional patterns are found in Conceição L. However, diatoms, dinoflagellates, and micro flagellates are more common than cyanobacteria, and anoxigenic autotrophic purple bacteria dominate at and below the halocline within the central sections of Conceição L., particularly during prolonged periods of weak tidal flushing and stagnation of bottom waters (Knoppers et al. 1984; Odebrecht and Caruso 1987; Odebrecht 1988).

The more rapidly flushed lagoons, Cananéia-Iguape and the Patos Estuary, are governed by autotrophic flagellates and centric and pennate diatoms (Tundisi et al. 1973; Abreu et al. 1992; 1994a,b). The striking feature in Piratininga L. is the succession between phytoplankton and the macroalgae. The macroalgae commence proliferation in winter and attain senescence in fall., as shown by the annual cycle of biomass dry weight (Fig. 3.9). The subsequent degradation of the macroalgae provides a nutrient pulse for the sustenance of phytoplankton, i.e., chlorophyll a (Fig. 3.9) until growth is resumed by the macroalgae. Similar successional patterns have been observed in the Peel-Harvey Estuary in western Australia, and the Venice Lagoon in Italy. The regeneration of nutrients occurs largely at the sediment-water interface (Sfrizo et al. 1988; Carneiro et al. 1993 1994).

Other primary producers that should be considered are mangroves and emergent macrophytes. Mangroves cover 930 km^2 along the southeastern coast, but are largely absent in the lagoons. The southernmost limit of mangroves in Brazil is Praia do Sonho, Santa Catarina (Schaeffer-Novelli 1989) at latitude 28°53'S, and thus no mangroves are found in Río Grande do Sul, including the Patos-Mirim Lagoon system. The only lagoon along the southeastern coast that has extensive mangrove forests is Cananéia-Iguape L., which is not a choked lagoon but is flushed twice a day by tides with a range of 1 m and exhibits salinities in the range 10–35. The Cananéia-Iguape mangroves cover 52 km^2 (Herz 1991) and consist mostly of *Rhizo-phora mangle* L., *Avicennia schaureiana* Stapf and Leech, and *Laguncularia racemosa* GFW Meyer (Cintrón and Schaeffer-Novelli 1992). The choked lagoons of Santa Catarina and Río de Janeiro, on the other hand, are characterized by occasional scrubby patches of mangroves along the borders of the lagoons. The reason for this is the lack of tidal range in the lagoons because of their choked characteristics, and also the rather low salinities in many lagoons but hypersaline conditions in Araruama L. (Oliveira 1959).

The predominant marginal vegetation in most of the low-salinity choked lagoons of Río de Janeiro consists of the macrophyte *Typha dominguensis* Pers. The distribution, density, and primary production along the lagoon margins are controlled by salinity, frequency of marginal inundation, freshwater run-off, and groundwater supply. The most dense and productive patches are encountered in the vicinity of riverine mouths. Studies of macrophytes, including their impact upon aquatic primary lagoon production, have been restricted to Maricá L. and Guarapina L. (Couto 1989). Although the marginal macrophyte vegetation may be important as a filter, these wetlands typically comprise less than 10% of the lagoon's surface area.

3.11 Primary Production

Measurements of primary production for complete annual cycles have been carried out in Flora L. and Urussanga L. (Carmouze et al. 1991), Guarapina L. (Moreira and Knoppers 1990; Knoppers and Moreira 1990), Barra L. (Carmouze et al. 1993), Cananéia-Iguape L. (Tundisi et al. 1973, 1978), and Patos L. Estuary (Proença 1990; Abreu et al. 1992, 1994b). Some daily and weekly measurements at some stages of the annual cycle have been carried out in Araruama L. (Knoppers et al., in press), Piratininga L. (Carneiro et al. 1993, 1994), and Conceição L. (Knoppers and Odebrecht, unpublished data). The annual cycles of primary production and mean daily rates of primary production and respiration in the lagoons have well-defined patterns (Table 3.4, Fig. 3.10). The phytoplankton-based coastal lagoons are, in general, characterized by a unimodal annual cycle, with highest production rates during the austral summer and lowest during the winter. The annual cycle measured for Barra L. was influenced by a dystrophic crises and a fish kill (Carmouze et al. 1993). The annual net primary production rates for the lagoons of the state of Río de Janeiro were 324–810 g C $m^{-2} yr^{-1}$. The highest daily primary production rates were recorded during the exponential growth phase by cyanobacteria in summer in Guarapina L. and Barra L., and by macroalgae in

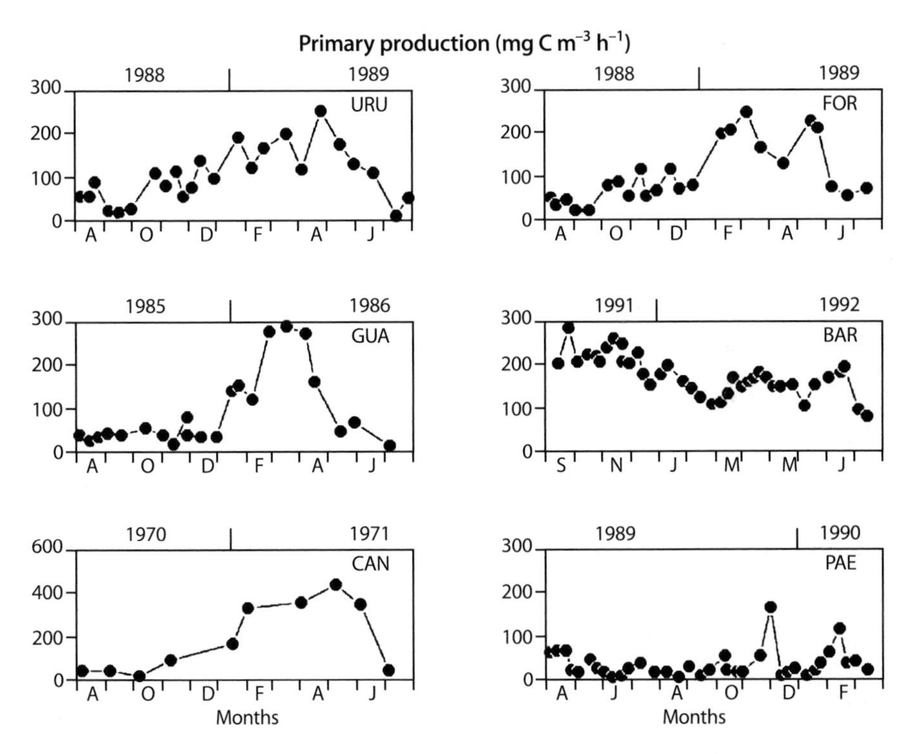

Fig. 3.10. Seasonal distribution of primary production (mg C $m^{-3} h^{-1}$) in Urussanga L. (*URU*), Flora L. (*FOR*), Guarapina L. (*GUA*), Barra L. (*BAR*), L. Cananéia-Iguápe (*CAN*) and the Patos L. Estuary (*PAE*)

Table 3.4. Net coastal lagoon primary production and respiration, and the dominant primary producers

Lagoon	Net *PP* (g C m^{-2} d^{-1})	*R* (g C m^{-2} d^{-1})	*n*	Study period	Primary producers	Methods	References
Araruama	0.18 ± 0.01 0.41 ± 0.10	0.43 ± 0.09 0.78 ± 0.24	34	Spring 1993 Fall 1995	Algal mats Algal mats	In situ chambers Diss. oxygen (*DO*)	Knoppers et al. (unpubl.)
Urussanga	1.28 ± 0.82	1.28 ± 0.60	25	1988–1989	Phytoplankton	Diurnal curve-CO_2	Carmouze et al. (1991)
Fora	1.25 ± 0.82	1.18 ± 0.59	25	1988–1989	Phytoplankton	Diurnal curve-CO_2	Carmouze et al. (1991)
Guarapina	0.89 ± 0.59	0.68 ± 0.44	17	1986–1987	Phytoplankton	Light/dark (L/D) in situ bottle-*DO*	Machado and Knoppers (1988)
Barra	2.22 ± 0.92 1.92 ± 0.84	2.18 ± 0.65 2.16 ± 0.61	4242	1992–1993 1992–1993	Phytoplankton Phytoplankton	Diurnal curve-*DO* Diurnal curve-CO_2	Carmouze et al. (1993)
Piratininga	1.15 ± 0.19	–	3	Summer 1989	M-phytobentos	Diurnal curve-*DO*	Carneiro et al. (1993)
Cananéia	0.35	–	12	1972	Phytoplankton	L/D in situ/^{14}C	Tundisi et al. (1978)
Conceição	1.59 ± 0.80	–	3	Summer 1984	Phytoplankton	Diurnal curve-*DO*	Knoppers and Odebrecht (unpubl.)
Patos Estuary	0.56 ± 0.68	–	46	1988–1989	Phytoplankton	L/D In situ/^{14}C	Abreu et al. (1994)

Piratininga L. Lower primary production rates were measured in Cananéia-Iguape L. with 125 g C $m^{-2} yr^{-1}$ and in the L. Patos Estuary with 50 g C $m^{-2} yr^{-1}$.

The choked lagoons with the least efficient tidal flushing (Table 3.2) in general exhibit the highest rates of primary production (Table 3.4). The exception is Araruama L., which is dominated by cyanobacterial algal mats; pelagic primary production was no more than 10% of the benthic primary production. Maximum total daily primary production rates were very low as compared to the other choked lagoons, but the rates were similar to those for the Patos L. Estuary. Both systems are particularly prone to wind forcing, which is pronounced because of the long fetches of the two lagoon systems (Herz 1977; Kjerfve et al. 1996) and seemingly inhibits primary production, perhaps because of strong mixing of the water column and resuspension of fine-grained bottom material.

Primary production and respiration rates in the lagoons were determined by different methods, but all measurements were performed in situ (Table 3.4). In the phytoplankton-based lagoons, the methodology varied from light/dark bottle incubation techniques, using either dissolved oxygen or radioactive bicarbonate ^{14}C measurements, to the more time-consuming in situ diurnal curve method, measuring dissolved oxygen and carbon dioxide. Primary production of the cyanobacterial algal mats in Araruama L. was determined by incubation of the sediment surface with benthic chambers; macroalgal primary production in Piratininga L. was determined by the diurnal curve method. The diurnal curve method in conjunction with incubation techniques and carbon dioxide measurements is the most appropriate methodology for these lagoons, for it enables the assessment of production and respiration of both the pelagic and benthic components, and also the whole system metabolism for conditions where many different autotrophs prevail (Reyes and Merino 1991; Knoppers 1994).

Abreu et al. (1994a,b) applied an interesting technique in the Patos L. Estuary, where particulate and dissolved phytoplankton production were measured by comparing ^{14}C uptake rates from size fractionated filtration and the acidification bubbling method (ABM). Production rates estimated by the ABM were equivalent to the sum of particulate and dissolved carbon production, simultaneously determined by the filtration technique. *DOC* released by phytoplankton represented approximately 22% of total dissolved and particulate carbon uptake.

The primary production estimates in the lagoons refer to the permanently aquatic component of primary production and therefore exclude the primary production of bordering wetland vegetation. Estimates of mangrove production from litterfall in southeastern Brazil, including Cananéia-Iguape L., ranged from 1–5 g C $m^{-2} d^{-1}$ on a mean annual basis (Kjerfve and Lacerda 1993). Studies in Guarapina L. and Maricá L. revealed that the above-ground biomass production of *Typha dominguensis* PERS ranged from 800 g dry weight $m^{-2} yr^{-1}$ under mesohaline conditions to 2 400 g dry weight $m^{-2} yr^{-1}$ under oligohaline conditions (Couto 1989).

Information on the contribution by the below-ground production is not accurate enough for these systems. Of particular interest is the impact of emergent macrophyte vegetation upon the aquatic primary production of the lagoons. It has been documented that leachates and oxidizable particulates from mangrove forests and emerged macrophytes may stimulate aquatic primary production in the vicinity of these habitats (Day et al. 1988). The impact by *Typha dominguensis* in Maricá L. and Guarapina L.

is not important. Signatures of $\delta^{13}C$ have shown uniform values of organic matter in surface sediments between the central and marginal parts of the lagoons, resembling those of phytoplankton. The macrophyte fringe seems to serve as a physical and biological filter for biogenic matter transferred from the drainage basin to the lagoon. However, some export of material from the vegetation belt may occur during sporadic inundation and wash-out events during the passage of meteorological fronts. Because of the small areal extent of the marginal wetlands, their impact on primary production of the lagoons is minor, with exception for the Cananéia-Iguape system.

3.12 Pelagic and Benthic Metabolism

In most phytoplankton-based estuaries and coastal lagoons, the relation between sediment oxygen consumption and pelagic primary production varies between 25 and 50% (Nixon 1982), and heterotrophic metabolism seems to govern these systems on an annual basis (Smith and Atkinson 1994). Measurements of production and respiration of the pelagic and benthic compartments exist for Barra L., Guarapina L., Urussanga L., Flora L., and Araruama L. (Machado and Knoppers 1988; Carmouze et al. 1991; Carmouze et al. 1993; Knoppers et al. 1996).

The four phytoplankton-based lagoon systems of the state of Río de Janeiro are characterized by:

1. a negligible benthic primary production as compared to the pelagic primary production;
2. benthic respiration measuring 25–40% of whole system respiration during winter to 30–55% during summer, indicating that the pelagic compartment oxidizes a slightly higher fraction of organic matter compared to the benthic compartment on an annual basis;
3. marked seasonal shifts between autotrophy and heterotrophy, with autotrophy dominating during the summer and heterotrophy during the winter;
4. net autotrophy and net heterotrophy being equal on an annual basis.

In contrast, Cananéia-Iguape L. is more controlled by marine sources and is probably also influenced by the metabolism and export of materials from adjacent mangrove forest (Adaime 1985). The hypersaline and carbonate-rich Araruama L. is also different, being oligotrophic, and with pelagic primary production measuring, at most, 10% of the benthic primary production. The total community metabolism was clearly heterotrophic (Knoppers et al. 1996).

3.13 Nutrient Sources and Primary Production

Nutrient sources to sustain pelagic primary production in coastal lagoons include input from the atmosphere, rivers, groundwater, and the sea, as well as internal recycling of nutrients between the pelagic and benthic compartments. In addition, nitrogen fixation is a source of nitrogen, and nitrogen is lost due to denitrification, export to the sea, and accumulation in sediments (Nixon 1982; Seitzinger 1988; Knoppers 1994).

Complete mass balance studies, including all sources and sinks, are still lacking for coastal lagoons (Smith and Atkinson 1994), including the lagoons of southeastern Brazil. In some lagoons, however, nutrient loading from rivers and nutrient release rates from the sediments have been measured, although information on the riverine load is available mainly for the dissolved inorganic fractions. To what extent the load of the dissolved and particulate organic fraction after oxidation within the lagoons represents a significant new nutrient source for the sustenance of primary production has yet to be established.

Mean annual areal loading of the Brazilian lagoons from fluvial input of dissolved inorganic nutrients is substantial (Table 3.5). For example, the nutrients introduced at the head of Patos L. by the Guaiba River, including the discharge from Porto Alegre (population 1 500 000), fertilizer plants, and agricultural run-off, accounts for 86% of the total riverine nutrient load to the system. The estimated transient time from the head to the mouth of Patos L. is 20 days (Herz 1977); nutrients are modified during the 250 km transit from head to mouth of the lagoon and particulate organic matter in suspension is oxidized (Niencheski and Windom 1994). In contrast, riverine loading of the smaller lagoons of the state of Río de Janeiro is mainly affected by domestic effluents (Knoppers et al. 1991). The drainage basins of Piratininga Itaipu L., Flora L., and Araruama L. are, for the most part, urbanized and are largely exempt from industrial and fertilizer runoff. The impacts of areal nutrient loads differ widely because of differences in lagoon water volumes. The most striking dilution effect is encountered in Araruama L., where the areal nutrient loading is very small in spite of substantial amounts of domestic effluents.

In general, the areal dissolved inorganic nutrient load sustains 1–10% of pelagic primary production in the smaller lagoons, irrespective of natural or cultural impacts. In the Patos L. Estuary, the dissolved inorganic nutrient load contributes as much as 15% of the primary production, mainly because the production here is much lower as

Table 3.5. Mean annual areal load of dissolved inorganic nitrogen (*DIN*) and phosphate (*DIP*) and percentage supply of the demand of primary production. Primary production (*PP*) of Piratininga L. refers to macroalgae in summer only, and the data for Patos L. were obtained from a mass balance study by Niencheski and Windom (1994). The remaining data are from Knoppers et al. (1991)

Lagoon	Areal load *DIN* ($mM\ m^{-2}\ d^{-1}$)	*DIP*	N/P	*n*	*PP* demand (%) *DIN*	*DIP*	Comments
Araruama	negligible		2.4	17	<1	<1	Hypersaline
Urussanga	0.05	0.021	2.4	16	≈1	2.1	Quasi natural
Fora	0.43	0.055	7.8	15	2.7	5.6	Some effluents
Guarapina	0.89	0.040	22.3	53	8.0	5.7	Includes the input from the adjacent L. Padre lagoon
Itaipu	3.21	0.213	15.1	12	–	–	*PP* not available
Piratininga	1.48	0.147	10.1	35	15.0	24.0	*PP* in summer only
Patos estuary	1.38	0.553	2.5	8	8.0	125.0	Includes the input from Patos L. proper

compared to the Río de Janeiro lagoons. Nixon and Pilson (1983) and Knoppers (1994) showed that the fluvial supply of dissolved inorganic nutrients to most estuaries and lagoons, subject to only moderate cultural eutrophication, sustains 10–30% of the average annual primary production. Higher contributions are found in individual lagoon cells and during some stages of the annual cycle. They are related to the diversity and locality of point and non-point sources and also meteorological events, as in the case of Araruama L. and other lagoons, such as Nichupte in Mexico, Venice in Italy, the Peel-Harvey in western Australia, and Ebrié in Ivory Coast (Hodgkin and Birch 1982, 1986; Carmouze and Caumette 1985; Sfrizo et al. 1988).

In phytoplankton-based coastal lagoons, the benthic supply and release of *DIN* generally sustains 10–30%, and release of *DIP* (orthophosphate) 5–30%, of the annual demand by pelagic primary production (Knoppers 1994). In some estuaries and lagoons, benthic release may supply 50%, and sometimes 100% at certain stages of the annual cycle (Nixon 1982; Zeitzschel 1980). In general, the *DIN* : *DIP* release rate from the benthic interface fluctuates somewhat but remains below the classic Redfield ratio, suggesting preferential release of *DIP* over *DIN* (Table 3.6). This is in part attributable to nitrogen loss by denitrification (Seitzinger 1988). It seems, however, that the slight nitrogen limitation imposed by sedimentary release may be compensated by nitrogen fixation in organic-rich coastal lagoons.

Measurements of nutrient release rates from the sediment-water interface have been made in Araruama L. (Knoppers et al. 1996), Flora L. (Belloto 1992), Guarapina L. (Machado 1989), Barra L. (Kuroshima 1995), Maricá L. (Fernex et al. 1992), and the Patos L. Estuary (Balzer and Niencheski, unpublished data). The release rates for ammonia and orthophosphate are summarized in Table 3.6. Complete annual cycles were determined in Guarapina L. and Barra L., whereas measurements in other systems were conducted only during the summer and winter. Flux estimates for Araruama L., Guarapina L., and the Patos L. Estuary include in situ benthic chamber and *in vitro* core incubations (Zeitzschel and Davies 1978). Flux estimates for other systems were

Table 3.6. Benthic nutrient release rates for several lagoons in southeastern Brazil

Lagoon	Benthic fluxes NH_4-N ($mM\ m^{-2}\ d^{-1}$)	PO_4-P	Study period	Method	Comments
Araruama	0.17 – 1.08	−0.03 – 0.03	Spring 1993 Fall 1995	In situ chambers	Knoppers et al. (in press)
Fora	0.03 – 8.80	0.01 – 0.09	Annual cycle	In situ pore water samples	Belloto (1992)
Guarapina	0.46 – 9.70	−0.20 – 0.06	Annual cycle	In situ chambers	Machado (1989)
Barra	21.5 7.9	1.02 0.50	Summer Winter	In situ pore water samples	Kuroshima (1995)
Maricá	7.65 – 21.0 4.04 – 20.5	0.04 – 0.09 0.12 – 0.36	Summer Winter	In situ pore water samples	Fernex et al. (1992)
Patos Estuary					Balzer and Niencheski (unpubl.)

calculated by Fick's first law from pore water collected using the in situ "Peeper" sampling technique (Fernex et al. 1992) and applying standard temperature-corrected diffusion coefficients. The results reflect seasonal variability of benthic nutrient fluxes, with the highest flux usually occurring during the summer when the primary production is enhanced. Assuming the Redfield ratio to establish the nutrient demand by phytoplankton primary production, the benthic supply of ammonia and orthophosphate sustains 10–30% of the primary production in the phytoplankton-based coastal lagoons on both a seasonal and annual basis. In the case of Araruama L., however, the microphytobenthos-based primary production was entirely sustained by ammonia and orthophosphate release in winter. In late summer, ammonia was preferentially released over orthophosphate, which covered only 5% of the demand by phytoplankton.

Considering that the relation between sediment oxygen consumption and pelagic primary production was 30–55% in the Brazilian lagoons, the pelagic compartment regenerates a somewhat higher fraction of nutrients than the benthic compartment. Some of the benthic oxygen consumption includes chemical oxygen demand, at least for the oxidation of sulfide to sulfate. Most of the lagoon sediments are anoxic a few centimetres below the surface.

3.14 Trophic State

The trophic state (TS), or degree of nourishment of coastal lagoons, is an indicator that can be used to as a first step to assess the process of eutrophication and also to develop management policies. There are many available methods for TS ranking (Lambou et al. 1983).

The simplest measure of TS-ranking is the average annual stock, or peak value, of chlorophyll a, total nitrogen (*TN*), and/or total phosphorous (*TP*). The choice depends on which nutrient is limiting for primary production and thus should be selected (Rast and Holland 1988). However, the extrapolation of the trophic state concept, which was originally developed for phosphorous-limited temperate lakes (Vollenweider 1968; Vollenweider and Kerekes 1982), presents problems when applied to nitrogen-limited, tidally dominated estuaries (Rast and Holland 1988; Vollenweider 1992). For example, using nitrogen as the TS indicator requires information on denitrification because it has a compensatory effect on eutrophication (Golterman and Oude 1991).

Complications may arise in tropical coastal lagoons because of the presence of continuous plant growth throughout the year and because of spatial and seasonal variability in nitrogen and phosphorous limitation. Examples of lagoons with such characteristics are Patos L. and Araruama L., as well as Ebrié in Ivory Coast. Thus, in comparing systems, chlorophyll a remains the only useful, practical, and simple measure of TS ranking (Golterman and Oude 1991).

Based on the mean annual chlorophyll stock (Table 3.3), the TS-ranking system (Rast and Holland 1988) indicates that Flora L., Urussanga L., Guarapina L., Barra L., Maricá L., Marapendi L., and Conceição L. are eutrophic and that Itaipu L., Cananéia-Iguape L., and the Patos L. Estuary are mesotrophic. Hypertrophism is found in Piratininga L. and oligotrophism in Araruama L. In Araruama L. and the Patos L. Estuary, some portions of the lagoons attain higher trophic states during parts of the year, although usually for only short periods of time. In Araruama L., cyanobacterial algal mats proliferate where the depth is shallower than 3 m, but the contribution of these mats is not included in the total chlorophyll a stock estimate. This is an additional problem in applying simple TS ranking to systems with different autotrophic populations. Some coastal lagoons may, at the same time, support phytoplankton, benthic micro- and macroalgae, macrophytes, and algal mats (Knoppers 1994). Schäffer (1988) applied different TS indices to 38 coastal lagoons and lakes in Río Grande do Sul. His results corroborate that comparative TS ranking of highly diversified coastal systems requires that the systems first be grouped according to their physical, geomorphological, and biogeochemical characteristics.

3.15 Human Impact

Many of the coastal lagoons of the microtidal southeastern coast of Brazil are subject to human impacts similar to Venice, Ebrié, Nichupte, and Peel-Harvey Lagoons (Sfrizo et al. 1988, Carmouze and Caumette 1985; Reyes and Merino 1991; Hodgkin and Birch 1986). Shore erosion, local rise of relative sea level, uncontrolled land use and drainage basin fertilization, deforestation, and urban-industrial expansion are of particular concern. Patos L. is the most prominent lagoon affected by all of these factors, but fortunately, the large size and storage capacity of the lagoon helps to counteract some of these human impacts. However, this is not true for the small coastal lagoons of the state of Río de Janeiro nor Conceição L., which are heavily impacted by domestic effluent discharge. Cultural eutrophication represents the primary water-quality problem in these lagoons. The nutrient loads correspond to those of other estuaries and lagoons subject to moderate cultural eutrophication (Nixon and Pilson 1983), and the lower N : P ratios of the effluent loads may drive some of the lagoons towards more severe nitrogen limitation. There is no information to support the concept that N-fixation by cyanobacterial may exert a compensatory effect.

Because of the lack of long-term interannual monitoring of trophic state indicators in all of the lagoons, eutrophication trends may be inferred only by indirect methods. Dystrophic crises, fish kills, excessive algal growth, nuisance odours from decaying organic matter, and the presence of pathogenic microorganisms during warm summer months are the main manifestations of eutrophication observed in some of the lagoons of Río de Janeiro and also in Conceição L. Some of these may also occur in the upper part of the Patos L. Estuary close to Porto Alegre and in the enclosed estuarine Saco da Mangueira near Río Grande (Persich et al. 1996).

All coastal lagoons along the southeastern coast of Brazil are being menaced by uncontrolled demographic expansion and increased sewage discharge. Estimates of augmentation of nutrient load due to population growth, from which to infer trends of eutrophication, have yet to be established. Moreover, such estimates may harbour problematic caveats (Golterman and Oude 1991). Independent of this, they are very difficult to calculate because of systematic lack of collection and documentation of reliable data by local, state, and federal government agencies in Brazil.

Statistical models based on comparative studies that link nutrient loading to standing stock of materials and hydraulic residence time (Vollenweider 1968) also serve to infer eutrophication trends. However, such models should not be used to extrapolate

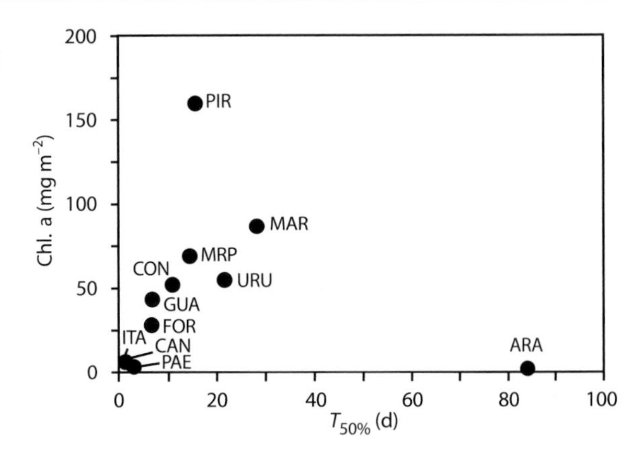

Fig. 3.11. Composite plot of the flushing half-life ($T_{50\%}$) and chlorophyll a for the lagoons

beyond the available range of data, especially when the maximum sustainable capacity of a system's metabolism is unknown. Before carrying out TS ranking, Knoppers et al. (1991) established a relationship between the standing stock of total phosphorous, chlorophyll a, and water turnover time for five phytoplankton-based coastal lagoons in Río de Janeiro. This relationship is quite robust (Fig. 3.11) and indicates that, with increasing water turnover time, the lower the standing stock of chlorophyll a is in these phytoplankton-based lagoons. However, the relationship is not reliable when a lagoon is dominated by either microphytobenthos (Araruama L.) or macrophytobenthos (Piratininga L.). Thus, in comparing trophic state and eutrophication of coastal lagoons, they should first be grouped according to the dominant autotrophic lagoon population.

Attempts to reconstruct eutrophication trends for the 20th century from analysis and dating of sedimentary cores in Piratininga L., Itaipu L., Maricá L., and Guarapina L. have not yielded consistent results. In order to eradicate malaria by improving drainage of freshwater wetlands adjacent to the lagoons and to increase marine flushing, engineering works were begun in the 1940s and 1950s that included the dredging of new channels between the lagoons and also new entrance channels to the South Atlantic Ocean (Oliveira 1959). These projects in Río de Janeiro greatly affected the hydrology of the coastal lagoons and their sedimentary depositional patterns.

3.16 Management Issues

How best to manage coastal lagoons does not have a simple solution, and the radical options that prevail (Hodgkin and Birch 1986) are difficult to digest politically. The only viable remedial option is to reduce and control the nutrient load, primarily by the management of watershed activities and the installation and maintenance of sewage treatment facilities, but possibly also, as suggested by Vollenweider (1992), a temporary "lagooning." In tropical coastal lagoons, reduction of the nutrient load is more difficult to monitor and accomplish, as it is necessary to manage the levels of all potentially limiting nutrients, not just a single limiting nutrient. Unfortunately, there is a general lack of information on the role of, for example, silicate and its synergistic interaction with nitrogen and phosphorous in controlling algal biomass in tropical coastal lagoons.

In Brazil, awareness by financial institutions of the need for environmental planning and clean-up is on the increase, and reasonable management plans have been proposed for many regions and ecosystems. The Brazilian environmental legislation (CONAMA) is exemplary. Also, a National Coastal Zone Management Plan was established recently (1988) to promote sustained protection of coastal ecosystems. The Coastal Zone Management Information System (SIGERCO) and a geographic information system (GIS) database already play important roles (Laydner 1996). However, solutions still need to be implemented, and they are likely to be very expensive and politically delicate – thus, slow and difficult to achieve. The ideal management strategy for tropical coastal lagoons, as for all ecosystems, is prevention rather than remedial action.

Acknowledgements

The manuscript is dedicated to Clarita (Clara Kjerfve) and Jojo (Johanna Knoppers). We would like to thank Dr. Werner Ekau for helping produce the graphics, Carlos Carvalho Bitton for helping compile the data tables, and Ursula Maria Neira Mendoza and Maria Eulália R. Carneiro for proof reading the references. We appreciate valuable comments on our manuscript by Drs. P.C. Abreu and C. Odebrecht and for providing us with unpublished data on Patos L. We also appreciate A. Schwarzbold for providing us with unpublished data. Over the years, our work on coastal lagoons has been supported by Conselho Nacional de Desenvolvimento Científico e Tecnológico (CNPq), National Science Foundation (NSF), National Geographic Society, Instituto Acqua through the PROLAGOS project, Financiadora Nacional Estudos e Projetos (FINEP), and Fundacão de Amparo a Pesquisa do Estado do Río de Janeiro (FAPERJ). This is contribution N° 1117 from the Belle W. Baruch Institute for Marine Biology and Coastal Research.

References

Abreu PC, Biddanda B, Odebrecht C (1992) Bacterial dynamics of the Patos Lagoon Estuary, southern Brazil (32°S, 52°W): relationship with phytoplankton production and suspended material Est Coast Shelf Science 6:621–635

Abreu PC, Granéli E, Odebrecht C, Kitzmann LA (1994a) Proença, and C Resgalla Jr, Effect of fish and mesozooplankton manipulation on the phytoplankton community in the Patos Lagoon Estuary, southern Brazil. Estuaries 17:575–584

Abreu PC, Odebrecht C, González A (1994b) Particulate and dissolved phytoplankton production of the Patos Lagoon Estuary, southern Brazil: Comparison of methods and influencing factors. J Plankton Res 16:737–753

Abreu PC, Hartmann C, Odebrecht C (1995) Nutrient-rich saltwater and its influence on the phytoplankton of the Patos Lagoon Estuary, southern Brazil. Est Coast Shelf Science 40:219–229

Adaime RR (1985) Produção do bosque de mangue da Gamboa Nobregua (Cananéia, 25° lat. S, Brasil.) PhD dissertation Instituto Oceanografico, Universidade de São Paulo

Alvarez-Borrego J, Alvarez-Borrego S (1982) Temporal and spatial variability of temperature in two coastal lagoons, CALCOFI Reports XXIII, pp 188–197

Atkinson MJ (1987) Low phosphorus sediments in a hypersaline marine bay. Est Coast Shelf Science 24:335–347

Belloto VR (1992) Regeneração de nutrientes nos sedimentos da Laguna de Saquarema, RJ Departamento de Geoquímica, Universidade Federal Fluminense, Niterói, RJ, Brazil

Bird EC (1994) Physical setting and geomorphology of coastal lagoons. In: Kjerfve B (ed) Coastal Lagoon Processes. Elsevier, Amsterdam, pp 9–39

Bruun P (1994) Engineering projects in coastal lagoons. In: Kjerfve B (ed) Coastal lagoon processes. Elsevier, Amsterdam, pp 507–533
Carlson RE (1977) A trophic state index for lakes. Limnol Oceanogr 22:361–369
Carneiro MER (1992) Ciclo anual do aporte fluvial e o estoque de matéria biogênica no sistema lagunar de Piratininga – RJ. MS thesis, Departamento de Geoquímica, Universidade Federal Fluminense, Brazil
Carneiro MER, Barroso LV, Ramalho NM, Azevedo Duarte C, Knoppers B, Kjerfve B, Kirstein KO (1993) Diagnóstico ambiental do sistema lagunar de Piratininga/Itaipu, RJ. Parte II: Hidroquímica. III Simpósio de Ecossistemas da Costa Brasileira subsídios a um gerenciamento ambiental, vol I, Mang uesais e Marismas, Academia Ciências, São Paulo, Brasil, pp 196–203
Carneiro MER, Azevedo C, Ramalho NM, Knoppers B (1994) A biomass de *Chara hornemannii* em relação ao compartamento físico-químico da Lagoa de Piratininga. Anaias Acad Brasil Cientifica 66:213–222
Carmouze, JP, Caumette P (1985) Les effects de la pollution organique sur les biomasses et activiés du fitoplancton et des bactéries hétérotrophiques dans le lagune Ebrié (Cote d'Ivoire), Revue Hydrobiologique Tropicale 17:175–189
Carmouze JP, Knoppers B, Vasconcelos P (1991) Metabolism of a subtropical Brazilian lagoon. Biogeochemistry 14:129–148
Carmouze JP, D'Elia CS, Farias B, Silva S, Azevedo S (1993) Ecological changes of a shallow Brazilian lagoon related to distrophic crisis. Verhandlungen Internationaler Verein der Limnologie 251:23–25
Cintrón-Molero G, Schaeffer-Novelli Y (1992) Ecology and management of new world mangroves. In: Seeliger U (ed) Coastal Plant Communities of Latin America, Academic Press, New York pp 233–258
Copeland BJ, Thompson JH, Ogletree WB (1968) Effects of wind on water levels in the Texas Laguna Madre, Texas, J Science 20:196–199
Costa CSB, Seeliger U, Kinas PG (1988) The effect of wind velocity and direction of the wind regime in the Lower Patos Lagoon Estuary, Ciência e Cultura 40:909–912
Coutinho R, Seeliger U (1986) Seasonal occurrence of benthic algae in the Patos Lagoon Estuary, Brazil. Est Coast Shelf Science 23:889–900
Couto ECG (1989) Produção, decomposição e composição química de *Typha dominguensis* Typhaceae na Lagoa de Guarapina, RJ. MSc thesis, Departamento de Geoquímica, Universidade Federal Fluminense, Brazil
Cromwell JE (1971) Barrier coastal distribution: A world survey. 2nd National Coastal and Shallow Water Research Conference, Abstract
Day Jr JW, Conners WH, Day RH, Ley-Lou F, Machado Navarro A (1988) Productivity and composition of mangrove forests at Boca Chica and Estero Pargo. In: Yáñez-Arancibia A, Day Jr JW (eds) Ecology of coastal ecosystems in the southern Gulf of Mexico: The Terminos Lagoon region. Instituto de Ciencias del Mar y Limnología, Universidad Nacional Autónoma de México pp 237–276
Delaney PJV (1963) Quaternary geologic history of the coastal plain of Río Grande do Sul, Brazil, Coastal Studies Institute, Louisiana State University, Tech Rep 7
Dias GTM, Silva CG (1984) Geologia de depósitos arenosos costeiros emersos – examplos longo do litoral Fluminense. In: Lacerda LD de, Araujo DSD, Cerqueira R, Turcq B (eds) Restingas; Origem, Estrutura, Processos. Universidade Federal Fluminense/CEUFF, Niterói, RJ, pp 47–60
Diegues AC, Oliveira ER, Moreira ACC, Marone E (1992) The vulnerability of principal Brazilian coastal ecosystems to climatic change and human impacts. In: The Environmental rmplications of global change, International Union for the Conservation of Nature and Natural Resources (IUCN), Gland, Switzerland, and Cambridge, UK pp 113–134
Dronkers J, Zimmerman JTF (1982) Some principles of mixing in tidal lagoons. Oceanol Acta Sp 107–117
Esteves FA, Ishii IH, Camargo AFM (1994) Pesquisas limnológicas em 14 lagoas do litoral do estado do Río de Janeiro. In: Lacerda LD de, Araujo DSD, Cerqueira R, Turcq B (eds) Restingas; Origem, Estrutura, Processos. Universidade Federal Fluminense/CEUFF, Niterói, RJ, pp 443–454
Fernandes LV, Marques Jr AN, Lira CA (1989) Estudos geoquímico da distribuição ertical de nutrientes na interface água-sedimento da Lagoa de Maricá – RJ. II Congresso Brasileiro de Geoquímica – Geoquímica Ambiental, Río de Janeiro, pp 339–350
Fernex F, Bernat M, Fernandes LV, Nepuceno Marques A (1992) Ammonification rates and ^{210}Pb in sediments from a lagoon under a wet tropical climate, Maricá, Río de Janeiro, Brazil. Hydrobiologia 242:69–76
Flores-Verdugo FJ (1985) Aporte de materia organica por los principales productores primarios a un ecosistema lagunar-estuarino de boca efimera, PhD dissertation, Instituto de Ciencias del Mar y Limnología, Universidad de Autónoma Nacional de México
Flores-Verdugo FJ, Day Jr JW, Mee L, Briseño-Dueñas R (1988) Phytoplankton production and seasonal biomass variation of seagrasses, *Ruppia maritima* L, in a tropical Mexican lagoon with an ephermeral inlet. Estuaries 11:51–56
Goltermann HL, Oude NT (1991) Eutrophication of lakes, rivers and coastal seas. In: Hutzinger O (ed) Handbook of environmental chemistry, vol 5A, Springer-Verlag, Berlin, pp 79–123

Guerra, AT (1962) O litoral Atlântico. In: Paisagens do Brazil. Instituto Brasileiro de Geografia e Estatistica (IBGE), Río de Janeiro, RJ
Herz R (1977) Circulação das águas de superfície da Lagoa dos Patos, PhD dissertation, Universidade de São Paulo, Brazil
Herz R (1991) Mangezais do Brasil. Special Publication, Instituto Oceanografico, Universidade de São Paulo, 233 pp
Hodgkin EP, Birch PB (1982) Eutrophication of a western Australian estuary. Oceanol Acta 5:313–319
Hodgkin EP, Birch PB (1986) No simple solutions: Proposing radical management options for an eutrophic estuary. Mar Pollution Bull 17:399–404
Keulegan GH (1967) Tidal flow in entrances: Water level fluctuations in basins in communication with seas. Corps of Engineers, US Army, Vicksburg, Tech Bull 14, 89 pp
Kjerfve B (1975) Tide and fair-weather wind effects in a bar-built estuary. In: Cronin LE (ed) Estuarine Research, vol 2, Academic Press pp 47–62
Kjerfve B (1986) Comparative oceanography of coastal lagoons. In: Wolfe DA (ed) Estuarine Variability. Academic Press, New York, pp 63–81
Kjerfve B (1994) Coastal lagoons. In: Kjerfve B (ed) Coastal lagoon processes. Elsevier, Amsterdam pp 1–8
Kjerfve B, Magill K (1989) Geographic and hydrographic characteristics of shallow coastal lagoons. Mar Geol 88:187–199
Kjerfve B, Knoppers B (1991) Tidal choking in a coastal lagoon, In: Parker BB (ed) Tidal Hydrodynamics. J Wiley & Sons, New York
Kjerfve B, Lacerda LD de (1993) Mangroves of Brazil. In: Lacerda LD de (ed) Conservation and sustainable utilization of mangrove forests in Latin America and Africa Regions, Part I – Latin America, ITTO/International Society for Mangrove Ecosystems, Okinawa, Japan pp 245–272
Kjerfve B, Knoppers B, Moreira P, Turcq B (1990) Hydrological regimes in Lagoa de Guarapina, a shallow Brazilian coastal lagoon. Acta Limnol Brasiliensia 3:931–949
Kjerfve B, Schettini CAF, Knoppers B, Lessa G, Ferreira HO (1996) Hydrology and salt balance in a large hypersaline coastal lagoon: Lagoa de Araruama, Brazil Est Coast Shelf Sciences 42:701–725
Knoppers B (1994) Aquatic primary production in coastal lagoons. In: Kjerfve B (ed) Coastal lagoon processes. Elsevier, Amsterdam, pp 243–286
Knoppers B, Moreira P (1988) The short term effect of physical processes upon nutrients, primary production, and sedimentation in Guarapina Lagoon (RJ), Brazil. Acta Limnol Brasiliensia 2:405–433
Knoppers B, Moreira P (1990) Matérial em suspensão e a sucessão de fitoplâncton na Lagoa de Guarapina (RJ). Acta Limnol Brasiliensia 3:291–317
Knoppers B, Opitz SS, Souza MP, Miguez CF (1984) The spatial distribution of particulate organic matter and some physical and chemical properties in Conceição Lagoon, Santa Catarina, Brazil, Arquivos Biologia e Tecnologia 27:59–77
Knoppers B, Kjerfve B, Carmouze JP (1991) Trophic state and water turn-over time in six choked coastal lagoons in Brazil. Biogeochemistry 14:149–166
Knoppers B, Landim de Souza WF, Landim de Souza MF, Gonçales ER, Vianna EC, Romanazzi A (1996) Benthic primary production, respiration and nutrient release rates in the hypersaline carbonate-rich lagoon of Araruama. Revista Brasileira de Oceanografía 44:155–165
Kuroshima KN (1995) Decomposição da matéria orgânica no sedimento da Lagoa da Barra – Maricá – RJ, MS thesis, Departamento de Geoquímica, Universidade Federal Fluminense, Niterói, RJ, Brazil
Lacerda LD de, Araújo DSD, Maciel NC (1993) Dry coastal ecosystems of the tropical Brazilian coast. In: van der Maarel E (ed) Dry coastal ecosystems Africa, America, Asia and Oceania. Elsevier, Amsterdam, pp 477–493
Lambou VW, Taylor WD, Hern SC, Williams LP (1983) Comparisons of trophic state measurements. Wat Res 27:1619–1626
Landim de Souza MF (1993) Distribuição espacial, sazonal e fontes fluviais de nutrientes na Lagoa de Araruama. RJ, MS thesis, Departamento de Geoquímica, Universidade Federal Fluminense, Niterói, RJ, Brazil
Laydner C (1996) The national coastal zone management programme In: Book of Abstracts: Socio-economic benefits of integrated coastal zone management, co-existence of economic development and ecosystem functioning: an international symposium, Bremen, Germany pp 27–28
Machado EC (1989) Desoxygenação e regeneração de nutrientes pelo sedimento da Lagoa de Guarapina, RJ, MS thesis, Departamento de Geoquímica, Universidade Federal Fluminense, Niterói, RJ, Brazil
Machado EC, Knoppers B (1988) Sediment oxygen consumption in an organic rich subtropical lagoon, Brazil. The Science of the Total Environment 75:341–349
Mandelli EF (1981) On the hydrography of some coastal lagoons of the Pacific coast of Mexico. In: Coastal lagoon research, present and future, UNESCO, Paris, pp 81–95

Martin L, Suguio K (1989) Excursion route along the Brazilian coast between Santos (state of São Paulo) and Campos (northern state of Río de Janeiro), International symposium on global changes in South America during the Quaternary: past-present-future, São Paulo, Brazil, 136 pp
Martin L, Dominguez JML (1994) Geological history of coastal lagoons. In: Kjerfve B (ed) Coastal Lagoon Processes. Elsevier, Amsterdam, pp 41–68
Martin L, Suguio K, Flexor JM, Bittencourt ACSP, Vilas Boas GS (1980) Quaternaire marin bresilien (littoral pauliste, sud-fluminense et bahianais). Cahiers Office du la Recherche Scientifique et Tecnique d'Outre-Mer, Serie Geologie 10:95–124
Martin L, Suguio K, Flexor JM (1988) Hauts nivaux marins Pleistocenes du littoral Bresilien. Palaeogeography, Palaeoclimatology, Palaeoecology 68:231–239
Mee LD (1978) Coastal lagoons. In: Riley JP, Chester R (eds) Chemical Oceanography 7. Academic Press, New York, pp 441–490
Merino M, Czitrom Jordán E, Martin E, Thomé P, Moreno O (1990) Hydrology and rain flushing of the Nichupté Lagoon system, Cancún, Mexico, Est Coast Shelf Science 30:223–237
Mesquita HS, Peres CA (1985) Numerical contribution of phytoplankton cells, heterotrophic particles and bacteria to size fractionated *POC* in the Cananéia Estuary (25°S 48°W), Brazil, Bol Instituto Oceanogr 33:69–78
Miyao SY, Nishihara L, Sarti CC (1986) Características físicas e químicas do sistema estuarino-lagunar de Cananéia Iguape. Bol Instituto Oceanogr 34:23–36
Miyao SY, Harari J (1989) Estudo preliminar da maré e das correntes de maré da região estuarina de Cananéia (25°S 48°W). Bol Instituto Oceanogr 37:107–123
Moreira PF (1988) Ciclo anual de nutrientes e produção primária na Lagoa de Guarapina, RJ. MS thesis, Departamento de Geoquímica, Universidade Federal Fluminense, Niterói, RJ, Brazil
Moreira PF, Knoppers B (1990) Ciclo anual de produção primária e nutrientes na Lagoa de Guarapina, RJ. Acta Limnol Brasiliensia 3:275–291
Muehe D (1984) Evidências de recuo dos cordões litorâneos em direção ao continente no litoral do Río de Janeiro. In: Lacerda LD de, Araujo DSD, Cerqueira R, Turcq B (eds) Restingas; Origem, Estrutura, Processos. Universidade Federal Fluminense/CEUFF, Niterói, RJ pp 75–80
Niencheski LF, Windom HL (1994) Nutrient flux and budget in Patos Lagoon Estuary. The Science of the Total Environment 149:53–60
Niencheski LF, Möller Jr OO, Odebrecht C, Fillman G (1988) Distribuição espacial de alguns parámetros físicos e químicos na Lagoa dos Patos – Porto Alegre a Río Grande, RS (verão 1986). Acta Limnol Brasiliensia 2:79–97
Nichols MM (1989) Sediment accumulation rates and relative sea-level rise in lagoons. Mar Geol 88:201–219
Nichols MM, Allen G (1981) Sedimentary processes in coastal lagoons. In: Coastal lagoon research, present and future, UNESCO, Paris, pp 27–80
Nixon SW (1982) Nutrient dynamic, primary production and fisheries yields of lagoons. Oceanol Acta 5:357–371
Nixon SW, Pilson M (1983) Nitrogen in estuarine and coastal marine ecosystems., In: Carpenter EJ, Capone DG (eds) Nitrogen in the marine environment. Academic Press, New York, pp 565–648
Odebrecht C (1988) Variações espaciais e sazonais do fitoplâncton, protozooplâncton, e metazooplâncton na Lagoa da Conceição, Ilha de Santa Catarina, Brasil. Atlantica 10:21–40
Odebrecht C, Caruso Jr F (1987) Hidrografía e matéria em suspensão na Lagoa da Conceição, Ilha de Santa Catarina, SC, Brasil. Atlantica, 9:83–104
Oliveira A, Kjerfve B (1993) Environmental responses of a tropical coastal lagoon system to hydrological variability: Mundaú-Manguaba, Brazil. Est Coast Shelf Science 37:575–591
Oliveira LPH de (1959) Limnologische Notizen über die Río de Janeiro-Lagunen. Archiv für Hydrobiologie 55:238–263
Pauly D, Yáñez-Arancibia A (1994) Fisheries in coastal lagoons. In: Kjerfve B (ed) Coastal lagoon processes. Elsevier, Amsterdam pp 377–399
Perrin P (1984) Evolução da costa fluminense entre as fronteras de Itacoatiara e Negra: preenchimentos e restingas. In: Lacerda LD de, Araujo DSD, Cerqueira R, Turcq B (eds) Restingas; Origem, Estrutura, Processos. Universidade Federal Fluminense/CEUFF, Niterói, RJ pp 65–73
Persich GR, Odebrecht C, Bergesch M, Abreu PC (1996) Eutrofizaçao e fitoplancton: comparacao entre duas enseadas rasas no estuario da Lagoa dos Patos. Atlantica 18:27–41
Phleger FB (1969) Some general features of coastal lagoons. In: Ayala-Castaneres A (ed) Lagunas Costeras. Universidad Nacional Autónoma de México, México, DF, pp 5–26
Proença LA (1990) Ciclo anual da produção primária, biomassa do fitoplancton e carbono organico particulado em área rasa da porção sul da Lagoa dos Patos, MS thesis, Universidade do Río Grande – FURG, RS, Brazil

Proença LA, Abreu PC, Odebrecht C (1988) Nutrientes inorganicos em água doce, meso-oligohalina e mixo-poli-euhalina no canal de acesso a Lagoa dos Patos, RS, Brazil. Acta Limnol Brasiliensia 2:57–77
Pritchard DW (1961) Salt balance and exchange rate for Chincoteague Bay, Chesapeake Sci 1:48–57
Rast W, Holland M (1988) Eutrophication in lakes and reservoirs: a framework for making management decisions. Ambio 17:1–12
Reyes M, Merino M (1991) Diel oxygen dynamics and eutrophication in a shallow well-mixed tropical lagoon (Cancun, Mexico). Estuaries 14:372–381
Schäffer, A (1988) Tipificação ecológica das lagoas costeiras do Río Grande do Sul, Brasil Acta Limnol Brasiliensia 2:29–55
Schaeffer-Novelli Y (1989) Perfil dos ecossistemas litorâneos brasileiros, com special ênfase sobre o ecossistema manguezal, Special Publication 7, Instituto Oceanografico, Universidade de São Paulo, 16 pp
Schwarzbold A (1982) Influencia da morfologia no balanço de substancias e na distribuição de macrófitas aquáticas nas lagoas costeiras do Río Grande do Sul. PhD dissertation, Universidade Federal do Río Grande do Sul, Porto Alegre, Brasil
Schwarzbold A, Schäffer (1984) A genese e morfologia das lagoas costeiras do Río Grande do Sul. Amazoniana 9:87–104
Seitzinger SP (1988) Denitrification in freshwater and coastal marine ecosystems: ecological and geochemical significance. Limnol Oceanogr 33:702–724
Sikora WB, Kjerfve B (1985) Factors influencing the salinity of Lake Pontchartrain, Louisiana, a shallow coastal lagoon: analysis of a long-term data set. Estuaries 8:170–180
Sfrizo A, Pavoni B, Marcomini A, Orio AA (1988) Annual variations of nutrients in the lagoon of Venice. Mar Pollution Bull 19:54–60
Smith SV, Atkinson MJ (1994) Mass balance analysis of carbon, nitrogen, and phosphorous fluxes in coastal water bodies, including lagoons. In: Kjerfve B (ed) Coastal lagoon processes. Elsevier, Amsterdamm, pp 133–155
Suguio K, Martin L, Flexor JM (1980) Sea-level fluctuations during the past 6 000 years along the coast of the state of São Paulo (Brazil). In: Mörner NA (ed) Earth Rheology, Isostasy and Eustacy. John Wiley and Sons, N.Y. pp 471–486
Suguio K, Martin L, Bittencourt ACSP, Dominguez JML, Flexor JM (1984) Quaternary emergent and submergent coasts: comparison of the Holocene sedimentation in Brazil and southeastern United States. Anais Acad Brasileira Ciencias 56:163–167
Tundisi J, Tundisi TM, Kuttner MB (1973) Plankton studies in a mangrove envrironment, VIII Further investigations on primary production, standing stock of phyto- and zooplankton and some environmental factors. Intern Rev Gesamet Hydrobiologie 58:325–345
Tundisi J, Teixeira C, Tundisi TM, Kuttner MB, Kinoshita L (1978) Plankton studies in a mangrove environment, IX, Comparative investigations with coastal oligotrophic waters. Rev Brasileira Biología 38:301–320
Vaulot D, Frisoni GF (1986) Phytoplanktonic productivity and nutrients in five Mediterranean lagoons. Oceanol Acta 9:57–63
Vilas Boas D (1990) Distribução e comportamento doas sais nutrientes, elementos maiores e metais pesados na Lagoa dos Patos – RS, MSc thesis, Fundação Universidade do Río Grande, Río Grande, RS
Vollenweider RA (1968) Scientific fundamentals of eutrophication of lakes and flowing waters, with particular reference to nitrogen and phosphorous as factors in eutrophication, OECD, Paris
Vollenweider RA (1992) Coastal marine eutrophication: Principles and control. In: Vollenweider RA, Marchetti R, Viviani R (eds) Marine coastal eutrophication. Science of the Total Environment, Supplement Elsevier, Amsterdam pp 1–20
Vollenweider RA, Kerekes JJ (1982) Eutrophication of waters: monitoring, assessment and control, OECD, Paris
Yáñez-Arancibia A (1987) Lagunas costeras y estuarios: cronología, criterios, y conceptos para una clasificación ecológica de sistemas costeros. Rev Soc Mexicana Hist Nat 39:35–54
Zee DMW, Lima ALC, Moreira MHR (1993) Some considerations about a coastal lagoon (RJ, Brasil) trophic level. In: Magoon OT, Wilson WS, Converse H, Tobin LT (eds) Proceedings of the eighth symposium on coastal and ocean management. New Orleans, LA, pp 2270–2282
Zeitzschel B, Davies JM (1978) Benthic growth chambers. Rapport P-v Reunion du Conseil International pour la Exploration du Mer 173:31–42
Zeitzschel B (1980) Sediment-water interactions in nutrient dynamics. In: Tenore KR, Coull BC (eds) Marine benthic dynamics. University of South Carolina Press, Columbia, SC pp 195–218
Zimmerman JTF (1981) The flushing of well-mixed tidal lagoons and its seasonal fluctuation. In: Coastal lagoon research, Present and Future. UNESCO, Paris, pp 15–26

Chapter 4

Nutrients and Suspended Matter Behaviour in the Patos Lagoon Estuary (Brazil)

Luis Felipe Niencheski · Maria da Graça Baumgarten
Gilberto Fillmann · Herbert L. Windom

4.1 Introduction

Estuaries are distinctive environments where continental runoff interacts with sea water. They are characterized by complex and usually strong circulation and this, along with the typically shallow depths of estuarine systems, leads to intense sediment-water interactions. Estuaries are generally also biologically productive.

Because human populations are concentrated in coastal regions, characteristics of estuaries can be influenced by man leading to changes in processes that occur there. A major impact of human activity is the release of contaminants and nutrients, the ultimate concentrations of which are controlled by the mixing of freshwater and seawater and by exchange between different components of the estuarine system (e.g., sediments, water column, biota).

Estuarine water quality is the result of an interplay of factors including material inputs, water movement and in situ biological and chemical processes. The changing chemical and physical conditions encountered in estuaries, along the salinity gradient (i.e., from freshwater to saltwater), may lead to changes in the solubility of substances resulting in either their removal from solution onto particles or their leaching from particles into solution. Chemical precipitation of substances may lead to the formation of particles which are preferentially removed from the water column during transport through estuarine systems, and biological processes may result in the exchange of materials, particularly nutrients, from particulate to dissolved form and vice versa.

The aim of this study was to describe basic chemical conditions of the estuary of the Patos-Mirim Lagoon system. The study is based on several cruises made during different environmental conditions, and involved the measurement of salinity, dissolved oxygen, suspended matter and nutrients. The main focus of the study was on the input of these substances to the estuarine system and processes affecting their behaviour.

The Patos Lagoon (10 360 km^2) and Mirim Lagoon (3 749 km^2) are located about 2 000 km South of Río de Janeiro and form the largest lagoonal system of South America (Fig. 4.1). The communication between these lagoons is through the 70 km long São Gonçalo Channel. The watershed emptying into these lagoons is about 201 626 km^2, of which 75% belongs to the Patos Lagoon. The only contact with the sea is through an inlet at the southern end of Patos Lagoon. Both lagoons are shallow (average deep about 6 m). Geomorphologically the southern region of the Patos Lagoon has the characteristics of a bar-built estuary with a 30 km wide upper limnic part which gradually, over 50 km, narrows into a 700 m wide access channel. All along

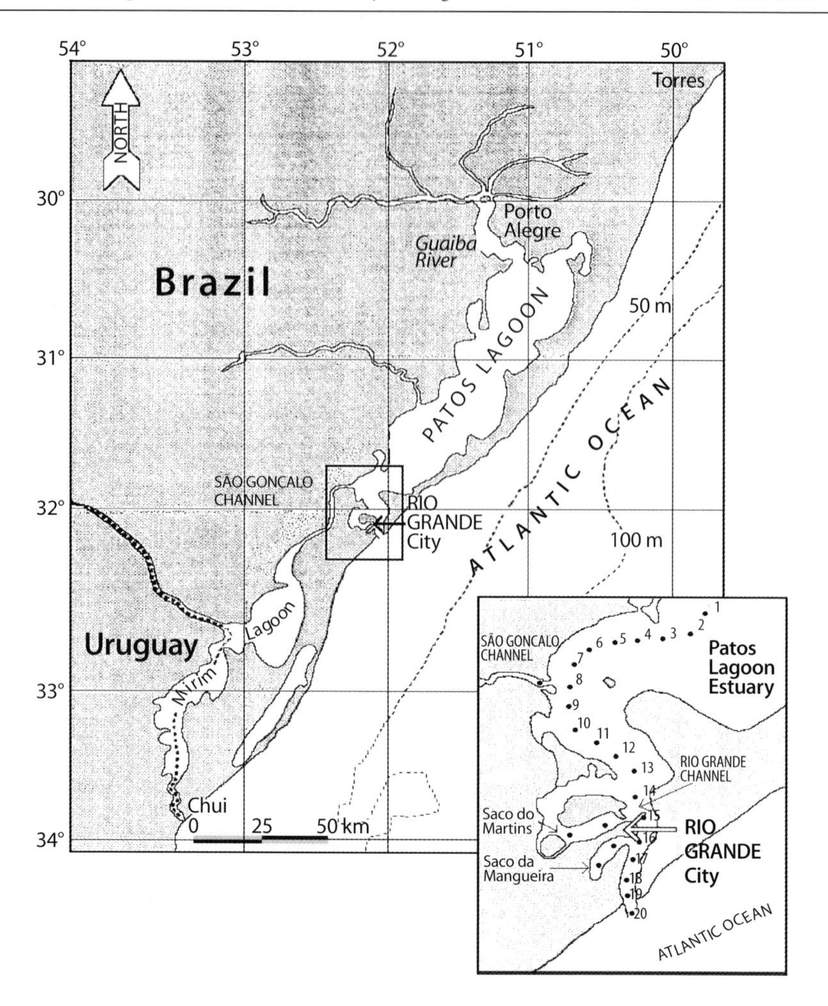

Fig. 4.1. Map of Patos-Mirim lagoon system showing station locations in the estuarine area

the estuarine area there is a navigation channel with average deep about 12 m. Almost 80% of this area is less than 2 m deep (Hartmann 1988). Because of this, wind action, waves and currents cause frequent resuspension of sediments. Additionally, the tidal range within the estuary is small (0.47 m) due to the low diurnal tidal amplitude in the adjacent South Atlantic (Möller et al. 1991).

Freshwater input to the estuarine region from Patos Lagoon originates from its main tributaries located in the upper region (Fig. 4.1). The Guaíba River, including the Jacuí, Sinos, Caí and Gravataí tributaries, supplies 86% of the average total freshwater input with most of the remainder coming from the Camaquã River. These rivers (along with inputs from the Southern Atlantic Ocean via the Patos Lagoon inlet canal) represent the major transport pathway of materials, such as nutrients and suspended matter, to the estuary. Nutrients originating from river inputs transits 250 km before reaching the estuarine region, and thus have a significant residence time within the lagoon during which biogeochemical process may result in their consumption, consequently reducing the inputs associated with freshwater to the estuary. In Patos Lagoon, the residence time depends more upon meteorological conditions (wind, rainfall, evaporation, etc.) than on tidal exchange because of the low tidal range (Möller et al. 1991). The time that water takes to travel from Guaíba River to the estuarine region is about 20 days (Herz 1977). In this area the annual rainfall exceeds annual evaporation. Vieira and Rangel (1988), using data collected over a twenty year period (1957–1977), calculated the annual rainfall to be 132 cm and the annual evaporation to be as ca. 90 cm.

There have been few studies of nutrients and suspended matter behaviour in this system. Vilas Boas (1990), however, suggests that the Northern part of Patos Lagoon may receive significant anthropogenic inputs as a result of rapid population growth and industrialisation along the northern region, particularly during the last few decades around Porto Alegre, the fifth largest Brazilian city. Another additional input to the estuary is that from São Gonçalo Channel, which has a discharge averaged over 30 year period of 700 $m^3 s^{-1}$ (Vieira and Rangel 1988). Concentrations of nutrients and suspended sediments in the São Gonçalo Channel input from Mirim Lagoon are the result of agricultural activities in its watershed (51 194 km^2) and untreated sewage from Pelotas city (300 000 hab.) which is located on the NE border of São Gonçalo Channel. Patos Lagoon Estuary also receives high nutrient inputs from Río Grande's urban area (200 000 hab.) and associated harbour and industrial activities (Almeida et al. 1993).

In spite of the constant short-term variability of salinity, dissolved oxygen, suspended matter and nutrients (Kantin and Baumgarten 1982; Kantin 1983; Niencheski et al. 1986; 1988; Baumgarten and Niencheski 1990) phytoplankton production and biomass present a clear seasonal pattern (Proença 1990) with maximum biomass values (10.56 mg l^{-1}) being related to the increase in light intensity, temperature and dissolved inorganic nutrients during spring and summer and low values during fall and winter (less than 2 mg l^{-1}) (Abreu 1992).

4.2 Materials and Methods

Water samples were collected in the southern part of the Patos Lagoon, during 8 cruises at twenty fixed stations within the estuarine mixing zone and one fixed station in São Gonçalo Channel in the main navigation channel and, during 15 cruises, covering four stations in semi-enclosed bays surrounding Río Grande City (Saco da Mangueira and Saco do Martins), and three stations in the Río Grande Channel (Fig. 4.1).The freshwater input from Patos Lagoon to the upper estuarine region was determined using data from Station 1 samples.

To obtain a better synoptic view of variability within the mixing zone, each of the 8 cruises were conducted during one day, using two vessels. The cruise dates were: 1) June 30, 1989, 2) August 30, 1989, 3) May 30, 1990, 4) June 30, 1990, 5) August 1, 1990, 6) August 30, 1990, 7) December 1, 1990, and 8) April 22, 1991. The 15 monthly cruises were carried out from May 1990 to July 1991.

Field teams worked concurrently so that samples could be collected and returned to the laboratory as quickly as possible. At each station, except in the semi-enclosed bays, vertical profiles of water temperature, salinity and conductivity were taken at

1.0 m depth intervals. Measurements were obtained using Yellow Springs Instruments Model 33 S-C-T meters. Each day, prior the sampling, all primary and backup instruments were intercalibrated.

In the main navigation channel, both surface and bottom water samples were collected using Van Dorn bottles (1.5 l) and placed into clean 1.0 l plastic bottles (nutrients) or into a 300 ml *BOD* glass bottles (dissolved oxygen). During the 15 monthly cruises only surface water sample were collected.

Samples for nutrient analysis were immediately filtered through cellulose acetate 0.45 mm filters and analysed immediately for ammonium. Filtered aliquots for nitrate, nitrite, silicate and phosphate determination were placed in individually cleaned polyethylene bottles and stored in a freezer until analysed ashore (Grasshoff et al. 1983).

Dissolved oxygen was analysed using the method of Grasshoff et al. (1983). Analysis of total suspended matter, ammonium, nitrite, nitrate, phosphate and silicate followed the procedures described by Aminot and Chaussepied (1983).

4.2.1 Data Reduction for Determining Estuarine Behaviour of Nutrients and Total Suspended Matter

Estuaries are often elongated and relatively shallow. The major interest in the spatial variability of materials has generally been in their distribution along the longitudinal axis of the estuary, in relation to the salinity gradient.

Advection-diffusion models have been used by many investigators to interpret estuarine chemical data referenced to salinity (e.g., Li and Chan 1979; Kaul and Froelich 1984). The distribution of a constituent in estuarine waters can be compared to salinity to determine whether a substance is:

1. conservatively transported through the estuary,
2. removed from the water column or
3. added to the water column via local inputs.

The only assumption required is that the concentrations of the constituent in the freshwater and oceanic end members are constant over the residence time of the estuary. For Patos Lagoon, this is assumed to be satisfied sufficiently to draw the conclusions presented in this paper.

In this study, data for dissolved nutrients and *TSM* were plotted against salinity for each cruise. Comparing concentrations of a substance relative to salinity, which is assumed to be conservative, along the salinity gradient provides a basis for judging the estuarine behaviour of the substance. For this purpose a line connecting the mean freshwater concentration to the concentration at highest salinity (i.e., the oceanic end members) is used to evaluate the deviation of concentrations, along the salinity gradient, from conservative mixing.

For cruises 1, 2 and 6, conducted in the main channel, where we do not have data from freshwater regions, we have estimated the freshwater end member (average and standard deviation) using historical data from Patos Lagoon (Vilas Boas 1990). This estimate is shown in figures using diamond symbols for estuarine nutrient distributions in the following sections.

4.3 Results and Discussions

4.3.1 Salinity

The penetration of seawater through the channel, responsible for the horizontal and vertical salinity distribution within the estuary, is more dependent on meteorological factors such as wind and precipitation than on tidal effects, which dominate many estuarine systems. Kjerfve (1986) attributes this feature to the large area of the estuary and the low elevation of the surroundings dune-oriented topography. Wind is thus the major factor controlling circulation and dispersion.

During three of the cruises (1, 3 and 7) the upper freshwater region of the estuary was well mixed (Fig. 4.2). From the middle of the estuary to the mouth, the water column was stratified. When rainfall is low and freshwater discharge is less important, as was the case during these cruises, water surface elevation at the head of Patos Lagoon is lower. This along with SW winds favours saltwater inflow, resulting in the observed stratification.

During cruise 2 (August, '89), the sampling region had higher salinity and was well mixed. Cruises 4 and 5 were carried out during periods of high rainfall when the estuary was completely fresh. Hartmann (1988) showed that the discharge from Patos Lagoon Estuary, during the winter, forms a low salinity plume that can penetrate 40 km across the shelf. The annual peak winter fresh water input exceeds 1500 $m^3 s^{-1}$ with extremes reaching 25000 $m^3 s^{-1}$ (Herz 1977). During cruises 6 and 8, the estuary was stratified except at either end (stations 1 and 20).

The stratification observed in cruises 1, 3, 6, 7 and 8 results in a modification of transport of material through Patos Lagoon Estuary which coupled with climatologic conditions favours removal and storage within the estuarine region.

4.3.2 Dissolved Oxygen

Dissolved oxygen in the estuary was close to 100% saturation during all cruises. Values decreased to 70% rarely. Because Patos Lagoon Estuary is shallow (approx. 50% of its area is less than 1 m depth) and windy (mean monthly wind velocity is 5.12 $m s^{-1}$ averaged over 30 years (Vieira and Rangel 1988), it is well oxygenated. Also, because of these features, it is unusual to find thermal stratification, except in the navigation channel (12 m depth).

4.3.3 Total Suspended Matter

During all cruises within the main navigation channel, observed total suspended matter (*TSM*) concentrations increased towards the sea. Similar patterns were found by Brockmann (1990) for the Elbe and Shannon estuaries. In these estuaries the cause of turbidity maxima is estuarine circulation and in Patos Lagoon Estuary the circulation is likely to be more dependent on meteorological conditions since wind than tidal events. There is no tide in Patos Lagoon Estuary because it's located close to an amphidromic point (region without tides) (200 km South) (Möller et al. 1991).

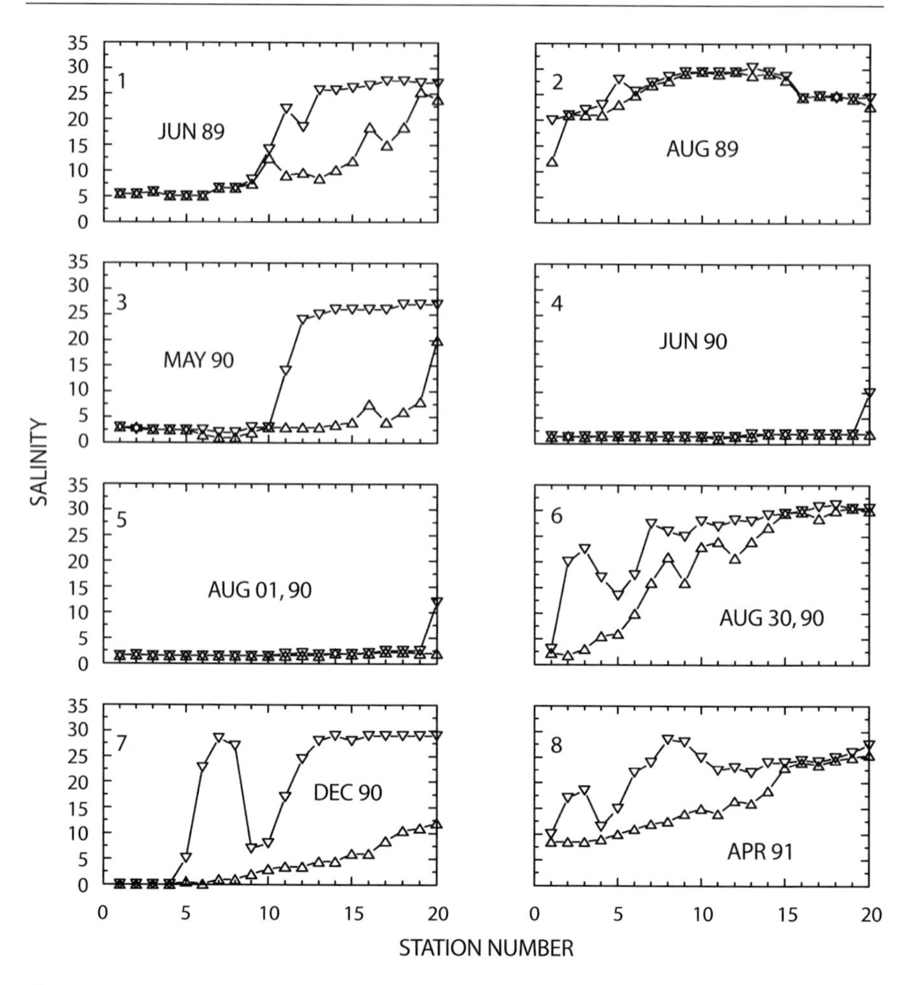

Fig. 4.2. Longitudinal (by station number) salinity observed during the eight cruises; Δ surface values; ∇ bottom values

Within the Patos Lagoon Estuary, the increase of *TSM* is due to the inflow of saltwater and the geomorphology of the Patos Lagoon Estuary that ends in a narrow access channel; conditions that favour resuspension of bottom sediments, which results in an average value of 50 mg l^{-1} (Niencheski and Windom 1994).

During two winter cruises (cruises 4 and 5), when salinity was low, the estuary had the lowest concentration of *TSM*. During this time, the *TSM* values might be expected to be high due to suspended materials discharged by the rivers. However suspended material input to Patos Lagoon by rivers occurs primarily in the north followed by desorption to sediments in that region.

Wind stress clearly exerts dominant control on *TSM* through resuspension in shallow areas of the estuarine zone. Winds from the South are responsible for the highest concentrations of *TSM*, above the average value of 30 mg l^{-1}.

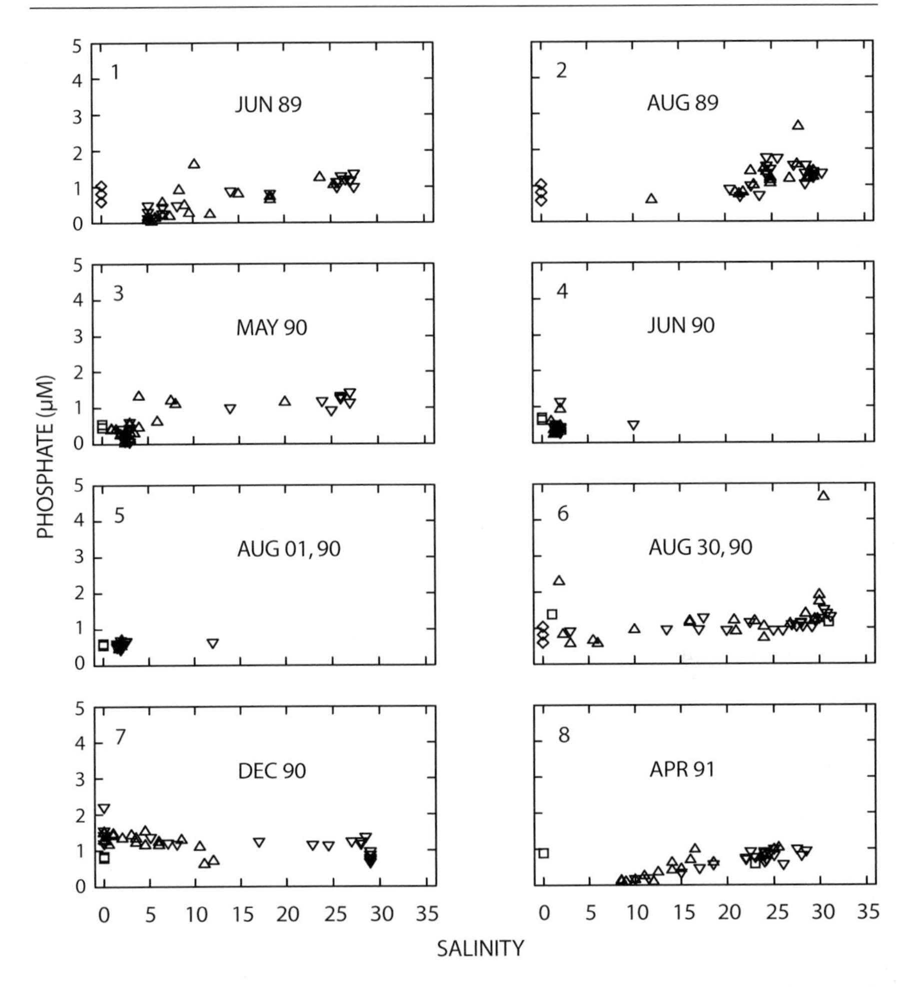

Fig. 4.3. Phosphate concentrations plotted against salinity; Δ surface values; ∇ bottom values; □ values from São Gonçalo Channel; ◇ average and standard deviation from freshwater end members

The observations discussed above suggest that water quality characteristics of Patos Lagoon Estuary is strongly dependent on meteorological conditions, which can vary from year to year. For example, rainfall exhibits irregular distributions between years in the Patos-Mirim Lagoon watershed, as do wind regimes.

4.3.4 Phosphate

Mixing diagrams for phosphate (Fig. 4.3), for the first (winter) and third (fall) cruises, suggest that this nutrient is removed from the water column in the lower salinity region (less than ca. 10) and released at salinities above ca. 10. For the second cruise, when salinity was high (between 20 and 30) throughout the study area, phosphate was enriched in the water column. For cruises 4 (winter) and 5 (winter) the phosphate concentration throughout the study area was dominated by the freshwater end-member and was relatively low.

During cruises 6 and 7, phosphate is generally stable throughout the estuary, increasing in the seaward direction in the cruise 6 and decreasing in a seaward direction during cruise 7. Finally, for cruise 8 (fall) phosphate was depleted in the low salinity region; increasing from virtually zero, at low salinities, toward the sea and exhibiting conservative behaviour.

As pointed out above, when Patos Lagoon is fresh, the concentration of both phosphate and *TSM* decrease. When the system has an established salinity gradient, the phosphate and *TSM* concentrations generally rise with increasing salinity.

Baumgarten and Niencheski (1990) showed that iron is presented in high concentrations in freshwater at the upper estuarine region. Iron compounds can react with phosphate and that are know to adsorb phosphate (Jonge and Villerius 1989). On the other hand, Niencheski et al. (1994) presented the carbonate content of samples range from 7.3–10% for those collected at the upper estuary and from 10–28% for those collected near the mouth of the estuary. This suggests that during transport to the upstream, some calcite formed at sea and in the estuary, dissolves and consequently some of the phosphate desorbs (Jonge and Villerius 1989). Although, the adsorption of phosphate by other suspended matter components, such as clay minerals and iron it seems to be more important, resulting in low concentration and consequently causing perturbation of the buffer phosphate mechanism, cited by Liss (1976). Presumably also occurs phosphate uptake by primary producers.

The higher salinity region presents phosphate concentrations higher than at the upper estuary, average $<2\ \mu M$, value considered normal in no impacted estuaries (Aminot and Chaussepied 1983). This suggests that a dissolved phosphate fraction is in equilibrium with an adsorbed fraction, with the latter acting as a buffer (Jonge and Villerius 1989) and indicating that a buffer mechanism operates under estuarine conditions.

The phosphate concentration found in the São Gonçalo Channel, during cruise 8, was comparatively larger than the estuarine values. It is obvious however, that at this time the input from the São Gonçalo had little affect on phosphate concentrations in the estuarine region. This is probably due to phosphate consumption by phytoplankton and/or to dilution of the input from São Gonçalo Channel by the much larger freshwater input from the northern part of Patos Lagoon.

On the estuarine semi-enclosed bays, highest phosphate values were observed in Saco da Mangueira during all cruises (Fig. 4.4). Data for Saco do Martins are scattered, but generally fit within the normal distribution observed for other stations in the study area. These two embayments have similar geomorphology and depth and receive untreated sewage inputs (Almeida et al. 1993). This suggests that the increased concentrations observed in Saco da Mangueira, in the South of Río Grande City, is not only due to natural inputs or sewage discharge but are probably due to industrial releases, especially from fertilizer complex, one of the largest in South America (annual produc-tion approximately of ca. 1.0×10^6 tons), located on its border (Baumgarten et al. 1995).

Taft and Taylor (1976) cited several studies concerning phosphorus cycling in coastal plain estuaries, where highest phosphate concentrations were observed in the sum-

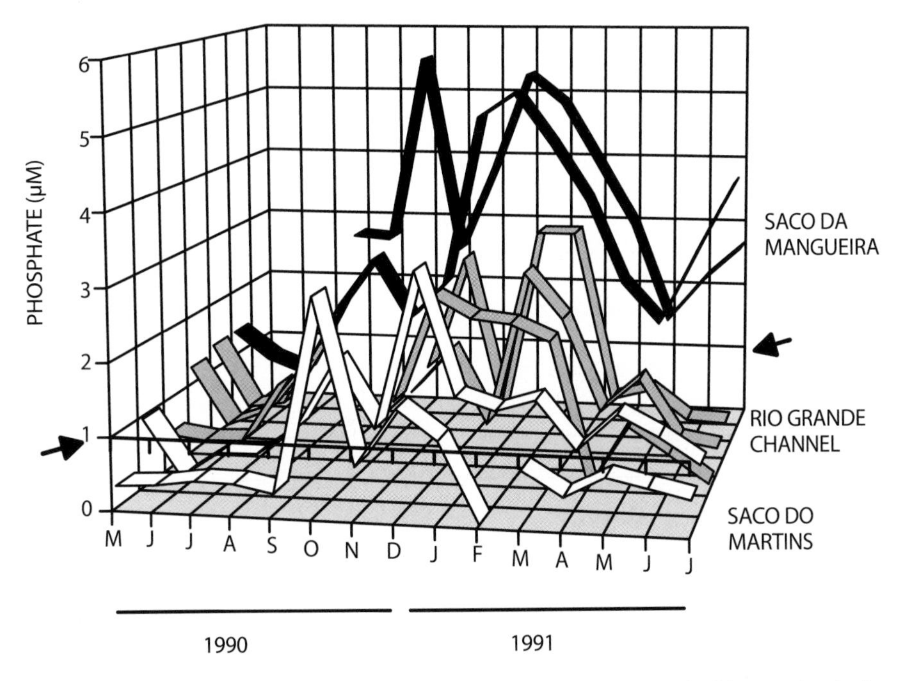

Fig. 4.4. Longitudinal (by month) phosphate concentrations observed during the fifteen cruises in the surrounding waters of Río Grande City. *Arrows* indicates the usual value found in no impacted estuaries

mer and the lowest concentrations observed during winter. The same pattern was observed for the surroundings waters of Río Grande City.

When these areas are dominated by freshwater, the phosphate concentrations seem to follow the same behaviour described for the upper estuary. In Saco da Mangueira, the phosphate concentrations depend on the action of salinity over the amount of phosphorus released by fertilizers factories. When this region is dominated by seawater, there is evidence to suggest that phosphate adsorption is depressed by increasing salinity. This may be a reflection of an ion exchange adsorption mechanism in which there is greater competition for ion exchange sites by other anions, e.g. Cl^-, SO_4^{2-}, Br^-, at higher salinities (Aston 1980). Liss (1976) presents that the decrease of magnitude of phosphate removal when salinity increase, may be due to blocking of anion exchange sites by ions such Cl^-, and SO_4^{2-} in the seawater.

4.3.5
Nitrate, Nitrite and Ammonium

Nitrate concentrations versus salinity during cruise 1 (Fig. 4.5) suggest production in situ by microbial processes (nitrification) in the upper estuarine region. For cruise 3, in situ production occurred throughout the estuary. For cruises 6 and 7, nitrate was generally conservative. However, during cruise 7 some remobilization occurred at high salinities. During cruise 8, nitrate concentration variations with salinity followed

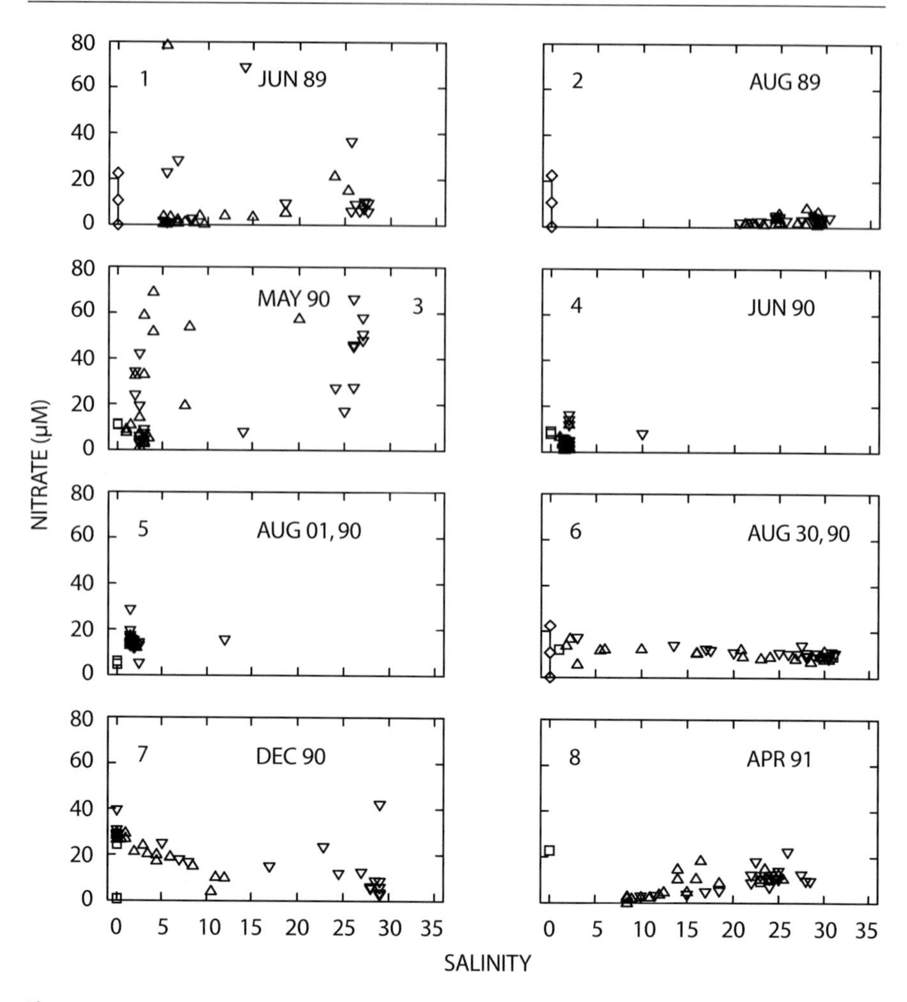

Fig. 4.5. Nitrate concentrations plotted against salinity. Δ surface values; ∇ bottom values □ values from São Gonçalo Channel; ◇ average and standard deviation from freshwater end members

closely phosphate concentration variations. Nitrite behaved similar to nitrate except for cruise 7 during which nitrite remobilization occurred.

Ammonium (Fig. 4.6) appears to be removed during cruises 1 and 3, at low salinity although in situ production appears at higher salinities in the first cruise. During the two 1990 winter cruises (June and August), the ammonium concentrations were elevated due to inputs from Patos Lagoon and São Gonçalo Channel. Also, surface waters near the mouth of the estuary had more elevated concentrations. This may be due to local atmospheric input resulting from the Río Grande's industrial park, which has been previously observed by Vilas Boas (1990).

During the sixth and seventh cruises, ammonium concentrations in the estuarine zone were greatly influenced by seawater penetration along the bottom which

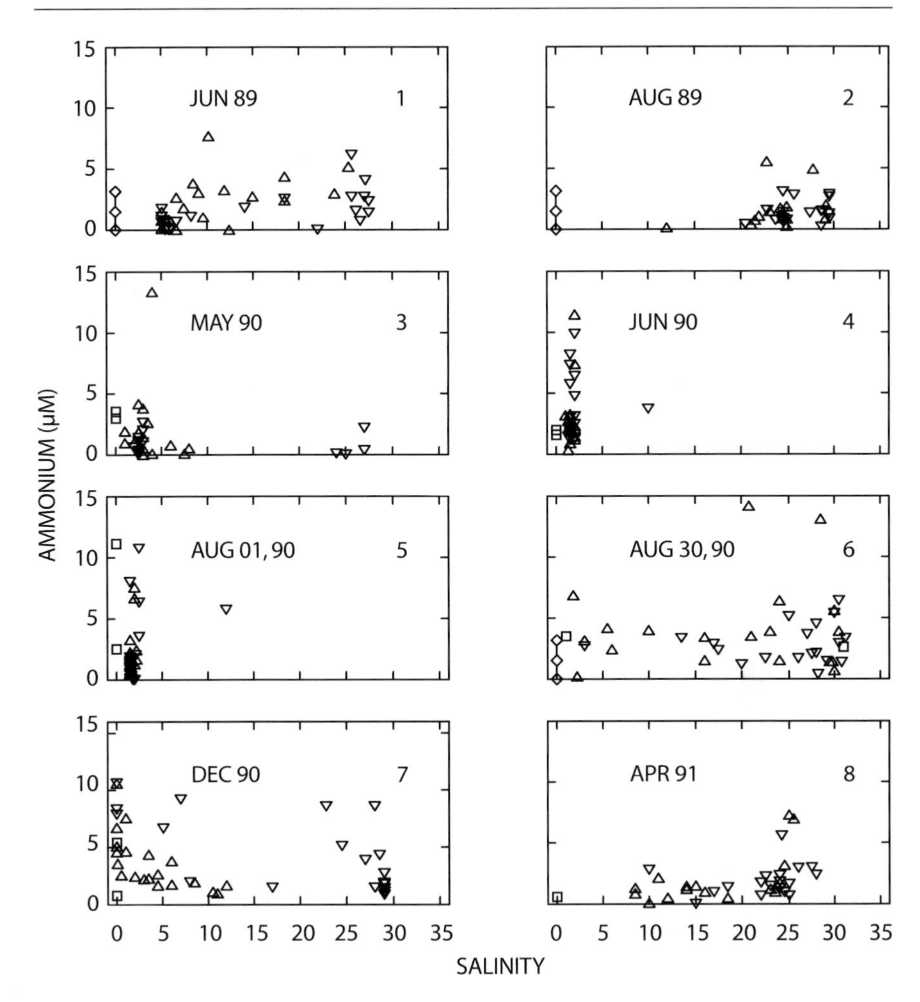

Fig. 4.6. Ammonium concentrations plotted against salinity. Δ surface values; ∇ bottom values; □ values from São Gonçalo Channel; ◇ average and standard deviation from freshwater end members

resuspended sediments and enhance sediment-water exchange. The ammonium concentration in the inputs from Patos Lagoon and São Gonçalo Channel were similar. Concentrations in the São Gonçalo Channel input from Mirim Lagoon are the result of agricultural activities in that watershed and untreated sewage from the city of Pelotas.

During cruise 8, ammonium was removed in the upper estuary and remobilized at higher salinities exhibiting variation versus salinity similar to that observed for phosphate, nitrite and nitrate. The period between cruises 7 and 8, is usually the season of high phytoplankton production in the Patos Lagoon Estuary (Kantin 1983) and this probably explains the depletion of dissolved phosphate and nitrogen in the upper estuary.

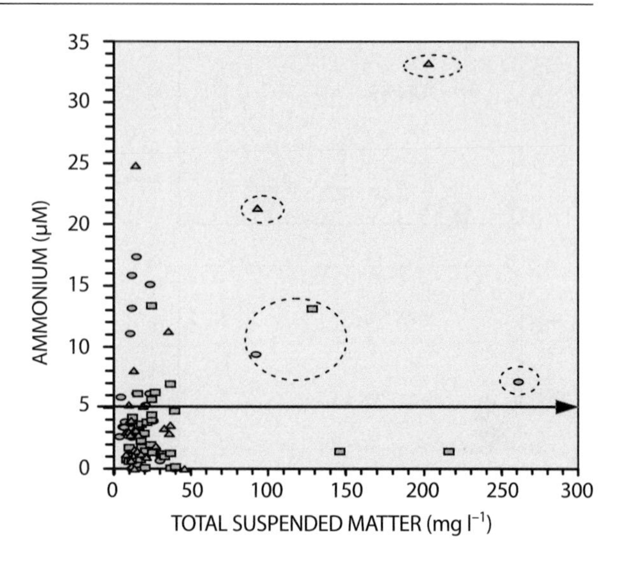

Fig. 4.7. Ammonium concentrations plotted against total suspended matter. The *points* inside the *circles* are the result of the strong south winds. The *line* indicates the maximum value from pristine estuaries

Like phosphate, nitrate and nitrite, ammonium concentrations are higher in samples from São Gonçalo Channel but the influence of this input is not observed at stations further along the longitudinal estuarine axis (i.e., salinity gradient) due to dilution from the greater input from the Northern part of Patos Lagoon.

Ammonium concentrations were higher in shallow waters due to mainly by two factors:

i. proximity of anthropogenic inputs, and
ii. resuspension of sediments rich in anthropogenic organic matter.

Figure 4.7 distinguishes these two features.

4.3.6 Silicate

The input from Patos Lagoon had an average silicate concentration of 175 μm, similar to that reported by Meybeck (1981) for the average discharge of 62% of the total world runoff. Although there is some variability in silicate concentrations along the salinity gradient, it generally decreases seaward in a conservative manner, except for the cruises dominated by the freshwater end-member (cruises 4 and 5) (Fig. 4.8).

In Patos Lagoon Estuary, diatoms are an important component of suspended matter (Hartmann 1988), suggesting dissolved silicate removal in the estuary. The silicate inputs from Patos Lagoon and São Gonçalo Channel appear to supply the uptake by phytoplankton with no apparent significant depletion.

Officer and Ryther (1980) presented some arguments that silicate is often the controlling nutrient in shifting production from diatoms to a flagellates. In Patos Lagoon Estuary silicate never appears to be limiting, thus providing favourable conditions for diatom production. Diatoms, which provide the major food source for zooplankton

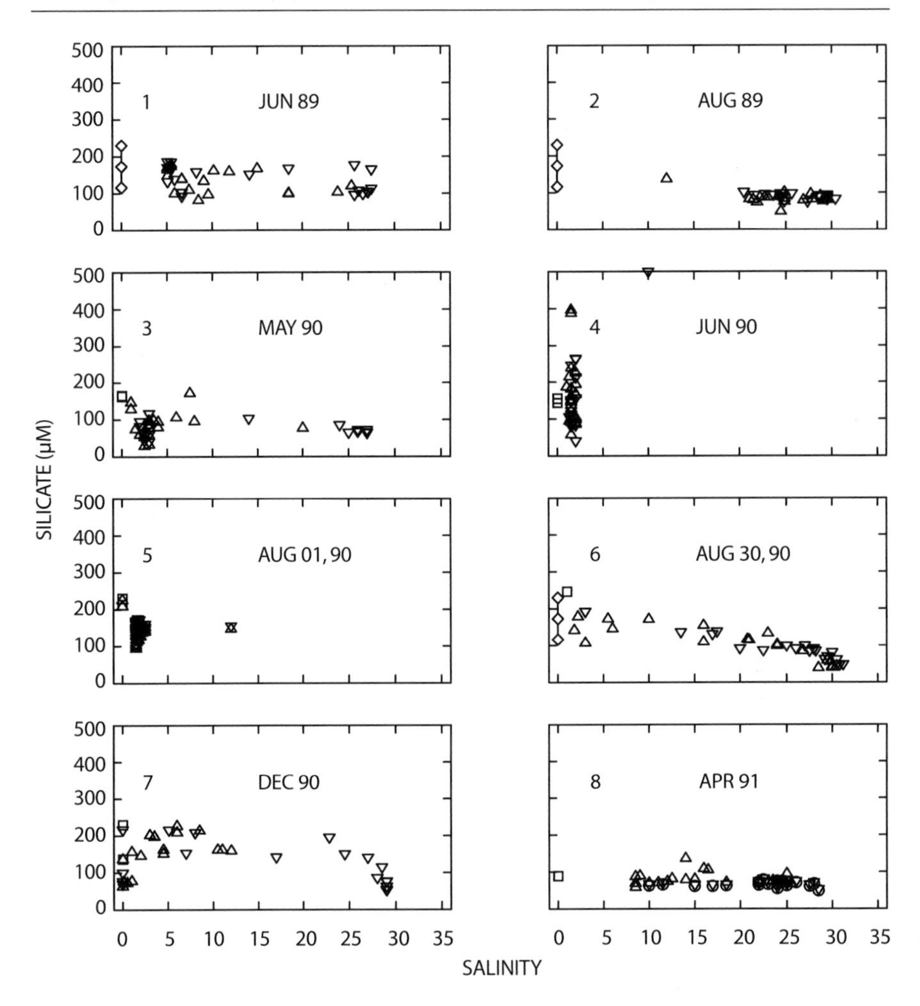

Fig. 4.8. Silicate concentrations plotted against salinity. Δ surface values; ∇ bottom values; □ values from São Gonçalo Channel; ◇ average and standard deviation from freshwater end members

and filter feeding fishes, supports commercial fish populations in the coastal zones, and are rarely a nuisance. Flagellate blooms, however, that generally result when silicate is limiting, are poor food for most grazers and are often a nuisance (oxygen depletion, red tides, etc.).

It is difficult to observe the influence of phytoplankton uptake because the sampling stations were spread over 50 km in the estuarine zone which receives inputs from several sources. Odebrecht et al. (1986) investigated variations of phytoplankton biomass in Patos Lagoon and observed high concentrations (>8 mg Chl-a l^{-1}) in the northern area, whereas lower values (<2.5 mg Chl-a l^{-1}) occurred in the estuarine region. They observed that nutrients were not limiting and therefore could not be the reason for the lower productions.

4.4 Conclusions

Nutrients concentrations in the Patos Lagoon Estuary exhibit small seasonal variability. From previous data (Kantin 1983; Niencheski et al. 1986; Vilas Boas 1990) annual nutrient patterns appear to vary somewhat from year to year. The complicated geography of the estuary results in complex hydrographic conditions. Also, in this environment there are several sources of nutrients, including surface freshwater drainage from land, direct waste discharge, fallout of particulates from the atmosphere and release during dredging. There are yearly variations in the rainfall and soil leaching, that cause variation in river input, which may propagate through the nutrient cycles.

Nutrient inputs to the estuarine region of Patos Lagoon from continental runoff is relatively small since the major 5 rivers that supply this system empty into the Northern region and the nutrients contained in the freshwater input are mostly consumed during transit, a 250 km path, to the estuarine zone.

Results of this study suggest that short term variability in nutrients and *TSM* in the main navigation channel of Patos Lagoon Estuary is primarily caused by resuspension of sediment and by the penetration of adjacent coastal waters along the bottom. In semi-enclosed bays of the estuary, winds play an important role in sediment resuspension. Nutrient input from Río Grande's urban, industrial and harbour activities is observed primarily in adjacent bays. However, atmospheric fallout near the industrial park, not studied as a part of this research, may explain some high surface water concentrations. The results also suggest that, at least on some occasion (i.e., in the fall) nutrients may in fact be limiting.

Acknowledgements

The authors would like to thank the CNPq/Brazil for a grant (202062/91) to Luis Felipe Niencheski during the study period and the Skidaway Institute of Oceanography, University System of Georgia, for all facilities provided.

References

Abreu PC (1992) Phytoplankton production and the microbial food web of the Patos Lagoon Estuary, southern Brazil. PhD thesis, Universität Bremen, Germany

Aminot A, Chaussepied M (1983) Manuel des analyses chimiques en milieu marin. Centre National Exploitation des Océans, Brest

Almeida MT, Baumgarten MGB, Rodrigues RMO (1993) Identificação das possíveis fontes de contaminação das águas que margeiam a cidade do Río Grande. Série "Documentos Técnicos-Oceanografía 06", FURG, Río Grande, 34 p

Aston SR (1980) Nutrients, dissolved gases and general biogeochemistry in estuaries. In: Olausson E, Cato I (eds) Chemistry and biogeochemistry of estuaries. Wiley, pp 233–262

Baumgarten MG, Niencheski LF (1990) O estuário da Laguna dos Patos: variações de alguns parâmetros físico-químicos da água e metais associados ao material em suspensão. Ciência e Cultura 42:390–396

Baumgarten MG, Niencheski LF, Kuroshima KN (1995) Qualidade das águas estuarinas que margeiam o município do Río Grande (RS): Nutrientes e detergente dissolvidos. Rev Atlântica 17:19–36

Brockmann U (1990) Nutrients and suspended material in Shannon and Elbe estuaries. In: Kausch H, Wilson JG, Barth H (eds) Biogeochemical cycles in two major european estuaries: the Shanelbe Project. Commission of the European Communities Water Pollution Research Report No15, pp 24–43

Grasshoff K, Ehrhardt M, Kremking K (1983) Methods of seawater analysis. 2nd edn, Verlag Chemie, Weinheim
Hartmann C (1988) Utilização de dados digitais do mapeamento temático para obtenção dos padrões de distribuição do material em suspensão na desembocadura da Lagua dos Patos. MS thesis, Instituto de Pesquisas Espaciais, São Paulo, Brazil
Herz R (1977) Circulação das águas de superfície da Lagoa dos Patos. PhD dissertation, Departamento de Geografia Universidade de São Paulo, São Paulo, Brazil
Jonge VN de, Villerius LA (1989) Possible role of carbonate dissolution in estuarine phosphate dynamics. Limnol Oceanogr 34:332–340
Kantin R (1983) Hydrologie et qualité des eaux de la region sud de la Lagune dos Patos (Brésil) et de la plataforme continentale adjacente. Thèses Doctorat D'Etat, Université de Bordeaux, France
Kantin R, Baumgarten MG (1982) Observações hidrológicas no estuário da Lagoa dos Patos: Distribuição e Flutuações dos sais nutrientes. Atlântica 5:76–92
Kaul LW, Froelich PN (1984) Modeling estuarine nutrient geochemistry in a simple system. Geochimica et Cosmochimica Acta 48:1417–1433
Kjerfve B (1986) Comparative oceanography of coastal lagoons. In: Wolfe DA (ed) Estuarine variability. Academic Press, New York, pp 63–81
Li Y, Chan L (1979) Desorption of Ba and ^{226}Ra from river-borne sediments in the Hudson Estuary. Earth Planetary Sci Lett 43:343–350
Liss PS (1976) Conservative and non-conservative behaviour of dissolved constituents during estuarine mixing. In: Burton JD, Liss PS (eds) Estuarine chemistry. Academic Press, New York pp 93–130
Meybeck M (1981) Pathways of major elements from land to ocean through rivers. Burton JD, Eisma D, Martin JM (eds) River inputs to oceans systems. UNESCO-UNEP SCOR Workshop, Rome, Italy pp 18–32
Möller OO, Paim PS, Soares I (1991) Facteurs et mécanismes de la circulation des eaux dans l'estuaire de Lagune dos Patos (RS, Brésil). Bull l'Instit Geol Bassin d'Aquitaine 49:15–21
Niencheski LF, Windom HL (1994) Nutrient flux and budget in Patos Lagoon Estuary. The Science of the Total Environment 149:53–60
Niencheski LF, Batista JR, Hartmann C, Fillmann G (1986) Caracterização hidrológica de três regiões distintas no estuário da Lagoa dos Patos-RS. Acta Limnol Brasiliensia 1:47–64
Niencheski LF, Möller OO, Odebrecht C, Fillmann G (1988) Distribuição espacial de alguns parâmetros físico-químicos na Lagoa dos Patos – Porto Alegre a Río Grande. Acta Limnol Brasiliensia 2:79–97
Niencheski LF, Windom HL, Smith R (1994) Distribution of particulate trace metal in Patos Lagoon Estuary. Mar Pollution Bull 28:96–102
Odebrecht C, Möller OO, Niencheski LF(1988) Biomassa e categorias do fitoplâncton total na Lagoa dos Patos, Río Grande do Sul, Brasil (verão de 1986). Acta Limnol Brasiliensia 2:367–386
Officer CB, Ryther JH (1980) The possible importance of silicon in marine eutrophication. Mar Ecology Progress Series 3:83–91
Proença LA (1990) Ciclo anual da produção primária, biomassa do fitoplâncton e carbono orgânico particulado em área rasa da porção sul da Lagoa dos Patos. MSc thesis, Universidade do Río Grande, Río Grande, Brazil
Taft JL, Taylor WR (1976) Phosphorus dynamics in some coastal plain estuaires. In: Wiley M (ed) Estuarine processes, vol 1, Academic Press New York, pp 79–89
Vieira EF, Rangel SRS (1988) Planicie Costeira do Río Grande do Sul Sagra, Porto Alegre
Vilas Boas DF (1990) Distribuição e comportamento dos sais nutrientes, elementos maiores e metais pesados na Lagoa dos Patos-RS. MS thesis, Universidade do Río Grande, Río Grande, Brazil

Chapter 5

Hydrographical Characteristics of the Estuarine Area of Patos Lagoon (30°S, Brazil)

Osmar O. Möller Jr. · Patrice Castaing

5.1 Introduction

The Patos Lagoon (Fig. 5.1a) situated in Southern Brazil, between 30 and 32°S is a very important water resource system for the Río Grande do Sul State. It serves as nursery ground for important commercial fish and shrimp species sustaining a fishery production of 182 kg $ha^{-1} yr^{-1}$ in average (Castello 1985). Major man activities are navigation and recreation. It also receives untreated domestic, agricultural and industrial sewage disposals produced by population living along the margins of the lagoon. The construction of two 4 km long jetties fixed the entrance and allowed navigation along the entrance channel. About 14 million m^3 of sediments where transported to the coastal zone in the first two years after the construction of these structures (Motta 1969). In recent years the channel has been constantly dredged to keep depth around 14 m.

Early studies (Rodrigues 1903; Malaval 1922) indicated that the circulation of this microtidal lagoon was driven by the NE-SW wind regime in absence of large river discharge. These aspects have also been discussed by Motta (1969), Herz (1977), Castello (1985), Costa et al. (1988) and Möller Jr. et al. (1991). Studies carried out through time series analyses and mathematical modelling by Möller Jr. et al. (1996); Möller Jr. (1996) demonstrated that:

a the wind drives the subtidal circulation of Patos Lagoon in the time interval of 3–16 days coincident with frontal system passages over this area;
b the lagoon responds to the local wind through the set up/set down mechanism of oscillation;
c the long period oscillations generated offshore by remote winds are attenuated as they propagate into the lagoon;
d subtidal exchanges between the lagoon and the coastal ocean are driven by the pressure gradient resulted from combined local and non local forcing effects with SW/NE winds driving landward/seaward flows;
e river flow dominates circulation during the seasonal strong flood period. The residual flow is directed to the ocean since it combines important contributions from the fluvial basin with the dominance of the NE winds.

If the basic circulation processes are, in a certain way, well established, the same is not true for mixing and salt transport mechanisms in spite of several studies conducted in the area (Castello 1976a–d; 1977a,b; 1978a,b; Hartmann 1984; Paim and Möller Jr. 1986).

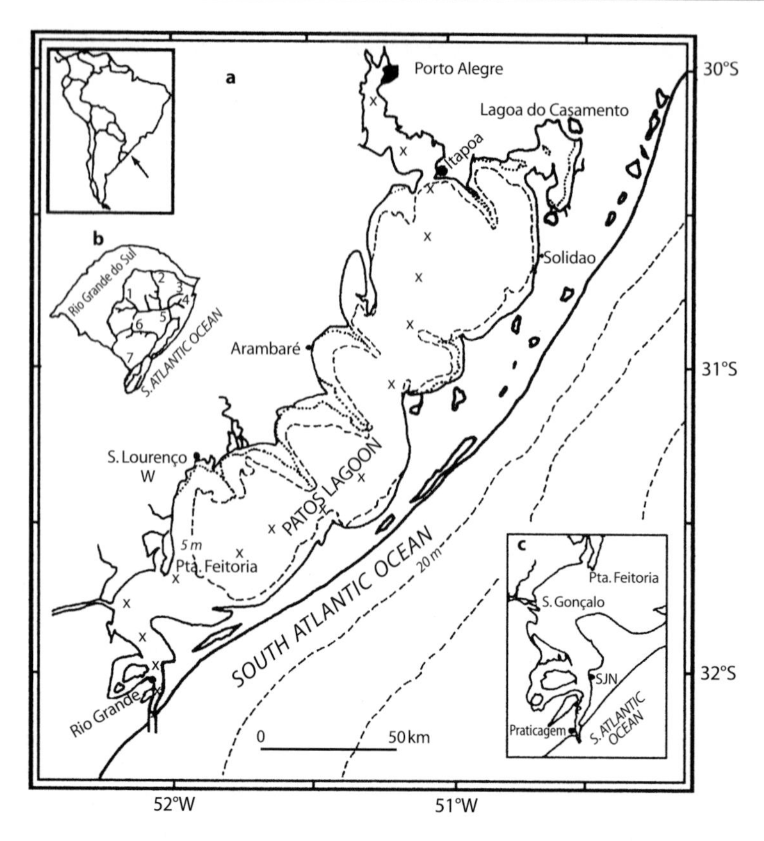

Fig. 5.1. a Location of the Patos Lagoon area. Crosses indicate hydrographical stations performed in 1988; **b** Hydrographical basin. Numbers *1–7* are Jacuí (*1*), Taquarí (*2*), Sinos (*3*), Caí (*4*), Guaiba (*5*) Camaquã (*6*) Rivers and Mirim lagoon basin (*7*); **c** Estuarine area with some of the localities mentioned in the text

Therefore, the objective of this paper is to discuss these aspects through hydrographical surveys which were carried out in two ways:

a longitudinal salinity and temperature profiles;
b short term time series of water level, salinity and wind and velocity vectors covering, at least one meteorological front passage.

5.2 Main Characteristics of the Area

The Patos Lagoon (Fig. 5.1a) is a choked lagoon located in the coastal plain of Río Grande do Sul state. Its typical dimensions are approximately the following: 250 km long, 40 km wide and 5 m deep. It is connected with the South Atlantic Ocean by a 20 km long, 2 km wide and 14 m deep inlet.

The climate of the region is temperate (IBGE 1977) characterized by the passages of meteorological fronts of polar origin. Associated to these fronts, are strong temperature variations, increase in rainfall rates and changes in the wind regime. NE winds are dominant over the year but during autumn and winter winds from west and southwest are also important (Herz 1977; Tomazelli 1993). Typical wind speeds are between 3 and 5 m s^{-1}.

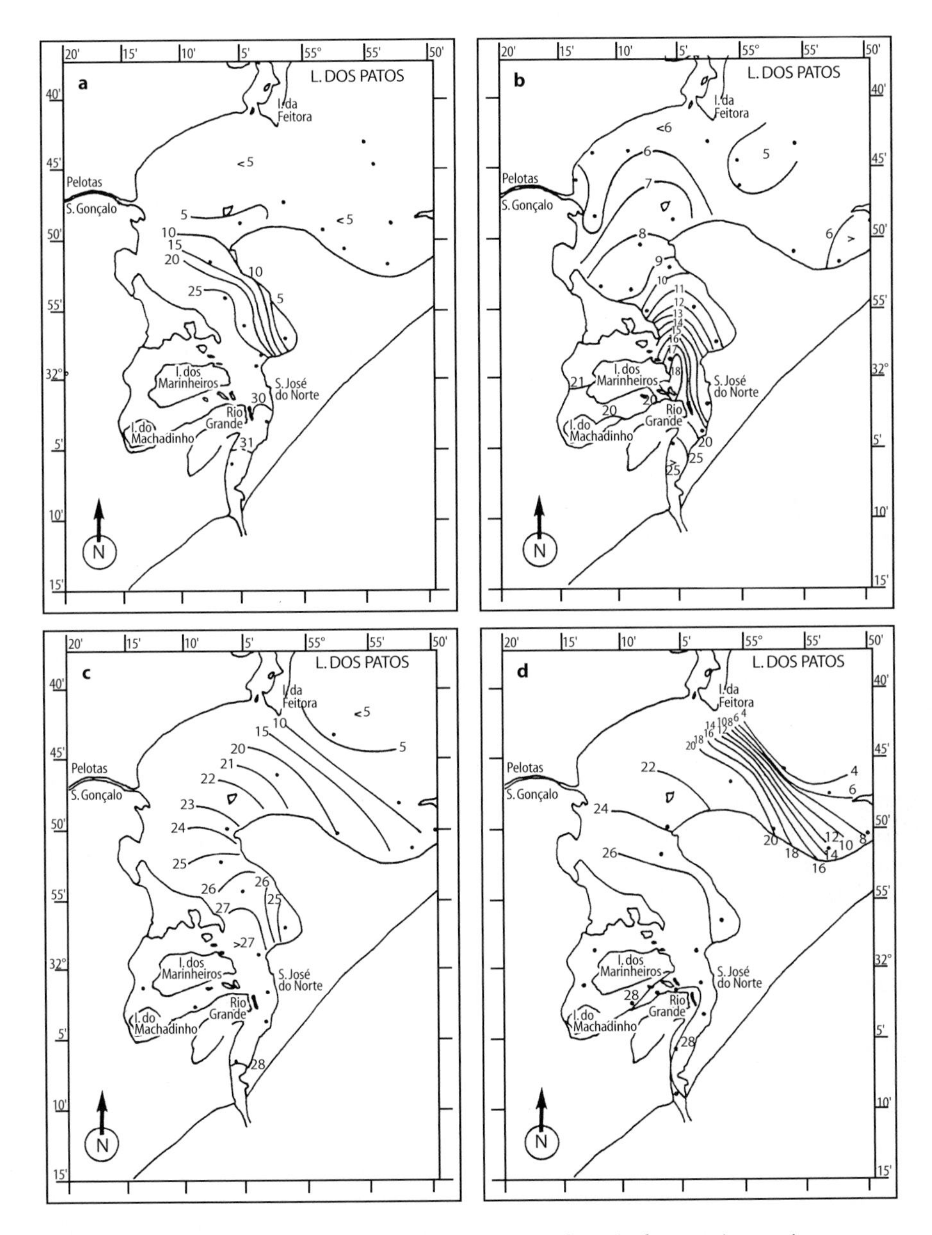

Fig. 5.2. Salinity (psu) distribution observed between 1976 and 1978 in the estuarine area in **a** summer, **b** and **c** autumn, and **d** spring, respectively

The Patos Lagoon is the only connection with the South Atlantic Ocean for a hydrographical basin that covers approximately 200 000 km^2 (Fig. 5.1b). The river discharge is characterized by strong floods during late winter/early spring season and low to moderate runoff during the rest of the year (Fig. 5.2). Large interannual variations are common (Paim and Möller Jr. 1986). The discharge of the Guaiba River, the most important tributary, may vary from 41–22 000 m^3 s^{-1} (Rochefort 1958) with an annual mean around 1 000 m^3 s^{-1} (Bordas et al. 1984). The Jacuí-Taquarí River system contributes with 85% of this value. Camaquã River is important only during flood periods when it may attain 5 000 m^3 s^{-1} (Herz 1977). No data at all exists for San Gonçalo channel (Fig. 5.1c), which connects Patos and Mirim Lagoons.

The tides are mixed and mainly diurnal reflecting the proximity of an amphidromic point for the M_2 tidal component (Vassie 1982). The mean range within the inlet is 0.32 m (Herz 1977). Friction provokes a liner attenuation as the tidal wave progresses landward (Möller Jr. 1996).

Two different geomorphological units can be distinguished along the lagoon: the lower (southern) and inner (northern) areas. They are naturally separated by the sandy banks located in the area around Ponta da Feitoria (Fig. 5.1c) where depths are around 1 m (Delaney 1965). The inner lagoon covers 90% of the total surface, has a mean depth of 5 m and is characterized by cuspate spits that had their origin associated with the wind wave field (Toldo Jr. 1994). The southern portion of the lagoon, also considered as the estuarine region (Closs 1962; Closs and Madeira 1968; Castello 1985), is dominated by very shallow zones (mean depth is 1.7 m) that are cut by natural and man made channels. The cross sectional area decreases exponentially from Ponta da Feitoria (7×10^4 m^2) towards the mouth (10^4 m^2) (Möller Jr. 1996). The estuary region is marked by the presence of several embayments.

The inner limit attained by salt water excursion is around Ponta da Feitoria. Figure 5.2 presents an overview of surface salinity distribution in the estuarine area based on data collected in different seasons: summer (Fig. 5.2a), autumn (Fig. 5.2b,c), and spring (Fig. 5.2d), among those reported by Castello (1976a–d; 1977a,b; 1978a,b). The strongest part of the longitudinal salinity gradient is always observed north of the entrance channel. Lateral salinity gradients are observed in the area where the cross sectional area increases. Mean annual salinity of the estuarine area is 13 psu (Castello 1985), with instantaneous values ranging from zero to 34 psu. Vertical salinity distribution varies from salt wedge type to a well mixed structure (Calliari et al. 1980; Möller Jr. et al. 1991).

5.3 Data Sampling and Treatment

5.3.1 Longitudinal Surveys

The longitudinal distribution of salinity and temperature were obtained in 1988 during a series of 12 cruises carried out along the entire lagoon as part of a multidisciplinary program named the Patos Lagoon Project. The sampling strategy consisted of 18 stations (Fig. 5.1a) located along the main estuarine channel and the deeper longitudinal axis of the lagoon and the Guaiba River. Salinity and temperature were measured with

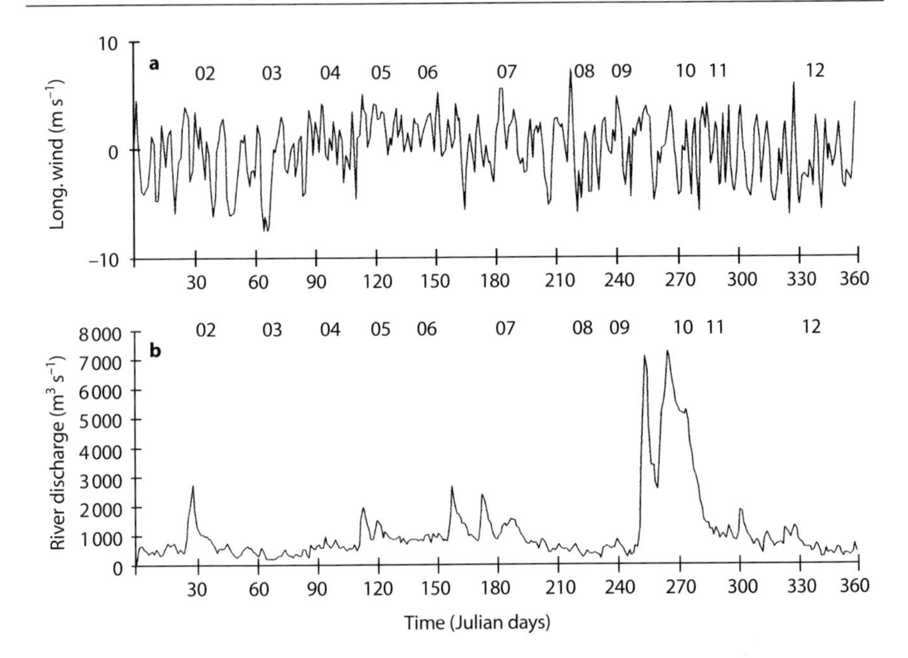

Fig. 5.3. a Low-pass filtered longitudinal wind velocity observed at S. Lourenço; **b** the river discharge of the Jacuí/Taquarí and Camaquã Rivers. Numbers (*02–12*) indicate the relative position of PLP cruises

a Yellow Spring thermo-salinometer calibrated before each cruise. The measurements were performed with depth interval of 2 m. Instantaneous current velocity data were also obtained, but the variations in current direction forced by tides and changes in the wind field during cruises (that lasted 2–3 days) make them difficult to interpret. The temporal position of each cruise with respect to the longitudinal wind component registered at São Lourenço and to river discharge (Möller Jr. 1996) is presented in Fig. 5.3.

5.3.2 Time Series Surveys

The time series of salinity, temperature, water level, current and wind velocities (Table 5.1) used for the estuarine classification and salt transport analyses were obtained in 1992 in a fixed point near Praticagem (Fig. 5.1c). The calculated cross sectional area that resulted from an bathymetric survey is 10 850 m². Because of the difficulties to install moorings in this area where fishing and shipping activities are intense, the measurements have been made from a boat. The same reason has prevented the boat to occupy the central axis of the channel, where maximum depth is around 18 m. Instead, the boat was aligned with a navigation buoy where depths are between 12 and 13 m. Cruises were scheduled in accordance with boat availability and have lasted for 110 h.

Measurements of vertical profiles of salinity, temperature and current velocity vectors were collected hourly with 2 m depth intervals. Salinity and temperature data were

Table 5.1. Characteristics of PLATES cruises and salt-transport component values[a]

	<u>	<s>	<h><u><s>	<HUS>	<h'u's'>	<hu"s">
PLATES I	−0.14	27.99	−47.84	−0.06	30.43	1.41
PLATES II	−0.27	21.86	−77.57	−0.97	39.88	4.3
PLATES III	−0.75					
PLATES IV	−0.4	11.35	−58.55	−0.6	51.18	0.81

[a] Units are, respectively: m s^{-1}; psu and kg m^2 s^{-1} for salt transport components.

obtained with the YS11 thermo-salinometer during the first three cruises; in the last one a Sensordata SD200 CTD was used. Sensors were calibrated before and after every survey. Current vectors were obtained with a hand held Sensordata SD4 current meter and with 3 self contained current meters installed at the depths of 3, 6 and 10 m in a cable sustained by the boat and heavily anchored to the bottom. A portable anemometer was used to obtain wind velocity and direction. Hourly water elevations at Praticagem gauge station (see Fig. 5.1c) were provided by the Office of Hydrography and Navigation (DHN). River discharge was calculated for the Camaquã and Jacuí-Taquarí River systems through the rating curve method using water level data provided by National Department of Waters and Electric Energy (DNAEE/MME).

Current velocity vectors were decomposed into their longitudinal and transversal components. Measured values were referred to the true north with the longitudinal axis following the orientation of the main channel. The same procedure was adopted to obtain longitudinal and transversal wind components. The angle of rotation, in this case, was equal to 37° to make the longitudinal axis parallel to the coastline.

The series were resolved in terms of tidal and subtidal components by fitting a fifth order polynomial curve through the original wind, water elevation and longitudinal current velocity data. Positive values indicate upstream flows and southerly winds. In the case of salinity only a single three points hanging filter was needed to remove the observed small time scale fluctuations. The exception was the part of the PLATES IV data where after hour 70 a strong tidal signal was recorded (see Fig. 5.9 for reference). In this specific case a linear trend was fit to remove this signal. Time and depth averages of properties were carried out following Kjerfve (1979). The non dimensional depth is defined as $\xi = (H - z) / H$, where H is the water depth and z is the depth at the level of measurement, so that $\xi = 1$ at surface and $\xi = 0$ at bottom.

5.4 Results

5.4.1 Longitudinal Salinity and Temperature Distribution

Longitudinal temperature and salinity variations along the lagoon axis are shown in Figs 5.4 and 5.5, respectively. In these figures day 1 corresponds to January 6 1988 which was the date of the first cruise, named PLP02.

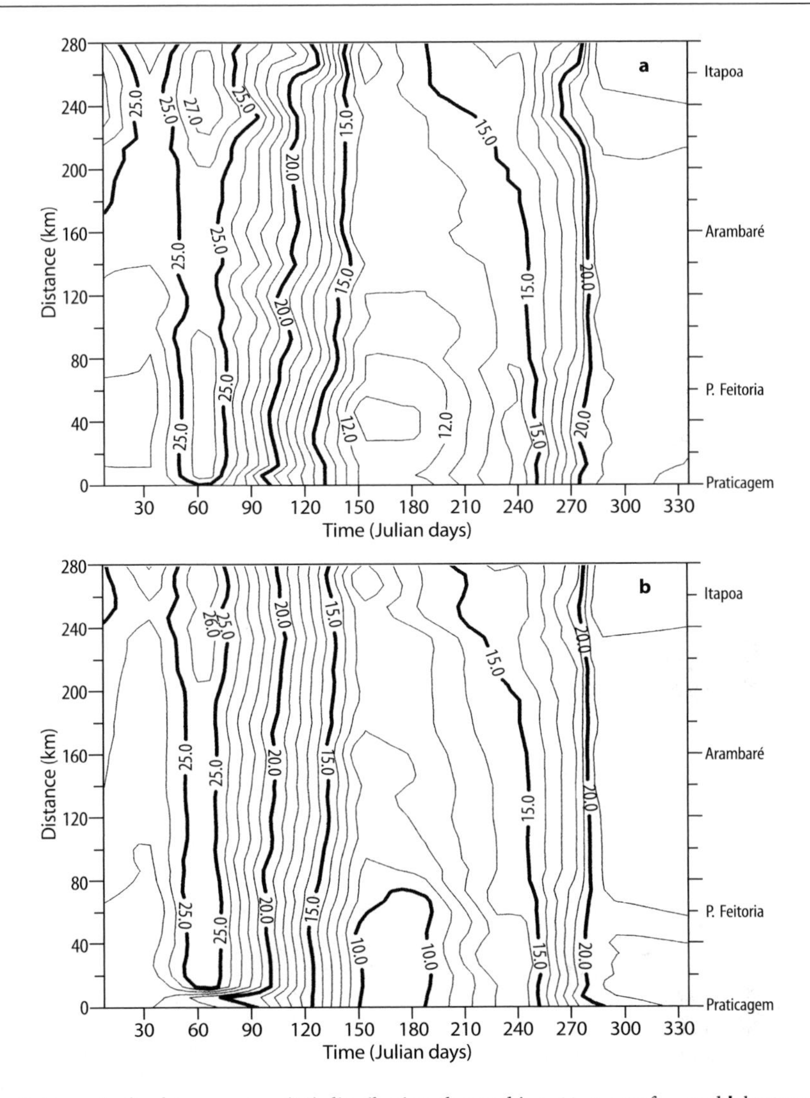

Fig. 5.4. Longitudinal temperature (°C) distribution observed in 1988 at **a** surface and **b** bottom

Temperature (Fig. 5.4a,b) shows typical seasonal pattern reaching values between 25 and 26° C in the Austral summer (Jan.–Mar., days 1 through 90). In winter temperature is around 10° C. In terms of longitudinal distribution small differences are found between northern and southern portions with temperature decreasing southwards. This is a latitudinal effect that can be also noticed through satellite images (Ghisolfi 1995). The water column tends to be homogeneous along the whole lagoon but a thermocline associated with entrance of saltier and colder coastal waters may be developed in the channel area.

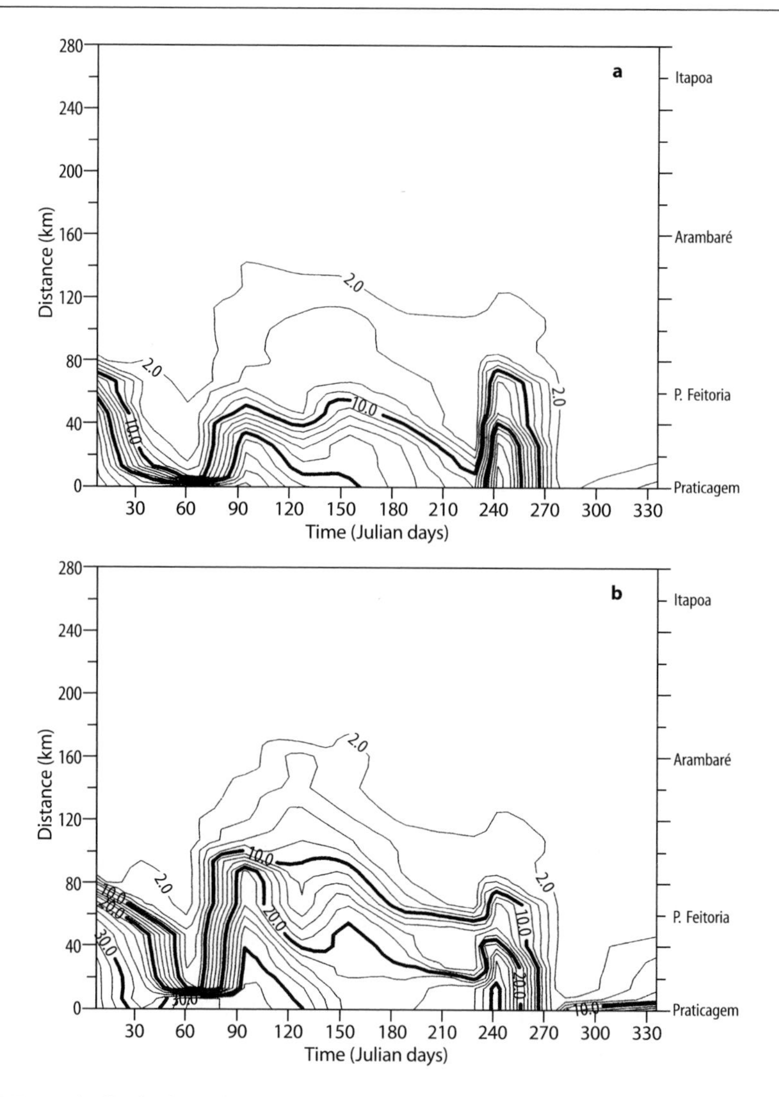

Fig. 5.5. Longitudinal salinity (psu) distribution observed in 1988 at **a** surface and **b** bottom

The inner limit of salt penetration (Fig. 5.5a,b) can be as far as 160 km north of the lagoon mouth, however a marked front is always formed around Ponta da Feitoria (60–80 km from Praticagem) where the salinity gradient can reach 1 psu km^{-1}. Weaker longitudinal gradients are found in the entrance channel up to 40 km from Praticagem.

The most important salt intrusion is seen during autumn between days 90 and 150 that occurs in a period when SW winds at 4 m s^{-1} in average where predominant. It can be noted that the fluctuations of salinity verified between Praticagem and Ponta da Feitoria are not followed in the inner lagoon. In two occasions the displacement of the front has reached the inlet region:

a between days 30 and 60 due to the influence of river discharge and an intense period of NE winds;
b in spring (after day 270), when the salt water was almost totally expulsed from the lagoon during the late winter/early spring flood period, a salt wedge type of estuary was then formed in the near entrance area.

This is a very common situation in this period of the year (Castello and Möller Jr. 1978; Castello 1985).

The comparison between the surface (Fig. 5.5a) and bottom (Fig. 5.5b) situations illustrates that, for the cases already discussed, stratified conditions have prevailed which indicates the importance of river discharge in the circulation of this system. This general trend for stratified conditions can be better observed in Fig. 5.6a. The curves correspond to an exponential decrease towards the inner parts with larger differences between surface and bottom occurring in the first 80 km from the mouth. Farther northwards, the vertical gradients decrease and the water column becomes weakly stratified. It may be argued also that this trend in stratification can be due to the fact that sampling was carried out along the deeper areas of the navigation channel. However the same trend stands even for shallow zones where salinity variations up to 16 psu were observed in areas with maximum depths between 2 and 2.5 m (Castello 1976b; 1977a; 1978a).

The longitudinal distribution of the depth averaged salinity ($\bar{s}$) calculated for data obtained during the first ten cruises, i.e., excluding spring data (when salty water was flushed out of the lagoon) is shown in Fig. 5.6b. If this distribution is assumed to be the steady state response to the mean river discharge (936 $m^3 s^{-1}$) the effective longitudinal dispersion coefficient K_x can be calculated from the balance between advection and diffusion as:

$$K_x = \frac{\overline{u}\,\overline{s}}{\frac{ds}{dx}} ,$$

where $\bar{u}$ is given by the ratio of the river flow to the mean cross sectional area and ds/dx the longitudinal salinity gradient. This one dimensional approach was initially

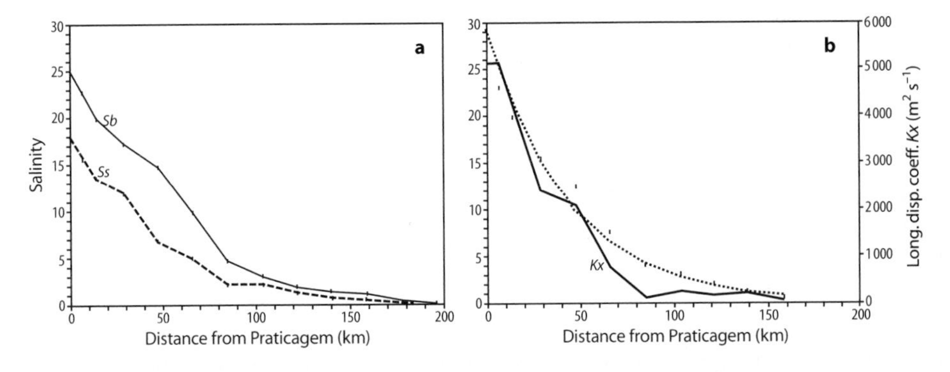

Fig. 5.6. a Mean surface (S_s) and bottom (S_b) salinity (psu) distribution along the longitudinal profile of the lagoon. **b** Mean depth salinities (psu) along the lagoon (*dots*); the exponential decrease in salinity calculated through regression analysis represented by the *dotted line* ($r^2 = 0.94$) and the variation of the longitudinal dispersion coefficient (K_x)

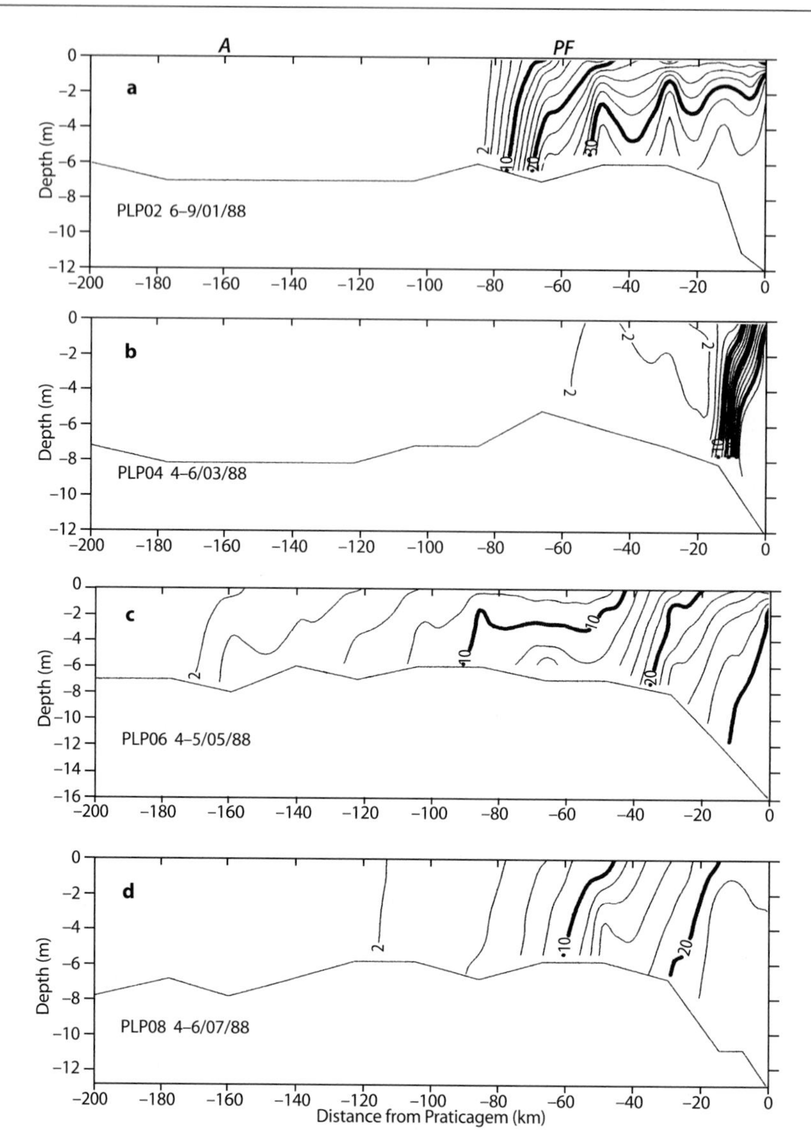

Fig. 5.7. Longitudinal profiles of salinity (psu) obtained during the cruises carried out in 1988: **a** January; **b** March; **c** May; **d** July. *A* and *PF* at the top of the Fig. 5.5. represent, respectively, Arambaré and Pta. Feitoria

proposed for salt balance studies in well mixed narrow estuaries but it is being applied as well for broad partially stratified systems (Fischer et al. 1979). The dispersion coefficient K_x (Fig. 5.7b) increases exponentially towards the mouth varying in the southern part of the lagoon from 2 000–5 000 $m^2 s^{-1}$. These are extremely high values compared to those presented by Dyer (1973), for instance. Comparable K_x values have been determined for the Gironde (Bonnefille 1978), Rotterdam Waterway (Winterwerp

1983) and the Delaware (Garvine et al. 1992) estuaries. The seaward increase of K_x is normally expected (Hamilton and Rattray 1988) however, in this case, it is mainly due the role played by the morphology of the lagoon.

In the same longitudinal section the estuarine zone may pass from a highly stratified to a partially stratified estuary (Fig. 5.7a) or it can behave as partially mixed estuary (cases 7b, 7c and 7d) with large fluctuations of the inner limit of the transition zone. In some of these cases, instantaneous two-layered reversed flows were observed with surface waters running seaward.

5.4.2
Temporal Variations of Water Level, Currents and Salinity

The series of four cruises carried out in the Praticagem area had the purpose of establishing the main mechanisms involved in the exchange processes between the lagoon and the coastal zone. This includes dynamical behaviour, mixing processes and salt flux parameters in the channel area.

In all examined cases the lagoon was subject to the influence of, at least, one meteorological front. During these events the wind rotates anticlockwise from NE to SW directions, returning to the NE position as the front migrates northwards (Stech and Lorenzzetti 1992). The way this system responds to this effect is shown in Fig. 5.8 that reproduces the data collected during PLATES I (6–11/04/92). A positive sea level elevation (Fig. 5.8b) reaches its maximum some 40 h after the establishment of the SW wind regime (Fig. 5.8a). This long period oscillation behaves like a progressive wave in the sense that the generated current velocity peaks (Fig. 5.8c) some hours before water level has attained its maximum. As the wind velocity decreases it no longer sustains the pressure gradient it generated at the coast and the flow reverses even before the establishment of the NE wind regime. The seaward velocities are stronger than the landward ones. The vertically stratified water column (Fig. 5.8d) observed in the first 15 h associated with a two-layered flow forced by tidal action, becomes well mixed when the current reverses throughout the entire column. This situation is also presented in Fig. 5.9 for the same period of time (6–11/04/92). It can be denoted that coastal waters exert the control on the maximum salinity values found within the lagoon.

Figure 5.9 presents the temporal variation of vertical longitudinal velocity and salinity profiles for the three series of measurements (PLATES I, II and IV) where mixing takes place within the lagoon. This wind forcing mechanism discussed above for PLATES I can be generalized to explain the situations found during PLATES II (18–23/05/92) and part of PLATES IV (5–10/10/92) cruises. The PLATES III (22–27/06/92) situation is not shown because the lagoon was subject to a river flow of 8 000 $m^3 s^{-1}$ and just behaved like a riverine system with mixing taking part in the adjacent coastal zone. Unfiltered data present maximum values around 2.5 $m s^{-1}$ for seaward flows observed during PLATES II and 3 $m s^{-1}$ in PLATES III.

During PLATES II (18–23/05/92) the situation is similar to that of the first cruise except that the SW wind regime with wind speeds between 6 and 10 $m s^{-1}$ was already established when measurements started. The freshwater discharged by Guaiba and Camaquã Rivers was 3 000 $m^3 s^{-1}$. vertically well mixed column has been formed and salinity decreased as the flow turned towards the coast. A second meteorological front reverses it at 100 h, forcing a return of mixed water into the lagoon.

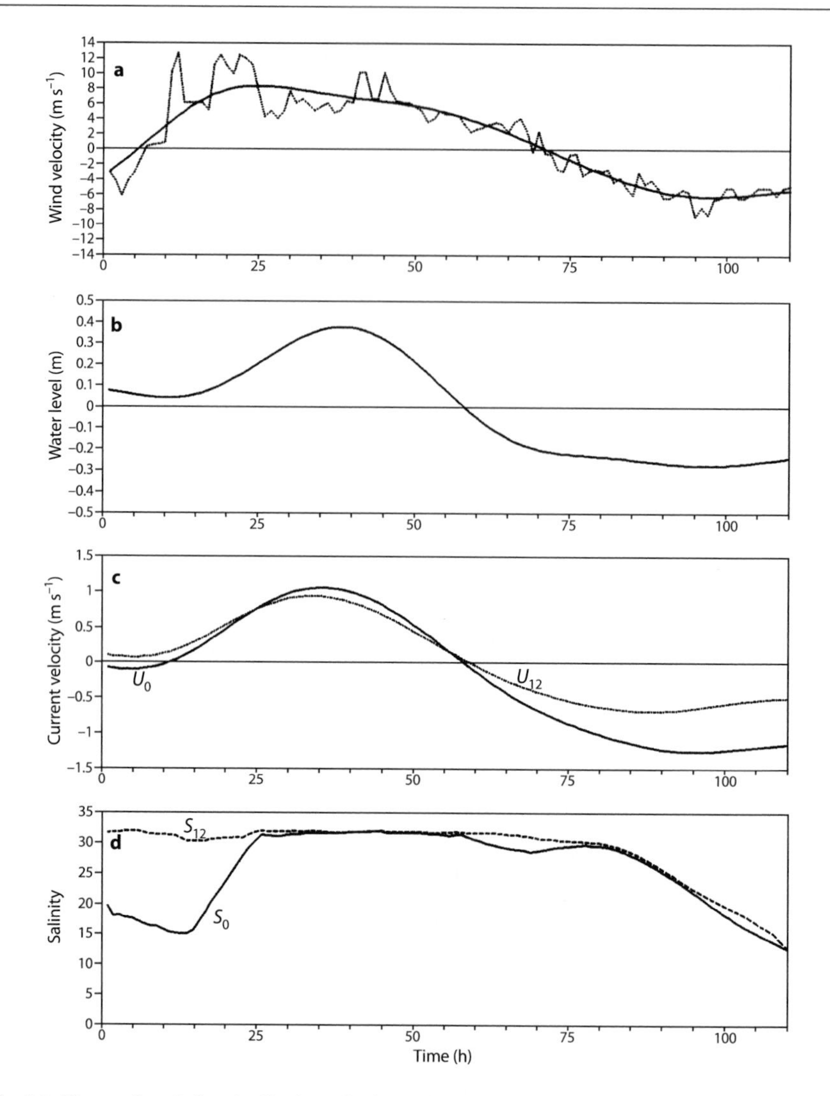

Fig. 5.8. Time series of **a** longitudinal wind velocity, **b** water level, **c** longitudinal current velocity component and **d** salinity for surface (U_0, S_0) and 12 m (U_{12}, S_{12}) obtained during PLATES I. Unfiltered data (*a*, *b*) are represented by *dotted lines* (data from Praticagem station)

During PLATES III survey, two meteorological fronts have passed over this area: on the second day with maximum wind velocities of 2 m s^{-1} and on the last day when SW winds reached 10–12 m s^{-1}. A weak current reversal was observed near the bottom related with the second front. However the freshwater persisted.

Some of the difficulties of working with data collected out of the deeper parts of the channel are shown in Fig. 5.9 for the case of PLATES IV. In order to demonstrate all possible situations that may occur in this area, the unfiltered salinity data are presented. In the first 60 h the response is similar to that observed for cases 1 and 2. A seaward

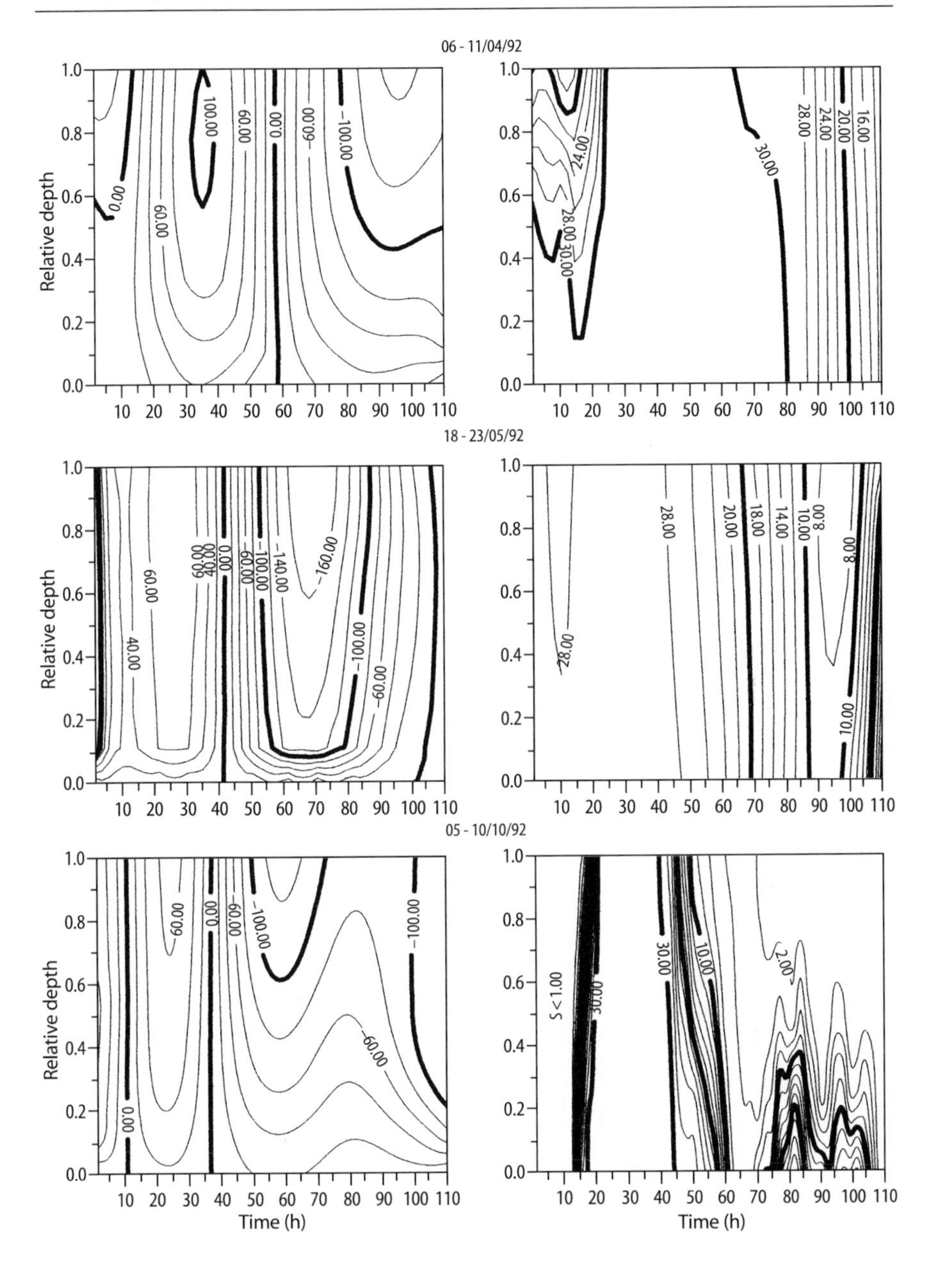

Fig. 5.9. Time series of longitudinal current velocity (m s^{-1}) and salinity (psu) profiles obtained during PLATES I, II and III

flow of freshwater is reversed by the SW wind (10–14 m s^{-1}) and 17 h later the section is vertically homogeneous. A weak stratification develops as the flow turns towards the sea. It holds until 65 h later, when a highly stratified (salt wedge) water column is formed.

It is not related to the mean flow, which is seaward as depicted from the velocity distribution. Rather these salinity oscillations are related with the tidal action that has reversed the flow in the deeper parts (17–18 m) of the channel and forced the entrainment of salt water into shallower depths. Following the 6 psu isoline one can observe that its vertical displacement has a velocity of 0.8 m h^{-1}. During this period the system is still under the action of light SW winds (2–3 m s^{-1}). The bending of the current velocity isolines may be the result of an incomplete filtering. A common feature among these three series is that well mixed conditions are forced by the advection of large amounts of coastal waters that displace the salinity front towards the inner parts of the lagoon.

5.4.3 Estuarine Classification of the Channel Area

The salinity and current velocity data presented above were used to classify the inlet according to the method proposed by Hansen and Rattray (1966). The method is based on the determination of the stratification and the circulation parameters. The stratification parameter is the ratio between the bottom (S_b) and surface (S_s) salinity difference with respect to the water column mean salinity S_O, $<S_b - S_s> / <S_O>$. The circulation parameter $<u_s> / <\bar{u}_f>$ the ratio of the surface mean velocity ($<u_s>$) to the water column mean velocity ($<\bar{u}_f>$). The brackets and bars indicate respectively time and depth averaged properties that were calculated following Kjerfve (1979). The diagram for cruises I, II and IV indicates the estuary type 1b, with a dominant seaward flow having little stratification where salt balance is driven by diffusion rather than gravitational circulation (Fig. 5.10). Two exceptions were observed:

a during PLATES I before the front passage (IBF, Fig. 5.10) when the situation was that of a partially stratified estuary, and
b at the end of the last cruise when a salt-wedge estuary was developed during a short period of time.

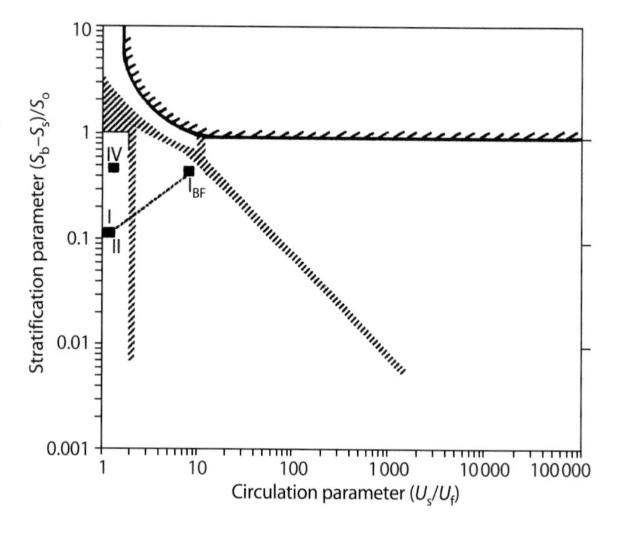

Fig. 5.10. The stratification/circulation diagram for the situations of PLATES I, II and IV. The index "*BF*" refers to PLATES I data obtained before the meteorological front passage

In this case the stratification parameter was 4.23, but it was not represented on the diagram because, as explained before, it is not related to the mean measured flow.

5.4.4 Longitudinal Salt Transport

Methods for estimating the principal factors affecting the salt balance of estuaries have been proposed by several authors (Dyer 1974; Officer 1976; Lewis and Lewis 1983). They are based on cross-correlations between flow components and salinity that may include longitudinal and transversal terms (Murray and Siripong 1988; Hughes and Rattray 1988) averaged over one or several tidal cycles. The same type of approach has been used for the determination of residual fluxes of suspended sediments and dissolved inorganic substances (Uncles et al. 1985; Dyer 1988; Pino et al. 1994). In spite of their short duration the data collected during cruises I, II and IV were used to establish the main factors involved with the longitudinal salt transport in the entrance channel of Patos Lagoon. Following Smith (1994) the instantaneous longitudinal velocity and salinity values can be given by:

$$u = \langle u \rangle + U + u' + u'' \;,$$
$$s = \langle s \rangle + S + s' + s'' \;,$$
$$h = \langle h \rangle + H + h' \;,$$

where: $\langle u \rangle$, $\langle s \rangle$ and $\langle h \rangle$ represent the time averaged velocity, salinity and depth; U, S and H are the tidal deviations; u', s' and h' are the low frequency deviations that are associated with wind forcing; u'' and s'' are the vertical shear effect which is a measure of the gravitational circulation. Deviations related to turbulence were disregarded.

The transport of salt (F) through the vertical water column of time averaged depth $\langle h \rangle$ can be given by:

$$F = \underbrace{\langle h \rangle \langle \bar{u} \rangle \langle \bar{s} \rangle}_{1} + \underbrace{\langle H \bar{U} \bar{S} \rangle}_{2} + \underbrace{\langle h' \overline{us} \rangle}_{3} + \underbrace{\langle h u'' s'' \rangle}_{4}$$

where the bar indicates depth average.

The calculated values for each one of these components are presented in Table 5.1. Two factors dominate this process: the advective term related to freshwater discharge (term 1) and the cross products associated between offshore forced oscillations and salinity (term 2). The tidal cross products and the shear stress are of minor importance. These terms do not balance each other resulting in a residual seaward salt transport for all the analysed cases which can be related to a decrease in the salt content in the inner part of the estuary (Dyer 1974).

It is clear that the difficulties to separate different signals in short time series may induce errors. This is evident in PLATES IV when the unbalanced salt flux may be tied with incomplete filtering and in the case of PLATES II where the series of measurements have started when the landward flow was already at its maximum. Nevertheless these results show the mean advective seaward flow and the deviations associated with the passage of meteorological fronts are the most important factors for salt transport in this part of the lagoon.

5.5 Discussion and Conclusions

Coastal lagoons are normally considered as well mixed type of estuaries. The wind action over shallow water bodies combined with low river discharge are the main factors that cause this situation. The longitudinal salinity distribution presented here indicate however that this is not the case of Patos Lagoon where stratification is, probably, the dominant feature. The considerable river runoff combined with a downestuary dominant wind field are the main reasons for this system to be stratified. The morphology of Patos Lagoon may also play its role because of the funneling in the lagoon section that tends to enhance the effect of river flow. The deep (14 m) inlet may contribute, sometimes, to flow separation in a two-layered vertical profile. In this case a lens of salt water can propagate near the bottom into the estuary, as described by Largier and Taljaard (1991) for microtidal estuaries, forming a highly stratified water column. Well mixed conditions were only found along the entrance channel after strong meteorological fronts or after a river freshet.

Stratification and destratification cycles in estuaries have been reported to be due: to the relative strength of river and wind action (Schroeder et al. 1990); to strong wind events (Goodrich et al. 1987) and to the river flow action (Smith 1985). In the case of Patos Lagoon it depends on the relative strength of the barotropic circulation induced by the wind field and river flow. Even during periods of high river flows (>4 000 $m^3 s^{-1}$) well mixed conditions can be attained in the inlet during the passage of a strong cold front that force large amounts of coastal waters to enter into the lagoon. However, velocity of winds in excess of 10 $m s^{-1}$ represents less than 1% of occurrences (Tomazelli 1993).

The salt balance is dominated by two terms: the residual seaward salt transport forced by river discharge and the advection of sea water caused by the passage of meteorological fronts. Diffusion is important only in the inner parts of the estuarine area where the salinity gradient is strong enough to balance river flow. Values of the longitudinal dispersion coefficient (K_x) between 100 and 500 $m^2 s^{-1}$, that are comparable with those obtained for many estuarine systems, were found only for the region near Ponta da Feitoria (some 50–60 km from the entrance) as indicated in Fig. 5.6b. Near the entrance, the salinity gradient is very weak and the requirement of balance between advection and diffusion would lead to unrealistically high values of K_x. This suggests that in fact the salinity distribution is dominated by advection in this area.

A basic mechanism for the formation of the estuarine zone is given by the pressure gradient generated between the coastal ocean and the lagoon by the action of SW winds. As the salty waters enter into the lagoon the mixing zone (denoted by the strongest part of the salinity gradient) is formed and displaced towards the inner parts of the lagoon as this process persists. Behind this area, the advection of unmixed coastal waters form a region with a very weak salinity gradient.

The time series indicate that for river flow exceeding 4 500 $m^3 s^{-1}$ only strong winds can force salt water into the lagoon. This intrusion is rapidly flushed out of the channel.

The data presented here can only give an insight into the main processes related with mixing and salt transport processes in the Patos Lagoon. 3D mathematical modelling and longer time series obtained at different sites are needed to understand these processes and their consequences within the lagoon.

Acknowledgements

The authors would like to acknowledge: the Interministerial Comission for Marine Resources – CIRM – and the Río Grande do Sul State Research Supporting Foundation – FAPERGS – (contract 90.1220.0) for the financial support given during the PLP and PLATES projects; the Marine Office of Hydrography and Navigation -DHN – and the National Department of Waters and Electrical Energy – DNAEE – for providing part of the data used in this study; to Drs. Frederico Isla, Peter Zavialov and one anonymous reviewer for criticising and improving the manuscript. This paper is part of a DSc. Thesis carried out by O. Möller at the University of Bordeaux I sponsored by the Brazilian National Council of Scientific and Technological Development – CNPq – under contract N° 200331-92.2.

References

Bonnefille R (1978) Residual phenomena in estuaries, application to the Gironde Estuary. In: Nihoul JCJ (ed) Hydrodynamics of estuaries and fjords. Elsevier Oceanographic Series 23, Amsterdam pp 439–482
Bordas MP, Casalas A, Silveira A, Gonçalves M (1984) Circulação e dispersão em sistemas costeiros e oceânicos, Caso da Lagoa dos Patos. Tech Report IPH/UFRGS, Brazil
Calliari LJ, Gomes MEV, Griep GH, Möller Jr OO (1980) Características sedimentológicas e fatores ambientais da região estuarial da Lagoa dos Patos. Proceedings XXXI Congresso Brasileiro de Geologia 2:862–875
Castello JP (Coord) (1976a) Projeto Lagoa, Relatório 1° Cruzeiro, Río Grande. FURG, BOA, Série Relatórios 1:1–23
Castello JP (Coord) (1976b) Projeto Lagoa, Relatório 2°/3° Cruzeiros, Río Grande. FURG, BOA, Série Relatórios 2:1–47
Castello JP (Coord) (1976c) Projeto Lagoa, Relatório 4°/5° Cruzeiros, Río Grande. FURG, BOA, Série Relatórios 3:1–48
Castello JP (Coord) (1976d) Projeto Lagoa. Relatório 6°/7° Cruzeiros, Río Grande. FURG, BOA, Série Relatórios 4:1–75
Castello JP (Coord) (1977a) Projeto Lagoa, Relatório 8°/9° Cruzeiros, Río Grande. FURG, BOA, Série Relatórios 5:1–56
Castello JP (Coord) (1977b) Projeto Lagoa, Relatório 10°/11° Cruzeiros, Río Grande. FURG, BOA, Série Relatórios 6:1–68
Castello JP (Coord) (1978a) Projeto Lagoa, II Transversal de Bentos e Relatório 12°/13° Cruzeiros, Río Grande. FURG, BOA, Série Relatórios 8:1–72
Castello JP (Coord) (1978b) Projeto Lagoa, Relatório 13°/14° Cruzeiros, Río Grande. FURG, BOA, Série Relatórios 9:1–72
Castello JP (1985) La ecología de los consumidores del estuario de la Lagoa dos Patos, Brasil. In: Yañez Arancibia A (ed) Fish Community ecology in estuaries and coastal lagoons, towards an ecosystem integration. DR (N) UNAM Press, Mexico, pp 386–406
Castello JP, Möller Jr OO (1978) On the relationship between rainfall and shrimp production in the estuary of Patos Lagoon (Río Grande do Sul, Brasil). Atlântica 3:67–742
Closs D (1962) Fora miníferos e tecamebas da Lagoa dos Patos. Bol Escola de Geologia, UFRGS 11:1–51
Closs D, Madeira LM (1968) Seasonal variations of brackish foraminifera in the Patos Lagoon, Southern Brazil. Publ Esp Escola de Geologia de Porto Alegre, UFRGS 15:1–51
Costa C, Seeliger U, Kinas P (1988) The effect of wind velocity and direction on the salinity regime in the Lower Patos Lagoon Estuary. Ciência e Cultura 40:909–912
Delaney P (1965) Fisiografia e geologia de superfície da planície costeira do Río Grande do Sul. Publ Especial Escola de Geologia, UFRGS 6:1–105
Dyer KR (1973) Estuaries: A physical introduction. J Wiley & Sons, London
Dyer KR (1974) The salt balance in stratified estuaries. Estuarine Coastal Mar Sci 2:273–281
Dyer KR (1988) The balance of suspended sediment in the Gironde and Thames estuaries. In: Kjerfve B Estuarine transport process. Univ of South Carolina Press, pp 135–145
Fischer HB, List EJ, Koh RCY, Imberger J, Brooks NH (1979) Mixing in inland and coastal waters. Academic Press, New York
Garvine RW, McCarthy RK, Wong KC (1992) The axial salinity distribution in the Delaware Estuary and its weak response to river discharge. Estuarine Coastal Shelf Sci 35:157–162

Ghisolfi RD (1995) Estimativas da velocidade superficial no Oceano Atlântico Sul Ocidental utilizando imagens seqüenciais do satélite AVHRR/NOAA MSc dissertation, CEPSRM, Universidade Federal do Río Grande do Sul
Goodrich DM, Boicourt WC, Hamilton P, Pritchard DW (1987) Wind-induced stratification in Chesapeake Bay. J Phys Oceanogr 17:2232–2240
Hamilton P, Rattray Jr M (1988) Theoretical aspects of estuarine circulation. In: Kjerfve B (ed) Estuarine transport process. Univ of South Carolina Press, pp 37–73
Hansen DV, Rattray Jr M (1966) New dimensions in estuary classification. Limnol Oceanogr 11:319–326
Hartmann C (1984) Material em suspensão e dissolvido no estuário da Lagoa dos Patos – Fase II. Technical Report, contract FURG/CIRM
Herz R (1977) Circulação das Águas de Superfície da Lagoa dos Patos. DSc thesis, Univ São Paulo, Brazil
Hughes FW, Rattray Jr M (1988) Salt flux and mixing in the Columbia River Estuary. Estuarine Coastal Mar Sci 10:473–493
IBGE (1977) Geografia do Brasil Região Sul. vol 5, 534 pp
Kjerfve B (1979) Measurements and analysis of water current, temperature, salinity and density. In: Dyer KR (ed) Estuarine hydrography and sedimentation. Cambridge University Press, Cambridge, pp 186–226
Largier JL, Taljaard S (1991) The dynamics of tidal intrusion retention, and removal of seawater in a bar-built estuary. Estuarine Coastal Shelf Sci 33:325–338
Lewis RE, Lewis LO (1983) The principal factors contributing to the flux of salt in a narrow, partially stratified estuary. Estuarine Coastal Mar Sci 16:599–626
Malaval MB (1922) Travaux du port et de la barre de Río Grande, Brésil. Paris Eyrolles Editeurs
Möller Jr OO (1996) Hydrodynamique de la lagune dos Patos: mesures et modélisation. DSc thesis, Université Bordeaux I
Möller Jr OO, Paim PS, Soares ID (1991) Facteurs et mecanismes de la circulation des eaux dans l'estuaire de la lagune dos Patos Bull Inst Géol Bassin d'Aquit 49:15–21
Möller Jr OO, Lorenzzetti JA, Stech JL, Matta MM (1996) The summertime circulation and dynamics of Patos Lagoon. Cont Shelf Res 16:335–351
Motta VF (1969) Relatório diagnóstico sobre a melhoria e o aprofundamento do canal de acesso pela barra do Río Grande. Tech Report IPH/UFRGS, Brazil
Murray SP, Siripong A (1988) Role of lateral gradients and longitudinal dispersion in the salt balance of a shallow, well-mixed estuary. In: Kjerfve B (ed) Estuarine transport process. Univ of South Carolina Press, pp 113–124
Officer CB (1976) Physical oceanography of estuaries (and associated coastal waters). John Wiley & Sons, New York
Paim PS, Möller Jr OO (1986) Material em suspensão e dissolvido no estuário da Lagoa dos Patos – Fase III. Technical Report, contract FURG/CIRM
Pino Q M, Perillo GME, Santamarina P (1994) Residual fluxes in a cross-section of the Valdivia River Estuary, Chile. Estuarine Coastal Shelf Sci 38:491–505
Rodrigues JAF (1903) As embocaduras das lagoas com aplicação à barra do Río Grande do Sul. Escola Polytechnica, São Paulo
Rochefort M (1958) Rapports entre la pluviosité et l'écoulement dans le Brésil Subtropical et le Brésil Tropical Atlantique. Travaux et Mémoires de l'Institut des Hautes Etudes de l'Amérique Latine
Schroeder WW, Dinnel SC, Wiseman Jr WW (1990) Salinity stratification in a river dominated estuary. Estuaries 13:145–154
Smith NP (1985) Numerical simulation of bay-shelf exchanges with a one-dimensional model Contrib Mar Sci 28:1–13
Smith NP (1994) Water, salt and heat balance of coastal lagoons. In: Kjerfve B (ed) Coastal lagoon processes. Elsevier Oceanographic Series 60, Amsterdan pp 69–101
Stech JL, Lorenzzetti JA (1992) The response of the South Brazil Bight to the passage of wintertime cold fronts. J Geophys Res 97:9507–9520
Toldo Jr EE (1994) Sedimentação, predição do padrão de ondas e dinâmica sedimentaria da anti-praia e zona de surfe do Sistema Lagunar da Laguna dos Patos, RS. DSc thesis CPGG, Universidade Federal do Río Grande do Sul, P Alegre
Tomazelli LJ (1993) O regime de ventos e a taxa de migracão das dunas eólicas costeiras do Río Grande do Sul, Brasil, Pesquisas 20:18–26
Uncles RJ, Elliot RCA, Weston SA (1985) Observed fluxes of water, salt and suspended sediment in a partially estuary. Estuarine Coastal Shelf Sci 20:147–167
Vassie J (1982) Tides and low frequency variations in the Equatorial Atlantic. Oceanol Acta 5:3–6
Winterwerp JC (1983) Decomposition of the mass transport in narrow estuaries. Estuarine Coastal Shelf Sci 16:627–638

Chapter 6

The Argentina Estuaries: A Review

M. Cintia Piccolo · Gerardo M.E. Perillo

6.1 Introduction

Despite of its long and extended coastline (5 700 km), the number of Argentine estuaries is relative small compared with other countries of comparable size. Mainly, the difference strives in the climate and the hydrological regime that characterize the country. Whereas the climate in the NE of Argentina is hot and humid with mean annual precipitation of 1950 mm and maximum of 2 437 mm; the NW is relatively dry and hot. In the center of the country the climate is mild with average precipitation on the order of 1 000 mm, and in the South is dry and cold with annual precipitation of 300 mm or less.

Based on the pluvial regime, Riggi (1944) defined a series of geohydrographic basins. According to him the rivers that flow into the Atlantic ocean can be classified as:

a Paraná-Uruguay River system,
b rivers that flows along the South of Buenos Aires plains,
c steppe basin of the Colorado River, and
d the Andean-steppe rivers that flow across Patagonia.

Each of these geohydrographic basins has its own physical characteristics but, when they are combined with the geomorphology and typical dynamics of the coast different types of estuaries are generated.

The regions with more precipitation are concentrated in the same river basin: Paraná and Uruguay Rivers basins that discharge into the Río de la Plata. Other examples are the rivers that belong to the Salado basin that discharge in Samborombón Bay (Fig. 6.1). These rivers flow in a region of low slope and mature relief; thus resulting in a meandering pattern. River discharge is significant and prevents the penetration of the saline water from the coastal ocean because it is opposed to a microtidal regime. Samborombón Bay is a shallow environment being greatly affected by the high sediment output from Río de la Plata (López Laborde and Nagy, this book; Framiñan et al. 1998).

Buenos Aires province has the largest number of rivers and small streams that discharge into the Atlantic ocean. Most of those located to the south of the Salado basin originate in the Tandilia and Ventania hill systems. In general, they are short streams that cross an extended region of dunes (2–8 km wide) before flowing into the coastal ocean. Dunes affect in two ways the fluvial discharge: on one side, they produce significant modifications in the stream flow as they force the streams to circulate on low relief areas increasing river length and generating sometimes shallow lagoons. Sec-

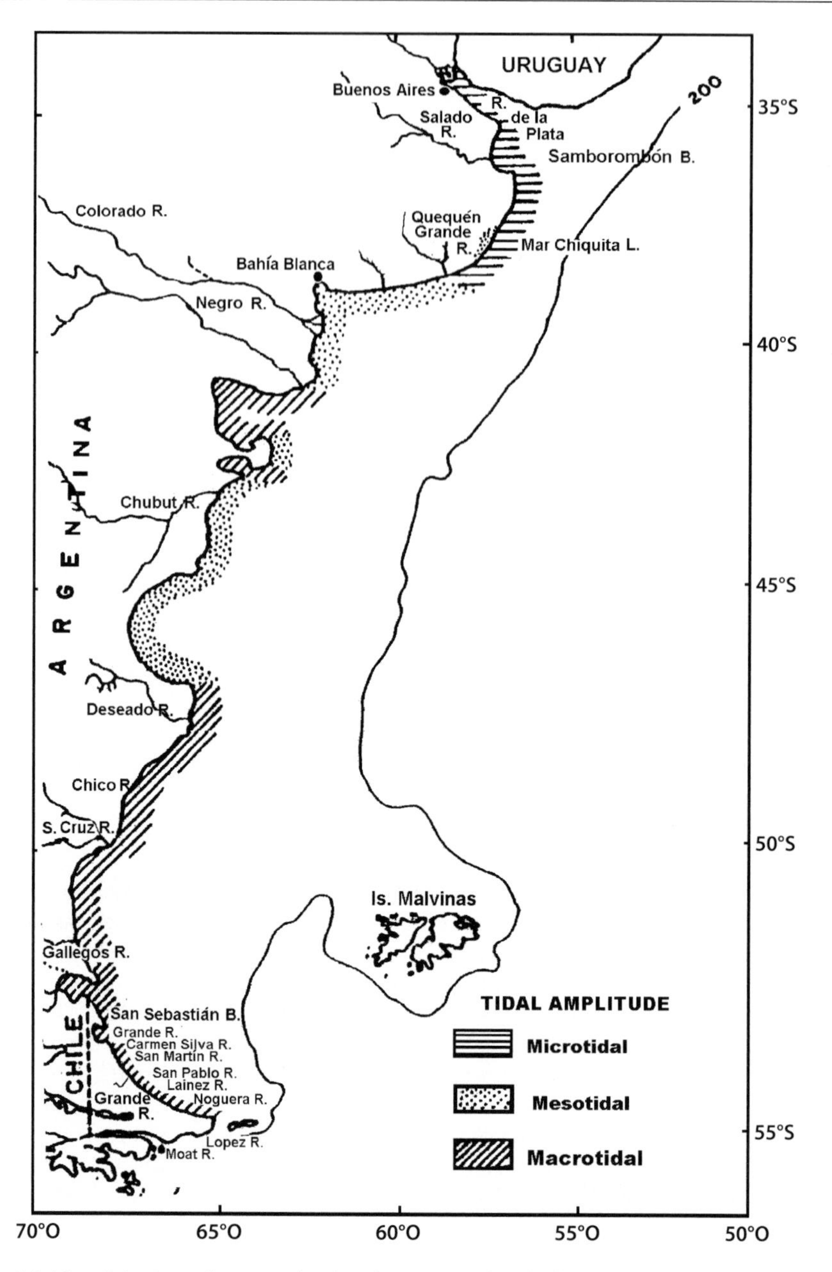

Fig. 6.1. Map of the Argentine coast showing the estuaries that discharge in the Argentine sea

ondly, the sandy soil induces high water infiltration. In both cases, there is a reduction of the stream discharge and; therefore, a reduction in the deepening and channel building capacity by the river.

To the south of Colorado River (Fig. 6.1) there are only ten major former fluvial valleys that discharge in the coastal ocean. All the Patagonian rivers have their origin in the Andean mountains and they flow across a high plateau with almost no tributaries. During the Holocene these rivers formed broad valleys that give an idea of their previous high runoff. At the present time, river discharge is low (and in some practically negligible) due to several decades of low precipitation and the building of dams along their courses.

Based on the hydrographic characteristics of the Argentinian rivers and the geomorphologic conditions in which they have developed, the main objective of the present article is to describe them from their geomorphological and physical points of view. The estuaries have been classified following the morphogenetic classification devised by Perillo (1995) (see Perillo and Piccolo, Chapter 1, this book) and presented in Table 6.1.

Table 6.1. Classification of Argentine estuaries based on the morphogenetic classification

PRIMARY	Former fluvial valleys	Coastal plain	Samborombón Salado Quequén Grande Quequén Salado Bahía Blanca Negro Chubut San Martín Carmen Sylva Grande
		Rias	Deseado Santa Cruz Coig Ewan or Brazo Norte San Pablo Lainez Irigoyen Noguera Bueno Bompland López Moat
	Former glacial valleys	Fjords	–
		Fjards	–
	River dominated	Tidal rivers	Río de la Plata
		Delta front	Colorado
	Structural		Gallegos???
SECONDARY	Coastal lagoons	Chocked –long inlet	Mar Chiquita
		Chocked –short inlet	San Antonio Bay Caleta Valdés
		Restricted	San Sebastian Bay
		Leaky	–

6.2
Río de la Plata Estuary

The Argentina coast starts in the Río de la Plata Estuary (Fig. 6.1), located at about 35°S on the Atlantic coast of South America. After the Amazon River, the Río de la Plata including its two major tributaries (Paraná and Uruguay Rivers) is one of the major waterways of the Southern Hemisphere as well as the second largest basin in Latin America (over 3 000 000 km²). Two major harbours and many cities are located on its shore from which Buenos Aires (Argentina) and Montevideo (Uruguay) are the largest and better known.

The Río de la Plata Estuary is 320 km long with a variable width between 35 and 230 km (Fig. 6.2). Its estuarine zone is estimated in 18 000 km². The northern coastline is 416 km long and the southern one 393 km (Framiñan and Brown 1996). It has an elongated shape whose width (b) varies exponentially with respect to its length (x) measured from its head (Paraná delta front) according to the following expression $b = 0.0885\, x^{1.402}$ $r^2 = 0.92$). Depth ranges between 2 and 18 m.

Many authors have described extensively its general hydrography (i.e., Glorioso and Boschi 1982; Boschi 1988; Framiñan and Brown 1996). The large inflow which average

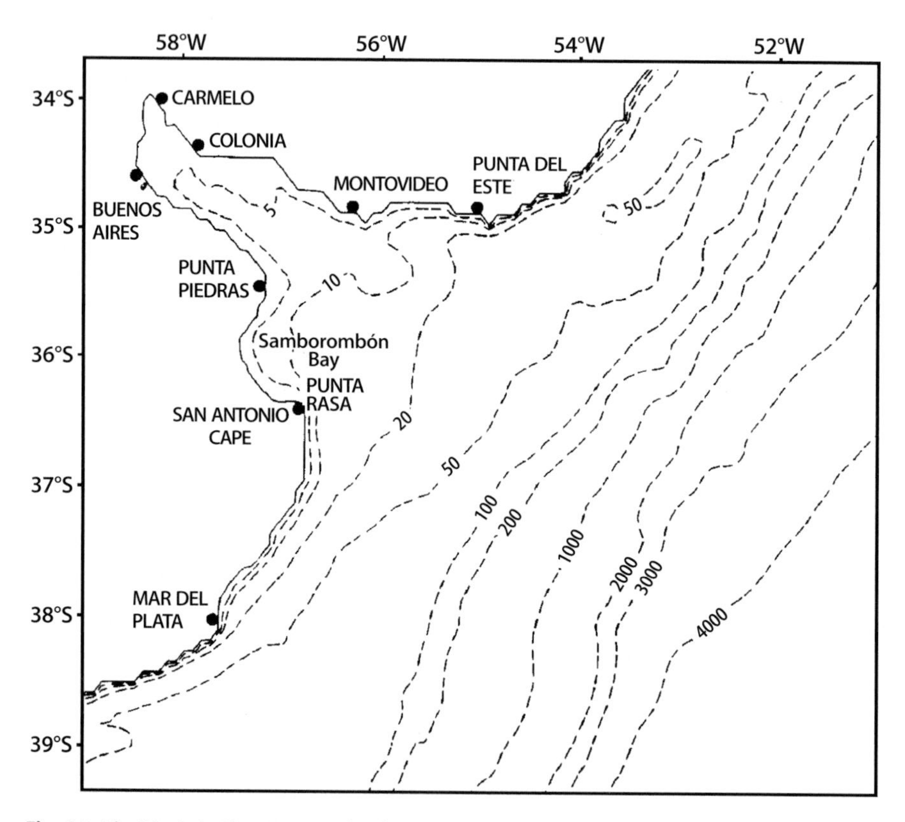

Fig. 6.2. The Río de la Plata Estuary, depth in metres

22 000 $m^3 s^{-1}$, governs the movements and mixture of river and marine waters being also controlled by wind and tidal action. Urien (1972), Gagliardini et al. (1984) and Boschi (1988) divided the Río de la Plata into three different zones:

i. the upper region (Playa Honda), from the head of the estuary to an imaginary line between Colonia (Uruguay) to La Plata (Argentina) cities (the narrowest cross-section, 40 km), with fluvial characteristics and affected by tides;
ii. an intermediate zone that extends up to the Montevideo-Punta Piedras line dominated by several shallow banks and estuarine fronts;
iii. an the wide outer region with typical marine features.

Sediment supply from the tributary rivers consist mainly of silt and clay. In the upper estuary sand is scarce and localised in banks, bars and beaches on the northern coast as it is provided mostly by the Uruguay River. This coast is rocky and sandy with many pocket beaches In the outer estuary a sand carpet extends from the inner continental shelf into the estuary. These are relict sands of the last Holocene transgression which invaded the estuary between 7 000 and 3 000 years B.P. The southern coast is nearly flat, with its maximum development in Samborombón Bay. The regional estuarine environment is mostly fluvial, but the mixing of ocean waters creates a gradual change from fluvial in the upper river to fluvio-marine and marine outer river (Urien 1972).

At the head of the Río de la Plata Estuary there is a very active delta. The geomorphological extension of the delta of the Paraná River exceeds its long-accepted limits as shown by sedimentological, morphological and stratigraphic studies (Parker and Marcolini 1992). The recognition of a subaqueous plain, a delta front and a prodelta, suggests that the delta extends to the outer zone of the Río de la Plata Estuary. Estimated growing rates for the delta is between 40 and 70 m yr^{-1}.

Tides are the physical processes most investigated in the estuary. The dimensions of the Río de la Plata are such that its natural period of oscillation is nearly that of the semidiurnal tide. The principal tide affecting the estuary is the semidiurnal lunar (M_2) with a period of 12.42 h. Due to Coriolis, the incoming tide is deflected towards the southern shore of the estuary (Balay 1961), yielding greater tidal amplitudes than along the northern shore. Maximum range is 1 m on the Argentina coast (Balay 1956; O'Connor 1991).

Much larger variations in sea level are induced by meteorological forcing. Winds that flow along the estuary produce the largest storm surges. Specially with winds from the S-SE during high tide, associated with stationary cyclones over the Río de la Plata can originate storm surges greater than 4 m with disastrous flooding on the coasts (Balay 1958; 1961). Several numerical models of tidal prediction in the Río de la Plata have been developed (Mazio 1990, 1991; O'Connor 1991) and Gagliardini et al. (1984) have studied the interaction between tides and the river flow using remote sensing images. Figure 6.3 shows the tidal amplitude and currents at two hours for high and low tide (O'Connor 1991).These results agree with those of Balay (1956; 1961) and Lanfredi et al. (1979). They also show that in Samborombón Bay there is a counterclockwise system of residual currents which flow seaward following the coastline together with fluvial drainage.

Surface salinity patterns are controlled by the wind field and, to a lesser extent by river discharge (Fig. 6.4). Both forces act upon the upper layer. Diluted shelf waters

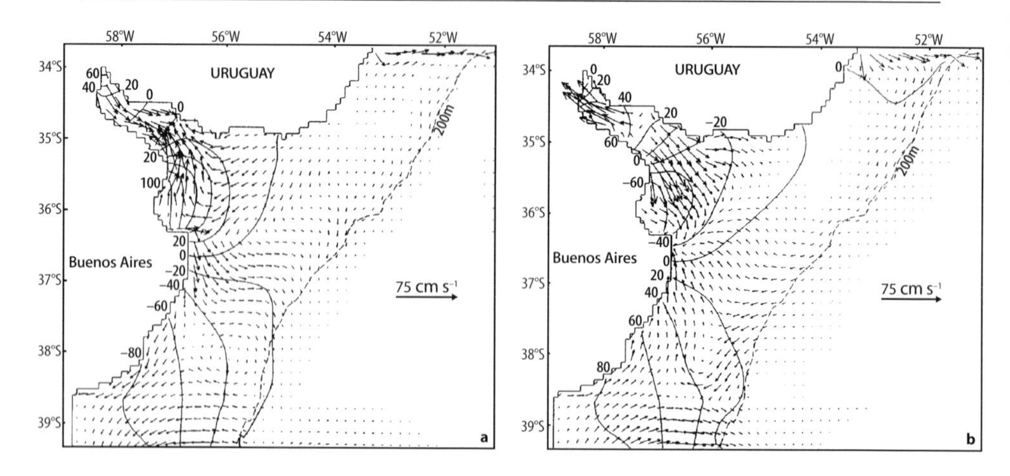

Fig. 6.3. The tidal amplitude in centimetres and tidal currents for the semidiurnal lunar M_2 tide **a** at cotidal hour four, low tide in Buenos Aires and **b** cotidal hour ten, high tide in Buenos Aires (after O'Connor 1991)

occupy the bottom layer, since the distribution is mainly controlled by topography (Guerrero et al. 1997). The estuary is characterized by two different salinity periods as a result of different forcing cycles: during fall-winter seasons, the estuary is characterized by the lowest wind influence and a maximum in river runoff resulting in a lower salinity regime; whereas during the spring-summer period, there are prevailing onshore winds and a minimum of riverine runoff inducing a higher salinity condition. The saline water movement is greater on the northern coast through the deep northern channel. In Samborombón Bay the saline water movement is quite restricted due to the shallowness of the area. In absence of wind the salinity follows the tidal circulation ranging from 0–35 psu at the surface and 15–36 psu at the bottom (Urien 1972).

On the other hand, water temperature varies between 8 and 27.5° C. The mean temperature field remains almost homogeneous for the warm (December–March) and cold (June–September) periods, both in vertical and horizontal scales (Guerrero et al. 1997).

As far as flora and fauna of the estuarine zone is concerned, a marked predominance of euryhaline marine species is evident and several fish species find refuge temporarily in the estuary, particularly Samborombón Bay, for the purpose of reproduction. Few fresh water species have been found and this only happens on the borelines. According to Boschi (1988) the biomass of fish vulnerable to the trawl obtained during different cruises show values between 60 250 and 141 800 t, wet weight, those of the spring being the largest. The estuarine ecosystem shows characteristics that are quite particular and do not agree with those of similar ecosystems in tropical and subtropical regions. No mangroves are present, whereas marshes grasses (*Spartina*) and other halophytic plants somehow contribute to detritus formation. No peneid shrimps or oysters which are very abundant in estuaries to the north of the Gulf of Mexico have been found (Boschi 1988).

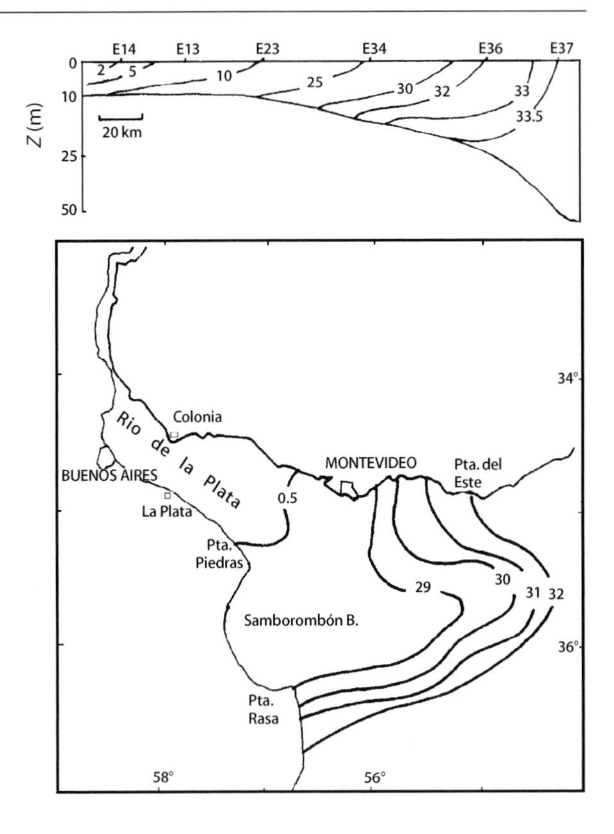

Fig. 6.4. Salinity distribution in the Río de la Plata Estuary (modified from Piccolo and Perillo 1997b)

6.3 Estuaries of the Buenos Aires Province Between Samborombón Bay and Bahía Blanca

Samborombón Bay is the southern limit of the Río de la Plata, between Piedras and Rasa points (Fig. 6.1). Samborombón and Salado Rivers and several small streams and channels flow into the bay. The rest of the streams that flow into the bay have a relatively short length and their basins are very small when compared with other argentine catchments. Unfortunately little is known about the streams that discharge into the bay, specifically about tidal dynamics and salinity input, although it is assumed that their influence is relatively small if any.

Further to the south there are a series a rivers and creeks that start at the Mar Chiquita Lagoon, some 120 km from Punta Rasa. These streams have their origin on the Tandilia and Ventania hill systems, having relatively small catchment basins and low discharges. Besides the low runoff, the development of estuaries on the Buenos Aires Province coast is highly dependent on the wide dune field that borders most of the coastline. Only the Mar Chiquita coastal lagoon, and Quequén Grande, Quequén Salado and Claromecó form estuaries. The rest of the streams arrive to the coast with very low runoff due to infiltration across the dunes and the low slope of the area. From this major estuaries, only Mar Chiquita Lagoon and the Quequén Grande River have been studied from the geomorphological and dynamical points of view. However, some work is presently being done at the Quequén Salado, and there are initial approaches at the Claromecó Estuary.

6.3.1
Mar Chiquita Coastal Lagoon

Mar Chiquita is the only coastal lagoon of Argentine that is chocked with a long inlet. It is located in the Buenos Aires province between 37°30'S–37°45'S and 57°37'W (Fig. 6.1). The lagoon has an elongated shape, with a general direction NNW-SSE. Its length is 25 km and its width varies between 500 and 4 500 m (Fig. 6.5). Mar Chiquita Lagoon has a mean area of 46 km^2 and a drainage basin of 10 000 km^2 approximately, changing according the total water volume of the lagoon (Fassano 1980). A chain of dunes developed over a Pleistocene wide barrier separates the lagoon from the ocean. Depth is variable, data obtained in 1972 indicated maximum values of 4.90 m during high tide to 1.80 m in low tide. However, López (1978) reported 1.5 m and 0.8 m respectively, with a volume of 36.8 hm^3. Naturally, the water level changes according to the groundwater run-off, the discharges of the small streams and the tides. However, the geomorphology varied considerably upon the construction of a bridge at the head of the inlet that induced a large shoaling of the inlet. Bathymetric information gathered in 1994 indicated that maximum depth are only 2 m at high tide (unpublished data).

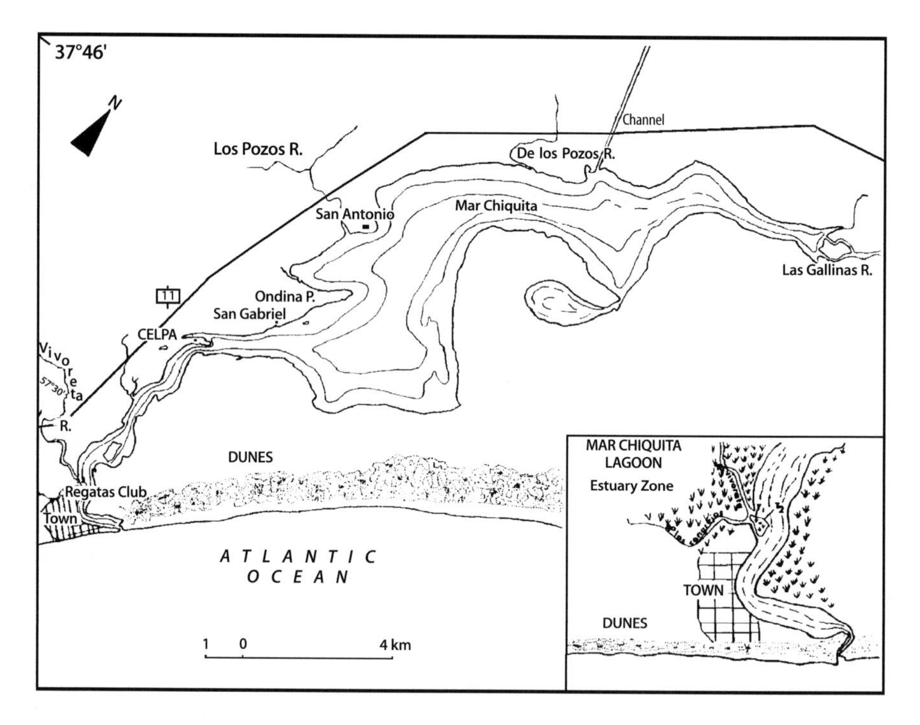

Fig. 6.5. The Mar Chiquita coastal lagoon (modified from Olivier et al. 1972)

The lagoon is connected to the sea by an elongated inlet channel, approximately 6 km and up 200 m wide. The inlet area is characterized by a small smoothly contoured seaward ebb delta highly variable, and therefore it was subject several studies (Isla 1980; 1984). Historical data and recent surveys indicate that the mouth of the lagoon has suffered changes in position and size with a northward pattern. The oldest information available, held by the Drainage Commission of the Buenos Aires province (January 1885), is reflected in the nautical chart of 1915, from which variations in location of the mouth are evident since 1908 (Fig. 6.6). After an initial stage, based on responses to effects of waves and related sedimentary processes, the mouth migrated approximately 100 m yr^{-1} (Lanfredi et al. 1987). The neighboring beaches are subject to a rapid erosive process, specially in the cross-shore component. Total obstruction of the mouth happened several times (1904, 1908, 1912, ..., 1933) being re-opened artificially (Isla 1980).

Several studies have been performed from a geological standpoint (i.e., Daus 1946; Schnack et al. 1980; Isla 1980, 1984), but scarce attention was paid to hydrographic conditions. This coastal lagoon is the first reservoir south of Samborombón that relates the inner river basins with the sea. Several fresh water courses discharge at the

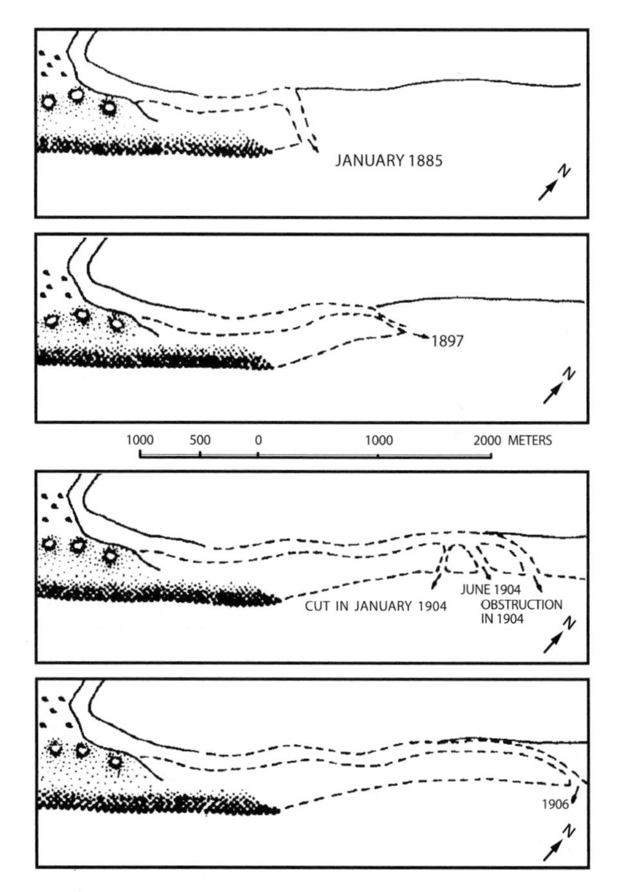

Fig. 6.6. Tidal inlet migration of the Mar Chiquita coastal lagoon between 1885 and 1908 (modified from Lanfredi et al. 1987)

lagoon. They are Arroyo Chico, De las Gallinas, Cangrejos, De los Huesos, Los Pozos, Grande, Dulce, Vivoratá streams (Fig. 6.5) and several channels discharge into the lagoon (Grondona 1975).

Significant salinity variations characterize the lagoon as it fluctuates from oligohaline to hyperhaline conditions, ranging from less than 5 psu to more than 30 psu (Alvarez et al. 1983). Hydrogeological investigations show the influence of the water table in relation to the lagoon. This groundwater contribution plays an important role for standard meteorological conditions with regards to the mean stored volume and it is particularly significant in prolonged dry periods. The water table rise associated with the rejected recharge effect inducing important variations in the lagoonal body (Fasano et al. 1982). Intrusion of the saline wedge varies according the ocean tide and the wind direction. Strong winds from the NE to SW direction during neap conditions prevent the entrance of marine waters into the lagoon. Winds from the N and NW directions aid the fast discharge of the lagoon water into the sea. On the other hand, strong winds from the sea during spring tides helps the marine water to enter into the lagoon several kilometres from the mouth (Reta et al. 1995).

Water temperature depends almost exclusively on the meteorological conditions (insolation, daily variations of the air temperature, etc.). There is approximately 2° C in water temperature variation between the mouth and the inner regions of the lagoon. The minimum and maximum temperatures registered in the period 1994–1995 was 9° C and 25° C, respectively (Martos et al. 1995). Significant air-sea-soil interaction processes in the area were observed. A diurnal energy balance study performed in the area showed that evaporation is the most important term in the heat balance equation, followed by the sensible and the soil heat fluxes. In the lagoon beach, the sensible heat and the soil heat flux have the same order of magnitude (Piccolo et al. 1995).

During calm conditions, tides are observed as far as San Gabriel (Fig. 6.5) With a lag time of 2.5 and 2.2 h from the mouth, for high and low tides respectively. The reduction in amplitude at San Gabriel is 78% with respect to the mouth (Table 6.2). The semidiurnal component (M_2) is the most important. Tidal currents inside of the lagoon are weak. Erosional capability has been observed by Lanfredi et al. (1987) in the entrance channel where the highest registered velocity was 140 cm s^{-1}. There is an evident relationship between the mouth and the outer banks. Due to the action of the storms, the banks supply sand to the mouth. The wave action is perceived in the first few metres of the channel; together with currents that transport sediments toward the interior of the lagoon, depositing them some hundred metres upstream. During the

Table 6.2. Tidal characteristics in the interior of Mar Chiquita Lagoon (after Lanfredi et al. 1987)

Station	Amplitude (m)	Time lag at HW (min)	Time lag at LW (min)
Mouth	1.38	–	–
Club regatas	0.96	30	72
CELPA	0.77	66	84
San Gabriel	0.30	138	132
San Antonio	0.05	–	–

ebb tide, sediments move along the bottom as bed load and the bank obstructs the back flow.

Mar Chiquita Lagoon is inhabitated by a dense population of two intertidal grapsid crab species. *Cyrtograpsus angulatus* and *Chasmagnathus granulata*. Besides frequently occurring larvae of the two dominating crab species, those of three other brachyurans (*Plathyxantus crenulatus, Uca uruguayensis, Pinnixa patagonica*) and of one anomuran (*Pachychelis haigae*) are also found occasionally. Caridean shrimp larvae also are found in the lagoon (Anger et al. 1994).

6.3.2 Quequén Grande Estuary

Quequén Grande River (Fig. 6.1) is one of the most important rivers of the SE Buenos Aires province (37°27'S–38°29'W). The river basin is located in a highly developed farming zone of Argentina and it has an area of 9 370 km². The river is 173 km in length. In the mouth of the estuary is found a vital harbour, Puerto Quequén and an important resort area, Necochea. The harbour is a very active commercial and industrial site, where many of the main products of Argentina are exported. In the last ten years the industrial and economical activities of the harbour have greatly increased and presently the estuary area in one of the most important economical zones of the province.

In the last 15 km the river has developed a meandering channel with cliffy banks cut on Pleistocene semiconsolidated sediment with high carbonate content. The width of the cross section is variable, between 150 and 200 m, reaching 400 m in the harbour area. Mean river discharge has varied along the years form 20 $m^3 s^{-1}$ (1918) to 5.25 $m^3 s^{-1}$ (1992), being the mean value 11.3 $m^3 s^{-1}$ (Sala 1975). Due to the harbour activities, the last 2 km of the estuary are kept with 12 m in depth by continuous dredging. However, further upstream the thalweg has a depth of 2–4 m.

Sediment in the estuary are composed of sand, clayey sand and silty clay. Sediment mean size decrease headward and it is function of the high energy conditions prevailing near the mouth of the estuary (wave activity, strong currents and tidal action) (Wright 1968). The bottom sediments within the estuary reflect the influence of the Pleistocene sediments of the adjacent area. However, at the head of the harbour and due to its particular dynamics, sediments are very fine and the conditions are highly reductive.

Salinity and currents are the major environmental factors which considerably vary in the estuary. Due to the tides, the salinity undergoes a daily variation of up to 23 psu at the harbour site of the estuary. Salinities greater than 30 psu are found in the first 2–3 km from the mouth of the estuary, meanwhile 1 psu are found at 15 km from the sea. However, maximum salinity intrusion varies from 15 km in high tide to 10 km at low tide. Mean temperatures varied between 7° C and 25° C along the year.

Vertical salinity profiles within the harbour area clearly show a strong halocline at about 1–2 m depth that is present along the tidal cycle. The upper layer is mostly freshwater with increased salinity towards the mouth due probably to upward salt transport by entrainment. The lower layer is vertically homogeneous reaching up to 33–35 psu.

Current velocity profiles taken at different stations for at least one tidal cycle both within the harbour and at the river show that the tidal influence is uncoupled between both layers. On the upper one, the flow is seaward practically during all the tidal cycle. The currents for the bottom layer are very small (<10 cm s^{-1}), specially near the harbour-head step, and having random directions (Cuadrado and Perillo 1996). It is apparent that the tide enters on the bottom and actually push up the upper layer which is flowing seaward.

Lanfredi et al. (1988) have studied the variations of the mean sea level in Puerto Quequén. They analysed 64 years of hourly tidal heights and obtained the long term trend of 1.6 mm yr^{-1}. The tide is predominantly semi-diurnal, mixed. Mean tidal range is 0.89 m. Yearly maximum tides were analysed by Piccolo and Perillo (1997a) for the period 1958–1994. The highest values were registered in 1962 and 1979 with 3.1 m. The most important astronomical components are the M_2, O_1 and K_1. Restricting the analysis to the period 1989–1993 and to the three main constants. Table 6.3 shows significant annual variations in amplitude and phase.

Piccolo and Perillo (1997a) have also studied the hourly departures of observed tides from predicted ones (storm surges). They presented maximum values of 1.5 m and −1.66 m in the period 1989–1994. Spectral analysis of the storm surges shows significant energy peaks in 60 d, 21 d, 10 d, 5.6 d and 12 h (Fig. 6.7). These periods indicate that the fluctuations are produced by meteorological processes in macro, synoptic and

Table 6.3. Amplitude and local phase angles for the principal diurnal

Year	M_2		K_1		O_1	
	A (m)	*G* (°)	*A* (m)	*G* (°)	*A* (m)	*G* (°)
1989	0.4378	178	0.1738	74.7	0.1755	3.7
1990	0.4387	75.1	0.1725	72.2	0.169	265
1991	0.4401	333	0.1607	69.5	0.1642	169
1992	0.4	249	0.1434	258	0.1372	266
1993	0.3626	165	0.1452	78.7	0.1366	3.4

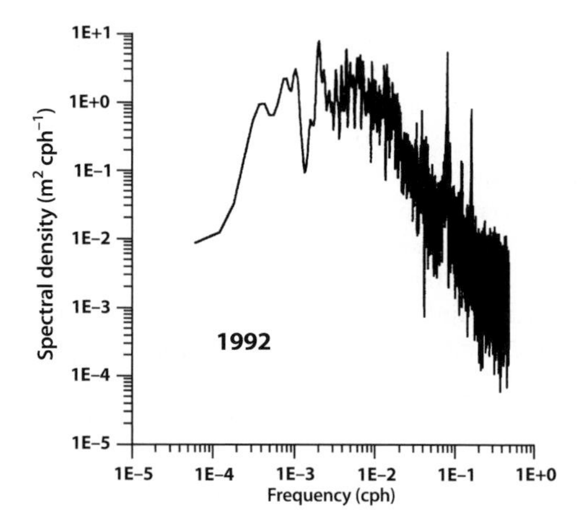

Fig. 6.7. Spectral density of the daily storm surges at Puerto Quequén (1992)

microscales. The 10 d peak corresponds to the frequency of storm passage for the study zone and the 5.6 d characterizes the synaptic scale. These periods are typical of high and mid latitudes. The 12 h peak corresponds to the local wind circulation: the sea breeze. Therefore, the sea level variations of the Quequén River Estuary is mostly due to the effect of the meteorological forcing.

Furthermore, tidal records also show the presence high frequency oscillations. The periods for these seiches are between 3 and 25 min, although the later is predominant. Even though normal oscillations have amplitudes of the order of 10 cm, commonly appear oscillations of up to 1.5 m. Normally the oscillations increase in size suddenly and unexplained, but inducing serious consequences for the navigation of large cargo vessels. At least two vessels departing from the harbour with full load hit the bottom due to the occurrence of sudden large oscillation, fortunately without damage. The resonance periods for the harbour calculated using Merian's and Helmholz's equations give theoretical values of 3.3 and 6.4 min, which are well below the typical 25 min. Most probably the seiches are produced by internal waves generated at the interface.

The foraminiferal and the camoebian biocenosis of the Quequén River Estuary has been studied by Boltovskoy and Boltovskoy (1968) and Wright (1968). They found *Miliammina fusca*, *Elphidium excavatum*, *Rotalia beccarii* ex gr. *Parkinsoniana*, *Bulimina patagonica* (*f. glabra*), *Trochammina inflata* and small calcareous Miliolidae. These species can tolerate very large changes in salinity (about 5–28 psu). The Quequén River Estuary is an interesting area for ecological studies not only on foraminifera, but for many other group of benthos and plankton.

6.3.3 Bahía Blanca Estuary

The Bahía Blanca area has suffered an accelerated urban expansion due to the presence of a major petrochemical industrial park, a thermoelectric plant, and an important deep-water harbour system from which most of the agricultural products of the country are exported. All of them are placed on the northern margin of the Principal Channel of the Bahia Blanca Estuary (Fig. 6.1). Industrial wastes (mainly oil derivates, pesticides, heavy metals, etc.) and untreated sewage discharges from coastal localities generate increasing contamination problems. Therefore, the distribution pattern of chemical parameters in a large populated and industrialized area of this estuary has received special attention (Freije et al. 1981; Sericano and Pucci 1982; Orozco Storni et al. 1984; Lara et al. 1985). The levels and ranges of heavy metals in estuarine areas not associated with anthropogenic activities have also been studied in water and sediments (Villa and Pucci 1985, 1987; Zubillaga and Pucci 1986; Villa 1988). Whereas the geomorphologic characteristics and hydrography of the Bahía Blanca Estuary has been extensively studied (Sequeira and Piccolo 1985; Serman 1985; Piccolo et al. 1987; Perillo et al. 1987a,b; Perillo and Sequeira 1989) and a chapter of this book is devoted to this estuary; therefore only a brief summary is presented here.

Bahía Blanca Estuary is a geomorphological complex environment derived from a former Late Pleistocene-Early Holocene delta complex. Tidal flats and channels with subordinate islands and few salt marsh areas are the main features of the estuary. These channels are partly closed by modified ebb deltas. The estuary has a general triangular shape with the major channels trending in a NW-SE direction. Dominant sedi-

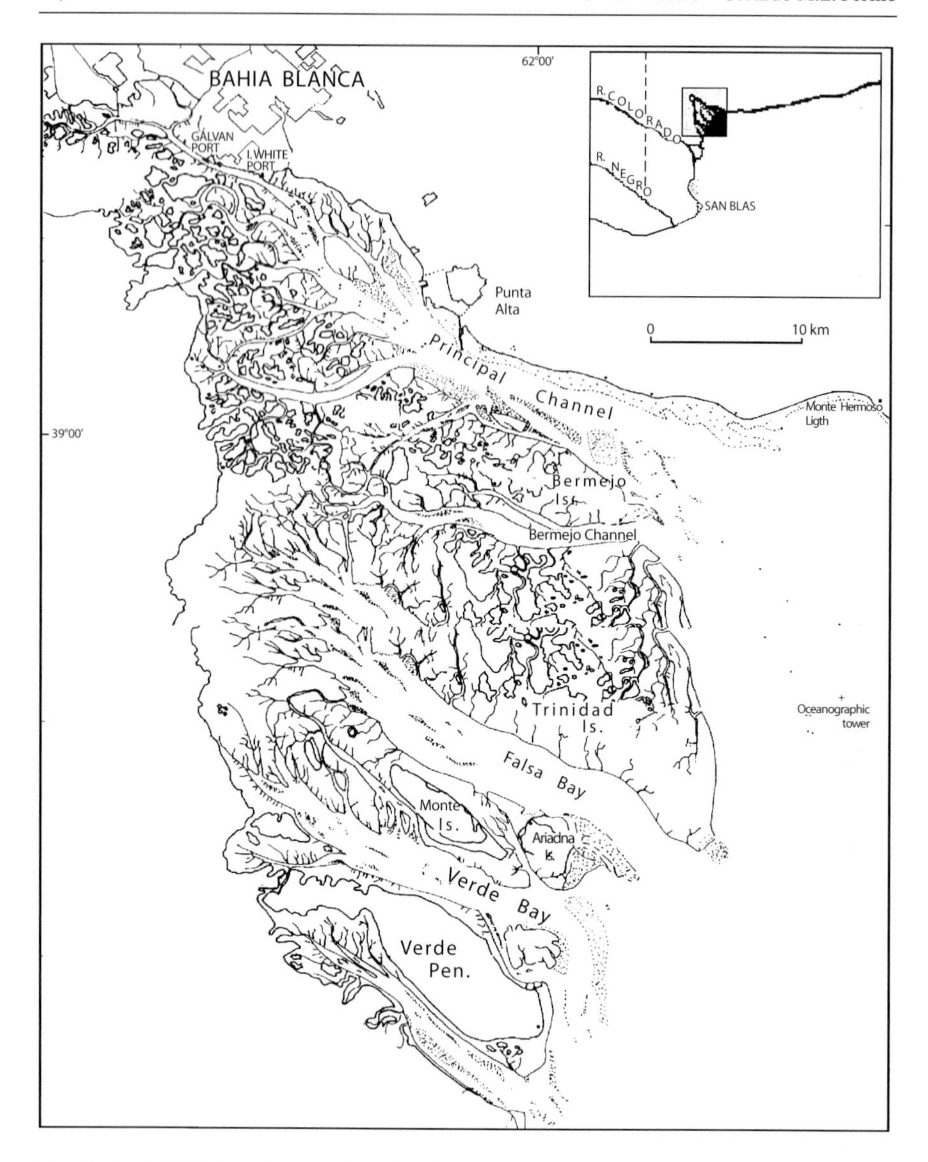

Fig. 6.8. The Bahía Blanca Estuary (modified from Piccolo and Perillo 1990)

mentology is based on silty clays on the flats and sand in most of the deeper parts of the channels. The latter give place to development of large migrating sand waves.

Based on the salinity and temperature distribution alone, the Bahía Blanca Estuary (Fig. 6.8) can be divided in two sectors (Piccolo and Perillo 1990). The inner one, from the mouth of the Sauce Chico River to Ingeniero White, can be classified as a partially mixed estuary during normal runoff conditions, but with a strong tendency to become sectionally homogeneous during low runoff. The outer reach is sectionally homogeneous. The boundary between both reaches is transitional depending on the river discharge. Fully turbulent conditions resulting in Estuarine Richardson Numbers less than 2 characterize the mixing regime. According to Hansey and Rattray (1966), the estuary is classified as Type 1a. However, when stations located across the different sections of the estuary are considered, the southern flank presents well defined Type 1a characteristics, meanwhile the northern flank, Type 2a one.

The residual circulation shows a marked difference in the direction of the mass transport. On the deeper parts of the sections (northern flank) the flow reverses with depth, being headward near the bottom. The net transport is completely landward on the shallower parts. This behaviour, added to the evaporation processes, produce a concentration of salt in the inner portion of the estuary and over the tidal flats, resulting in salinities larger than those observed in the inner continental shelf when these areas are flooded. In the southern flank of the Principal Channel the asymmetry of the tidal current is mainly due to the extensive tidal flats bordering it. Perillo and Sequeira (1989) have also shown the same pattern for the middle reach of the Bahía Blanca Estuary which can be explained by the fact that the volume of water crossing a point in the channel adjacent to the tidal flat during flood does not necessarily returns through the same point at ebbing tide, thus mass conservation is not attained for specific cross-sections.

A quasi-stationary tidal wave provides the main energy input to the Bahía Blanca Estuary. However, the complex geometry of the estuary and the wind produce deviations from the estimated values from theoretical estimations. The calculated damping coefficient is small thus resulting in an hypersynchronous estuary where convergence is the dominant factor producing a marked increase in the tidal amplitude. Unfortunately the number of tidal stations with long and continuous records is small and all of them located on the northern shore of the Main Channel. Therefore with these data it is very difficult to assess the actual influence of the tidal flats on the tidal wave. The approach followed here to estimate the convergence and friction effects only used mean tide conditions. One can imagine that the convergence effect ought to disappear as soon as the tide overcomes the channel banks and runs freely over the extensive tidal flats. To the authors' knowledge no theory has been developed yet that takes into consideration large tidal flats.

In a region where the wind blows on average at 16 km h^{-1} with average maximum of 58 km h^{-1} and gusts of over 100 km h^{-1} (Piccolo 1987) its influence over the tide is very important. The deviations of the real tide in relation to the predicted reaches up to 4 m. This type of differences seriously affect the general circulation of the estuary specially when it is compared with the tidal range that varies from 2–4 m and the amplitude-to-depth ratio is relatively large (Perillo and Piccolo 1991).

Heat exchange across the sediments of the tidal flats of the estuary were analysed by Piccolo and Dávila (1993) during one year. Thermistors were installed at 0.05 and 0.15 m below the sediment surface, and at 1 and 10 m above the mud flat. Assuming that soil temperature varied as a sinusoidal function of time, thermal diffusivity (k) was determined from a wave amplitude-depth relationship resulting in a mean value of 0.7×10^{-6} $m^2 s^{-1}$. The magnitude and rate of heat exchange during tidal inundation is dependent upon the relative temperature of the mud and the incoming water layer. The net result of tidal inundation during daylight hours, specially after noon, is a very sharp fall in temperature, the magnitude of which is closely related to the timing of tidal water movements. However, the largest fall in mud temperature occurred during a late afternoon flooding. Power spectra of air, water and soil temperature showed maximum peaks at 3.6, 4.3 and 5.3 h, respectively.

Tidal and wind records from two stations located one on an estuarine bank and another offshore the Bahía Blanca Estuary were analysed by Piccolo and Perillo (1989) to establish the atmospheric forcing on the subtidal fluctuations for a four month period. Although the residual tidal records for both stations have large coherences at all scales, the phase relationships may indicate that forcing is produced first at the left bank for time scales shorter than 16 days. It is then evident that the continental winds, which are strongly prevailing in the area, are the main factor to produce variations in the tides. Wind forcing is also influenced by the geomorphologic characteristics of the particular setting of both stations. The left bank of the estuary has faster response to wind forcing than the offshore one because the tide is channelized at least part of the time and the NW-SE winds are more effective in producing the water level set up/ set down.

Numerous studies have been performed about the different species and zones that characterize the estuary. A brief resume of some of the results will be presented here. Macrobenthic intertidal associations in the estuary are described by Elias (1985) presenting the first delimitation of infaunal communities in the mud flats. Physiographically the estuary corresponds to a salt marsh, where two communities are conspicuous: the "espartillar" (*Spartina* sp. – *Salicornia* sp.) and the cangrejal (*Chaemagnathus granulata*). Besides two communities were found: *Laeonereis acuta – etone* sp. and *Axiotella* sp. – *Ninoe* sp. Wagner et al. (1991) studied the intertidal communities in Ingeniero White harbour. At present, *Balanus glandula* and *Balanus a. amphitrite* inhabit that zone (Fig. 6.8). Seasonal pulses of *Balanus* larvae from Bahía Blanca are observed in annual cycles. *Balanus glandula* is dominating with high biomass values at the middle intertidal zone. On the other hand, Barria de Cao (1986, 1992) analysed the numerical abundance and seasonal cycles of tintinnines species in the estuary. She identified 25 species of *Tintinnina* (Protozoa, Ciliophora) belonging to 11 genera. Hoffmeyer (1983, 1994) studied the faunistic composition of the zooplankton of the inner part of the estuary. She identify 35 taxa corresponding to 8 phyla being the most common (*A. tonsa*, *Paracalanus parvus*, *Euterpina acutifrons* etc.)

The waters of the estuary are characterized by the presence of Scyphomedusae. Three species were found: *Chrysaora lactea*, *Aurelia aurita* and *Drymonema gorgo* in summer and efirae of the two former from October to January. These species are commonly found in the Brazilian waters. Their occurrence in higher latitudes may be due to influence of warm waters coming from the north or the heating of local waters which could maintain their populations temporarily during spring and summer (Mianzan 1989).

6.4 Estuaries of the Patagonia

Rivers in the Patagonia are allochthonous with source of water provided by the precipitation and/or the melting of the snow on the Andes. They flow across the arid and desert Patagonia region practically no tributaries are received. Several of the river are some of the largest in the country in valley size and river discharge such as the Colorado, Negro and Santa Cruz Rivers. Río Gallegos is probably the one that has the largest estuary after Río de la Plata and Bahía Blanca, although its freshwater input is reduced. Other estuaries such as Deseado and San Julián have recently become fully marine environments since no freshwater is being input there. Both San Antonio Bay and Caleta Valdés only receive some freshwater from groundwater since no surficial rivers flow into them.

The climate of Patagonia is semiarid to arid. It is characterized by strong westerly winds throughout the year. Typical mean speeds are 30–35 km h^{-1} (16–19 kts). The predominant winds are from W, SW and NW, which also are the directions of the strongest winds. The winds from the SW-NW quadrant have six times the frequency of all the other directions combined.

Unfortunately there are few studies related to the geomorphology Patagonian estuaries. The more important ones are for the estuaries of the: Colorado, Negro and Chubut Rivers (Fig. 6.1). There is a lack of studies about the dynamics of the estuaries of the southern part of Argentina; therefore, different and disperse processes in different time scales will be described here trying to update the present knowledge of them.

6.4.1 Colorado River Estuary

Colorado River sources (Fig. 6.1) are in the dry regions of the Central Andes mountains. The mean annual discharge is 148 $m^3 s^{-1}$, being the maximum and minimum mean daily values 1 053 $m^3 s^{-1}$ and 30 $m^3 s^{-1}$ (Evarsa 1995). Its maximum discharge occur in spring (October) because of the melting of the winter upper mountain snow, ending in December or January. Nevertheless, this values are presently changing since the end of 1996 because a dam was constructed some 340 km from the mouth that has modified both water and sediment load discharge. In fact, the name of the river came from its brownish color due to the high suspended sediment load.

As is explained in more detail in Perillo and Piccolo (this book), the Colorado River Delta extended originally from Bahía Blanca to Anegada Bay, some 200 km of the coastline, although there is no information about its extension offshore. The present day delta extends about 40 km along the coast and begins at 86 km from it. However, this delta has suffered important modifications in the last 100 yr both in morphology and channels of the delta (Fig. 6.9). According to Soldano (1947) in the delta there were four main discharges of the river into the ocean. From North to South their names were: Colorado Chico, Cauce Nuevo, Brazo Antiguo and Riacho Azul. Between the last two river channels there were several tributaries of the Riacho Azul that formed a microdelta. However, Weiler (1983) showed significant differences from the description of Soldano (1947).

The coast between Laberinto Point and Unión Island shows a light convex curvature along the North-South direction. Extended beaches and a low geomorphology are found in the river mouth. However, the region is characterized by a significant erosional processes of the old delta, that still have not been evaluated. According to Codignotto and Weiler (1980) model, the erosive process of the delta probably occurred during the Flandrian ingression that has its maximum penetration 6 900 yr ago and a short regression spatial period 400 years ago. Colorado River mouth, as well as in other river mouths in this part of the Argentina coast, shows a deflection towards the North due to littoral drift.

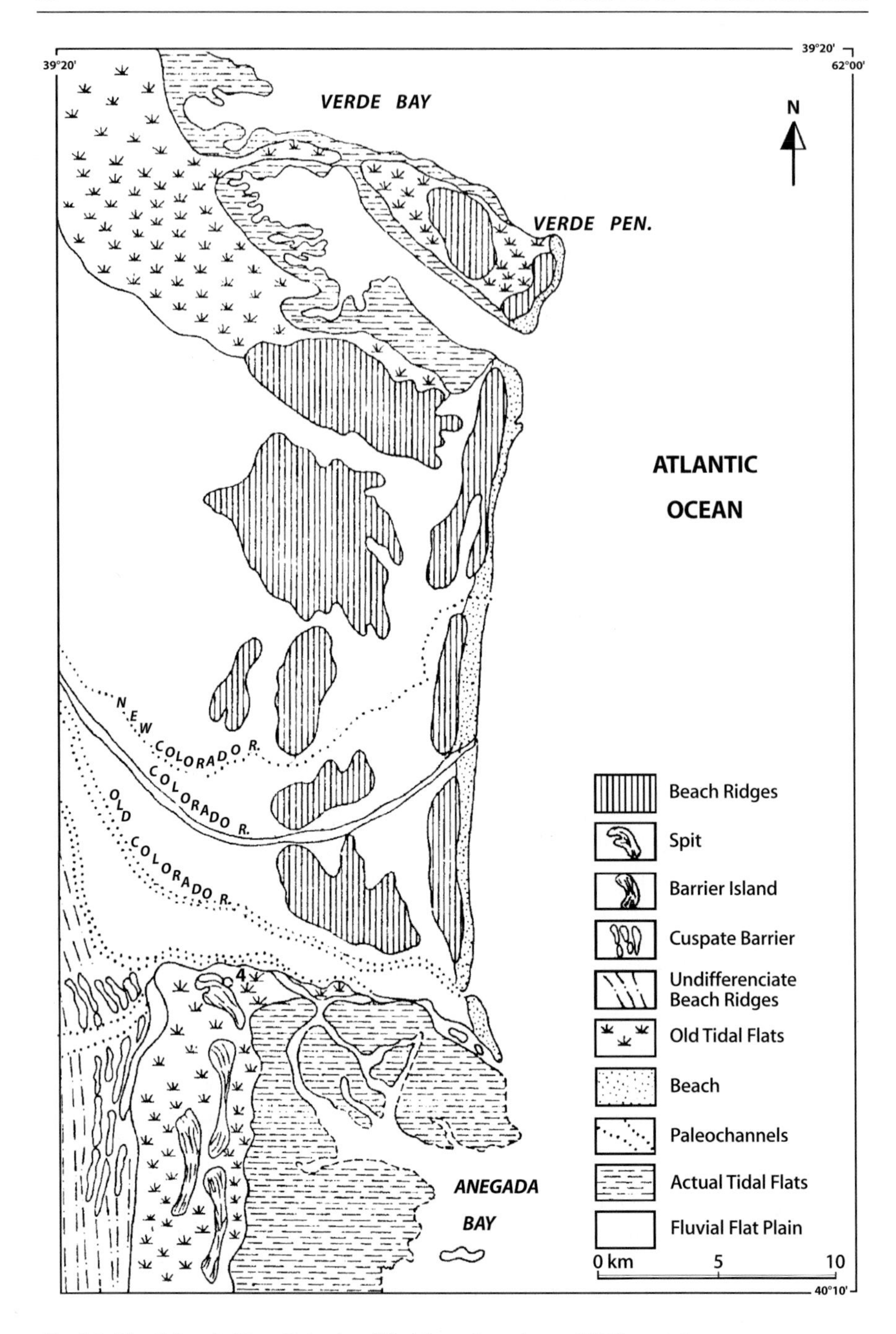

Fig. 6.9. The Colorado River Delta (modified from Gonzalez and Weiler 1983)

Mean tidal amplitude is 1.8 m and it penetrates into the river 60 km inland, approximately. Due to the construction of the dam (Casa de Piedra), the discharge has suffered a great reduction to its historical minimum. Therefore, an important sea water ingression into its channel is anticipated, exceeding what it was observed in 1984 (20 km from the coast). However, there is no physical data from the estuarine sector.

Nevertheless, Cuadrado et al. (1998) analysed temperature, salinity, suspended sediment and tidal current data obtained from the inner shelf offshore the Colorado River Estuary. They detected for the first time the influence of the Colorado River plume into the inner shelf. The distribution of the suspended sediment concentration in the surficial water showed two well defined plumes with values higher than 50 mg l^{-1}. One of them was at the North of the Colorado River exceeding 100 mg l^{-1} with a NNE direction. The other plume was located southwards, with concentrations of 80 mg l^{-1}. Unfortunately, to the authors knowledge little is known about the biology, tides, etc. of this estuary.

6.4.2 Negro River Estuary

Both the Colorado and Negro Rivers constitute the northern border of Patagonia. From economical and hydrological reasons, the Negro River valley is the most important of the region. The Limay and Neuquen Rivers are its main tributaries joining at about 550 km from the mouth and it does not receive any other tributary along the way. The mean annual river discharge is 858 $m^3 s^{-1}$, being its maximum and minimum mean daily 3 405 $m^3 s^{-1}$ and 75 $m^3 s^{-1}$, respectively (Evarsa 1995). In the estuary mouth current velocity was estimated in 3.83 $m^3 s^{-1}$ for a discharge of 550 $m^3 s^{-1}$ (INCYTH 1975). According to satellite data, the plume of the Negro River discharges towards the South.

Even though some geological surveys have been done in the lands around the estuary only one has deal directly with the geomorphologic characteristics and sediment transport processes occurring at the mouth of the estuary (del Río et al. 1991). At the mouth the estuary has a width of 1 km and flow along a valley of 12 km approximately; however, the river width varies between 500 and 800 m. Its depth vary between 5 and 10 m (Soldano 1947). Two banks are found in its mouth (Miguel at the SW and La Hoya at the NE, Fig. 6.10). They form an open ebb delta although they have different characteristics without the frontal shield.

The ebb delta presents a principal channel, which is dominated by ebb currents, and marginal ones which are dominated by flood currents. Both banks are formed by swash bars and shelves of fine sands with a high range of selection. Del Río et al. (1991) mention the presence of megaripples produced by the variations of the tidal currents during the ebbing tide. The authors also indicated that the ebb delta has an asymmetric form due to the littoral drift that characterize the area, which is from South to North (Aliotta 1983). The Miguel bank had grew about 2 500 m towards the East. On the other hand, La Hoya bank has a clear tendency of migration towards the coast.

Mean spring and neap tides at the mouth are 3.35 m and 2.44 m, respectively, and tidal influence are observed up to 66 km from the mouth (Soldano 1947). Naturally, ebb currents are more important in magnitude than flood ones. They reach values of 2.5 $m s^{-1}$ (del Río et al. 1991). The slack water of high water is 30 min and at low water is 17 min. Strong winds from the S, SW and W advance the ebb tides and delay the

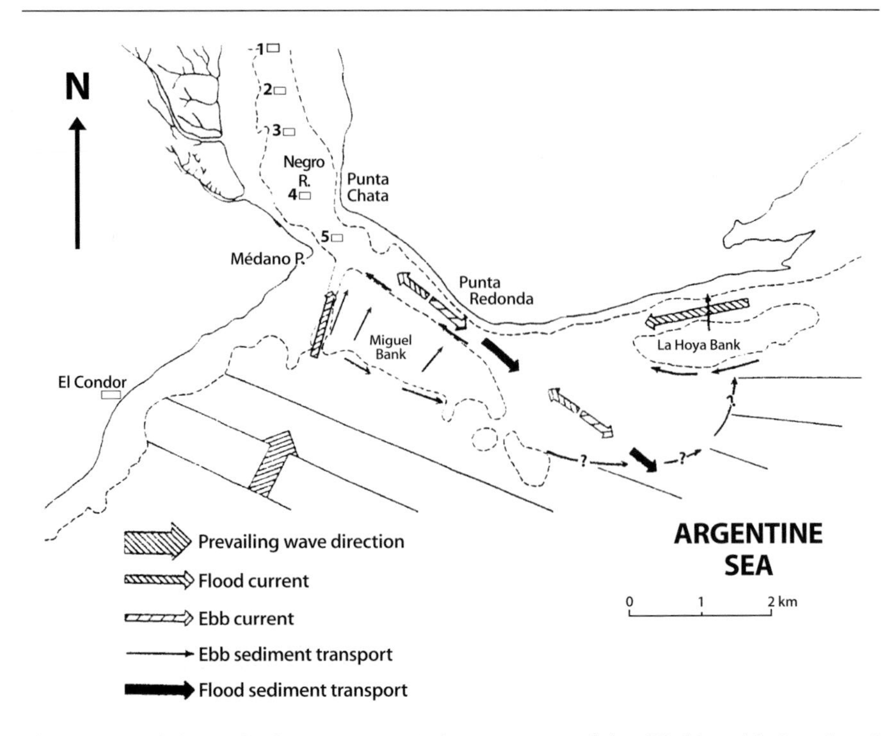

Fig. 6.10. Morphology and sediment transport in the Río Negro mouth (modified from del Río et al. 1991)

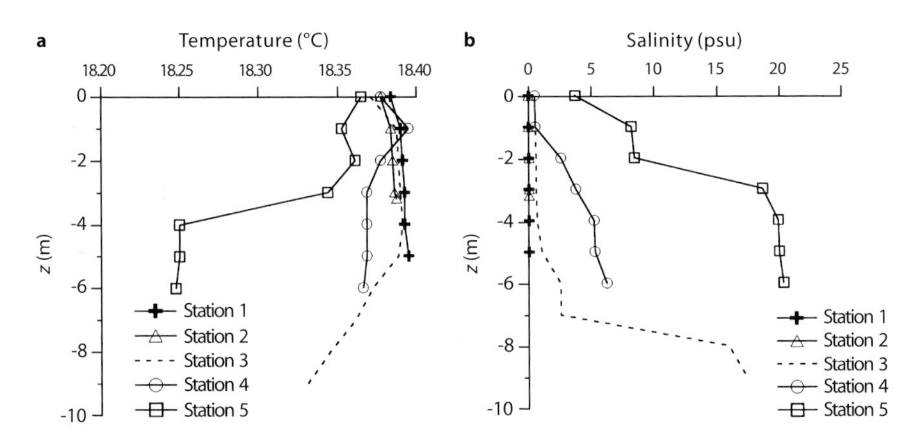

Fig. 6.11. a Temperature and **b** salinity profiles in the Río Negro Estuary in March 1998

high tides. Winds from the E and N produce the contrary effect. Although the prevailing winds are from the NW (INCYTH 1975).

Examples of the temperature and salinity profiles in the estuary for autumn are given in Fig. 6.11. Estuarine waters are thermally homogeneous and decrease towards the mouth of the estuary. Station 1 and 5 are 4 and 1 km from the mouth, respectively. The salinity profiles are vertical homogeneous with the presence of the fresh river waters only at station 1; however, a saline intrusion of marine waters was observed at depths from 6–8 m at station 3. The steep vertical salinity gradient was of the order of 12 psu m^{-1}. At stations 4 and 5, closer to the mouth of the estuary, although not at its mouth, the profiles show significant vertical salinity gradients. Further measurements in the estuary mouth would help to understand the penetration of the marine waters into the estuary due to tidal action.

6.4.3 Chubut River Estuary

The Chubut River is the most important watercourse of the Chubut province, in the Patagonia (Fig. 6.1). The river has been dammed at about 120 km from its mouth and passes through several of the major cities of the province. River water is used for agricultural, industrial and human consumption. Several industries are installed along the lower Chubut River, and a major fisheries-industrial park has being installed close to Rawson Harbour (located only 600 m inland from the river mouth). Dredging of the last 1 000 m of the navigation channel to accommodate medium size fishing vessels has been made several times. The present day industrial and municipal sewages are normally discharged into the river with very little or no treatment. Some work has been done on the chemical characteristics of the estuary (Orfila et al. 1987). High gradients of Cl^-, Na^+, SO_4^{2-}, K^+ and Mg^{2+} have been found in longitudinal surveys from the mouth to about 2 km upstream. On the other hand the longitudinal concentration of Ca^{2+} and carbonates were constant in the water column. Waters are rich in silica.

Considering the period 1990–1994, the annual average river discharge is 35.1 $m^3 s^{-1}$, and extreme mean daily values vary from 354–4.93 $m^3 s^{-1}$ (Evarsa 1995). The Chubut River belong to a wide basin denominated by Soldano (1947) the Chubut-Senguer basin. The meandering channel varies from 70–200 m in width, and averages about 2 m in depth. The river bed shows several sigmoidal bars constituted by medium to coarse sand. The bars divide the river into two channels, which are connected intermittently by small oblique interbar channels. From the harbour to the mouth, the bottom is mainly formed by very coarse sand and large patches of well-rounded gravels (Perillo et al. 1989). Tides are semidiurnal and have mean spring and neap ranges of 3.8 and 2.3 m, respectively.

Perillo et al. (1989) investigated the hydrography and circulation of the Chubut River under exceptionally low river discharge. Previous studies (Orfila and Scapini 1987; Orfila et al. 1987) showed that the frontal zone formed by the entrance of the tide into the estuary migrates with river discharge, but most of the time it is found at less than 2 km from the mouth. However, Perillo et al. (1989) observed it as far as 4.5 km from the mouth showing that the salt intrusion due to the tidal effect reached further inland than during normal river discharge (Fig. 6.12). Based on the classification of Hansen and Rattray (1966) the estuary pertains to the Type 1, with certain vertical stratification observed on the seaward side of the frontal zone. The time-mean salinity over the tidal cycle was calculated at different levels in the Chubut River Estuary. The station near the mouth shows a vertical profile characteristic of a partially mixed estuary with an upper layer of about $z / d = 0.25$ and a mixing zone of similar thick-

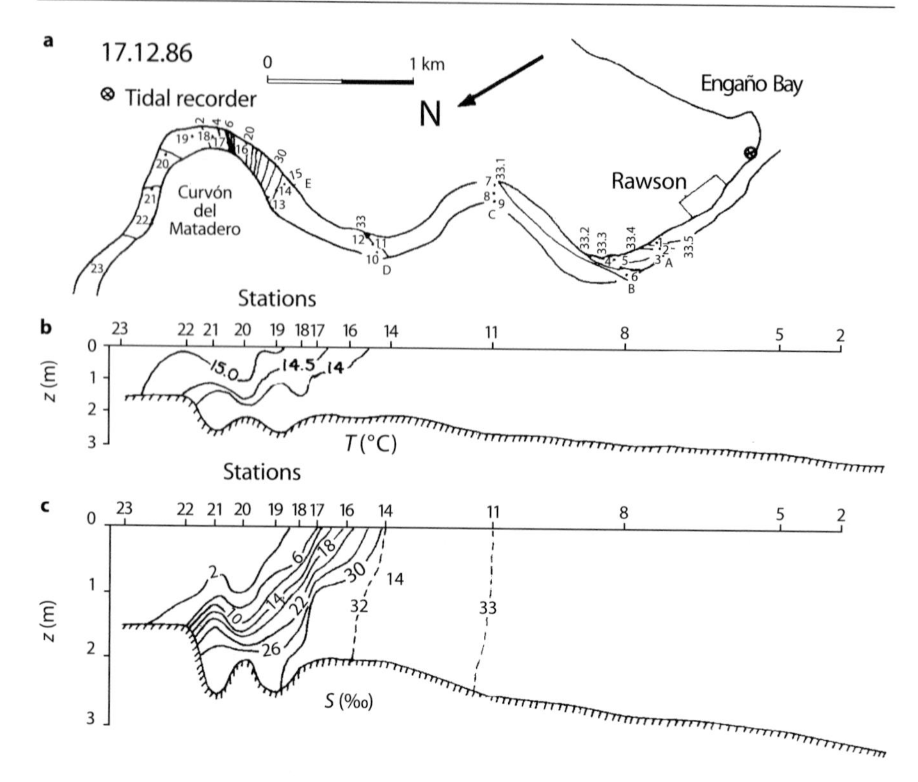

Fig. 6.12. Temperature and salinity profiles along the Chubut River Estuary, **a** location of the stations and distribution of the surface isohalines, **b** longitudinal profile of temperature, **c** longitudinal profiles of salinity (modified from Piccolo and Perillo 1997b)

ness. On the contrary, a station located in the inner estuary showed vertically homogeneous conditions with typical salinity values of fresh water.

Lateral salinity gradients are the result of the general morphology of the estuary. The presence of meanders and interchannel bars originate a secondary circulation driving fresh water near the surface towards the right margin and concentrating the saltier water on the left bank. On the other hand, wind effect is a major component of the circulation and mixing of this shallow estuary (Perillo et al. 1989). The longitudinal distribution of temperature and salinity shows a saline wedge. In the frontal zone the longitudinal gradients are 1.4×10^{-3} °C m^{-1} and 4.5 psu m^{-1} in temperature and salinity respectively. Salinities at the mouth of the estuary vary between 31.8 and 33.9 psu with almost no vertical stratification.

6.5 Estuaries of Southern Patagonia

From the Chubut River Estuary and along the 750 km of the Patagonian coast there are no other watercourse till the Deseado River Estuary in the Santa Cruz province (Fig. 6.1). The Santa Cruz coast have four rias, being the most important ones in Argentina (Castaing and Guilcher 1995). They are formed by the discharge into the ocean of the Deseado, Santa Cruz, Coig Rivers plus the San Julián; which used to receive fluvial discharge trough the Bajo de San Julián, (San Julián Lowland) but at the present time it is only a net marine embayment.

Only five rivers are found in 1800 km of the Patagonian coast. This fact shows the poor hydrological resources of this zone. Besides, the Deseado River discharge is limited since the streams from the Buenos Aires and Pueyrredón lakes at the Andes were captured by backward erosion towards the Pacific basin (Kühnemann 1963). The Santa Cruz River have a mean annual discharge of 697 $m^3 s^{-1}$, being the maximum and minimum mean daily values 2520 and 180 $m^3 s^{-1}$ (Evarsa 1995). Unfortunately, there is no data from the other watercourses, although it is known that Gallegos and Coig Rivers have smaller discharges than Santa Cruz.

Patagonian estuaries were formed by the inundation of the old river valley cut through the Tertiary sediments of Patagonian formation. The Deseado River has a general WSW-ENE orientation with a form of an elongated 40 km funnel. Its width vary between 2500 and 400 m within 18 km from its mouth. It has an irregular form due to the presence of islands, beach ridges and points that originate small bays (i.e., Uruguay, Concordia). Between Puerto Deseado and its offshore zone the maximum depths vary between 30 and 37 m, but towards its interior, its depth decrease in a short distance from 20 m to only 5 m (Piccolo and Perillo 1997b).

Even tough the mean spring and neap tides are 4.2 and 2.9 m, respectively, there are not large tidal flats except in very few places with restricted circulation. According to the nautical charts the maximum tidal velocities of high and low tide vary between 2.5 and 3 m s^{-1}. This indicates the strong dynamics and turbulence that characterize the circulation of its waters. Kühnemann (1963) describes the strong turbidity of the Deseado waters produced by the fluvial discharge of volcanic ashes (mostly white-gray clays) giving Secchi values that vary between 6 m at the mouth to 0.5 at 12 km inland from it. The author also mentions the presence of a turbidity front at its mouth. Pallares (1968) shows variation of temperature and salinity for a 15 days period at Cavendish Point, the harbour and Uruguay Bay (Fig. 6.13). Maximum and minimum temperatures are observed in January–February (14° C) and July–August (5° C), respectively. Meanwhile salinity variation is small no exceeding 1 psu.

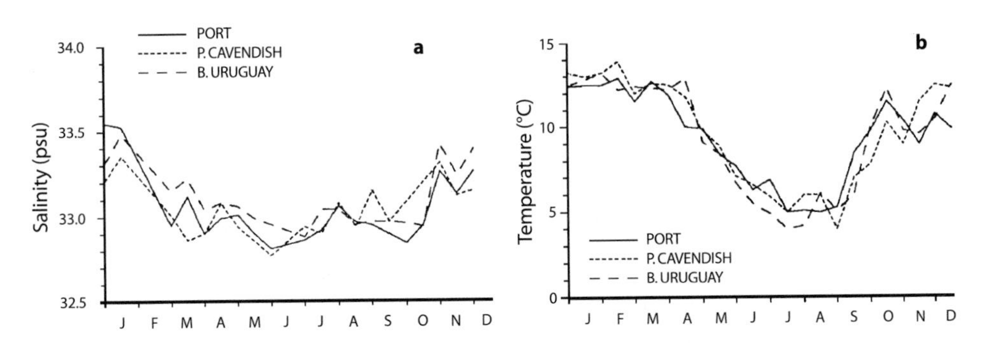

Fig. 6.13. Annual distribution of **a** salinity and **b** temperature in the Deseado River Estuary (modified from Piccolo and Perillo 1997b)

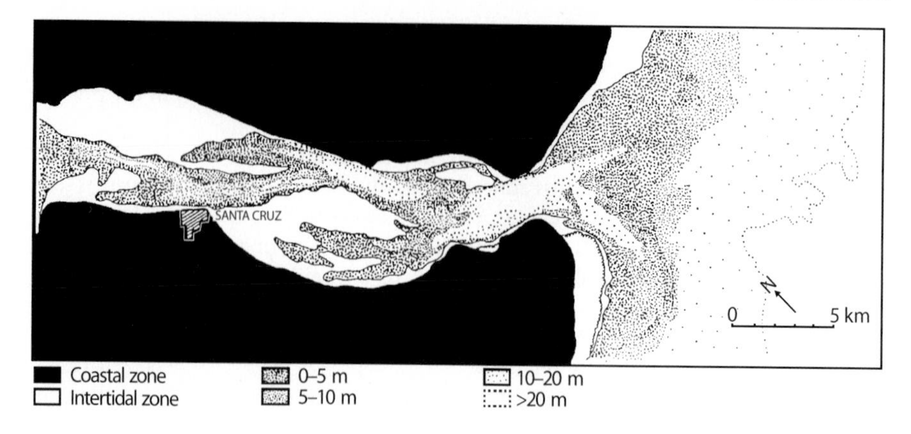

Fig. 6.14. Geomorphology of the Santa Cruz River Estuary. An ebbing delta and with two principal channels are present in its mouth (modified from Piccolo and Perillo 1997b)

The river valleys of Santa Cruz and Coig are very similar. They develop large tidal amplitudes (9.5 and 5.4 in spring and neap tides, respectively). The ebb delta of the Santa Cruz River (Fig. 6.14) presents two ebbing channels. The southward channel is the most active cutting the frontal shield. In both cases it is not observed marginal channels flood dominated, this fact may indicate that tidal intrusion into the estuary is as a water mantle, that will means the presence of bores. The formation of an ebbing delta is produced because the mouth of the estuary is narrow (2 km approximately) compared with its precedent valley. This originates to strong ebbing currents reinforced by the westerlies winds (Piccolo and Perillo 1997b).

Coig River Estuary is the most typical ria type estuary in the Argentina coast, but also is one of the least studied places. It is unknown if any salinity or temperature data was ever taken and there is no single current meter information. Although some 50 years ago there was a small town near its mouth, at present there is no single settlement along the whole river making the logistic to study it extremely difficult.

6.5.1 Río Gallegos Estuary

The macrotidal Río Gallegos Estuary is the southernmost Atlantic estuary on the South American continent (51°30'S, 69°W). Mean tidal range at the mouth (Punta Loyola, Fig. 6.15) is 9.5 m and 5.4 m, for spring and neap conditions respectively, and 12 km upstream at Río Gallegos City, corresponding heights are 9.6 m and 5.8 m. This estuary is also the third largest estuary in Argentina after the Río de la Plata and Bahía Blanca estuaries (Piccolo and Perillo 1997b). Its length, marked by tidal influence, is about 40 km and its width varies from about 8 km near its mouth and 800 m at the head. Depth also varies inland from about 25 m in the entrance channel to only few metres in the inland channels among the head tidal flats. The Chico River is the only tributary to the Río Gallegos in the study area and provides negligible freshwater input during the largest part of the year (Perillo et al. 1996).

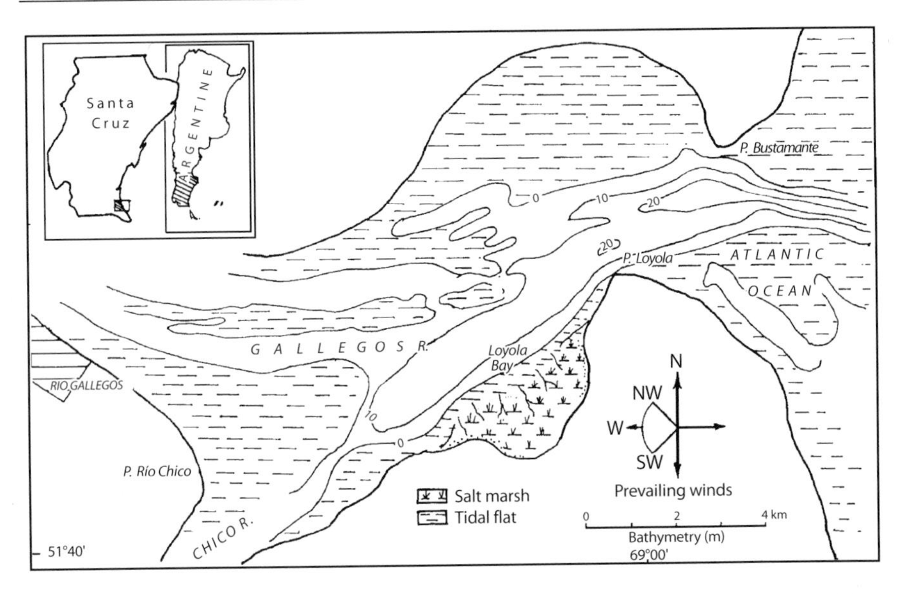

Fig. 6.15. Location map of Río Gallegos Estuary and the Loyola Bay intertidal complex (modified from Perillo et al. 1996)

Río Gallegos City is located on the southern margin of the estuary. Few studies have been performed about its geomorphologic and dynamic characteristics previous to 1994 (i.e., Pastor and Bonilla 1966). Since then some preliminary information has been gathered. The strong westerlies winds, originated by the circulation of the subtropical semipermanent high of the Pacific Ocean, penetrate into the Patagonia with velocities of 100 km h^{-1} and gusts of 200 km h^{-1} making the study of the estuary extremely difficult. Therefore, from April to August, the windy period, the region is subject to extreme low temperatures that causes significant problems in the city and in the navigation of the estuary.

The origin of the Gallegos Estuary, as it also the case for the Deseado and Santa Cruz ones, is associated with a Tertiary fault that runs along the northern border of the river, which may have been reactivated in the Pleistocene (about 88 000 years B.P.) (Pino 1995). Estimation of vertical movement is complicated, because sediment formations throughout the area are thick, horizontally bedded and with little variation in the vertical lithological sequence. The northern border of the estuary is composed of cliffs 80–150 m high. Whereas the southern bank is a relatively low lying area having the same stratigraphic characteristics of the northern bank in the subsurface. The surface of this southern block is covered by glacifluvial deposits. During the Kansas Glaciation, the Gallegos River defined the northern limit of continental glaciation. However, there are no clear indications of glacial reworking within the actual estuary.

Between the head of the estuary and Río Gallegos City, the estuary channel is almost occluded by extensive tidal flats and complex channels. Seaward of Río Gallegos City, the flats pass to a series of longitudinal banks whose crests become exposed during low tide (all conditions). These banks are also present at the mouth of the Chico

River (Fig. 6.15). Seaward of the confluence, a deep channel crosses the mouth of the estuary between Punta Loyola and Punta Bustamante, with a depth of over 23 m.

Between Punta Loyola and the mouth of the Chico River an extensive intertidal zone fills Loyola Bay. The area is backed by a raised shingle beach, while shingle spits and shoals are also observed on the upper intertidal salt marsh levels. The intertidal system compress a tidal flat and a salt marsh.

Although some areas of the Argentine coastline, south of Río Gallegos, are still scientifically unexplored, it is believed that the Loyola Bay salt marsh is the southernmost within continental South America. There are at least two other salt marshes in San Sebastian Bay and in the Río Grande Estuary (190 and 270 km south from Río Gallegos, respectively) and also along the southern coast of Magellan Strait, all in Tierra del Fuego.

The tidal flat occupies the lower intertidal zone. It varies in width from a few metres near Punta Loyola (Fig. 6.15) up to around 700 m during low tide, spring conditions, towards the mouth of the Río Chico. The sedimentology of the tidal flat is rather complex. The upper 0.5 m sediment layer (no data from below) is formed by rounded gravels (mean diameter 20–150 mm) with a matrix of clayey silt. The gravels are glacifluvial materials derived from the Rodados Tehuelches formation. This formation covers practically all of Patagonia, and the gravels form most of the beaches on the Argentine coastline to the south of Peninsula Valdés. In sectors of the flat, gravel shoals, 0.15–0.30 m relief, occur devoid of fine material (Perillo et al. 1996).

A gravel beach extends from Punta Loyola into the Loyola Bay with a local relief of about 3 m and a steep (8–15°) slope. The beach terminates in three spits that are encroaching upon the marsh, suggesting the dominant sediment transport in this area was from Punta Loyola inward due to ocean wave diffraction around it. The lower portion of the flat has an undulating surface, being entirely covered with minor creeks. These grooves are generally 0.5–1.0 m wide and 0.10–0.40 m deep, separated by relatively flat inter-channel areas ca. 0.5–2.0 m wide. Only a few larger channels, both in depth and width, cross the flats and these are directly branched from the largest channels on the marsh.

The salt marsh covers the upper portion of the intertidal environment. It displays the three typical levels of a salt marsh: lower, middle and upper. *Salicornia antigua* is the dominant plant species throughout the marsh, and *Franchicenia* cf. *microphylla*, *Festuca* sp., and *Griceda* cf. *ayantinensis* and are also present. At least three other plants were sampled, identification was not possible because the plants were not at suitable growth stage. *Salicornia* was more abundant throughout the lower and middle marsh areas and within the channel flanks of the upper marsh area, whereas *Festuca* and *Griceda* became progressively more abundant on moving from the middle to the upper-marsh areas (Perillo et al. 1996).

Within the marsh tidal channels of various dimensions are present, but larger channels are dominant. Channels within the tidal flat have little sinuosity, but as they penetrate further into the marsh, channel characteristics change dramatically. The lower marsh has the highest density (number of creeks/per unit area) of tidal creeks, but with the middle and upper segments containing only a slightly lower creek density. The drainage system is dendritic. Backing the salt marsh there is an area considered to be a former salt marsh identified only by the drainage pattern because the vegetation there is indicative of a supratidal environment and estuarine water may only inundate these upper reaches during less than once a year or less.

Perillo et al. (1996) found on the marsh a large number of salt pans with their long axis mostly oriented parallel to the main wind direction. They were able to discover a mechanism of tidal channel formation by the interconnection of these pans. This is a remarkable finding since it is normally regarded that salt marsh channels are inherited features from the tidal flats previous to their colonization by plants.

Further geomorphologic studies have been concentrated on the northern shore of the estuary, where the evolution of gravel spit were analysed within historical times by comparing maps and aerial photographs. Melo et al. (1998) discussed the effect of wind, tidal currents but specially the waves formed along the estuary water body as the most important agents in the evolution of the spits and cliff recession.

6.6 Tierra del Fuego Estuaries

In Tierra del Fuego island exists numerous but small estuaries. The more significant are the estuaries of Grande and Carmen Sylva Rivers which are of the coastal plain type. San Sebastian Bay has the typical characteristics of a restricted coastal lagoon because of the large spit that partly close its mouth and the influence of the San Martín River that discharge into the bay. In fact the San Martín River itself is a coastal plain inner estuary. Several studies have been done in the bay, but unfortunately to the authors knowledge little is known about the rest of the Tierra del Fuego estuaries.

Along the Mitre Peninsula (the tip of Tierra del Fuego) there are several small rivers that develop ria type estuaries. They are from north to south Ewan or Brazo Norte, San Pablo, Lainez, Irigoyen, Noguera, and Bueno. Few estuaries also flow to the Beagle Channel, these are Bompland, López and Moat Rivers. Most of these estuaries are located in such a remote area that there are no roads allowing for adequate transport of equipment to study them. Also, the Strait of Lemaire, between Mitre Peninsula and the los Estados Island is very dangerous navigation passage for the small boats needed to study these estuaries.

6.6.1 San Sebastián Bay

San Sebastián is a wide bay located in northern Tierra del Fuego having a semicircular shape partly closed by a long and narrow gravel spit. The bay is 55 km in length and 40 km wide (Fig. 6.16). The spit has a length of 17 km with width varying between 50 and 1 000 m (at the tip). The open mouth is about 20 km wide. Freshwater input into the system is provided by the San Martin River (there is no gauging of river discharge) at the southwestern part of the bay. Tidal range is 10 m and wind influence is from the west with the same characteristics described by Río Gallegos. Perillo (1995) suggested that the bay can be considered as a restricted coastal lagoon.

Wave action in the bay is two fold. The oceanic waves enter into the bay through the mouth and attack the southern coast up to the mouth of the river and along the outer face of the spit. Locally generated wave have the fetch of the bay and move seaward or towards the inner face of the spit. The spit also controls the circulation of the tidal currents. Basically the flood currents enter along the southern border of the mouth and circulate clockwise along the bay and inner face of the spit, ebbing on the

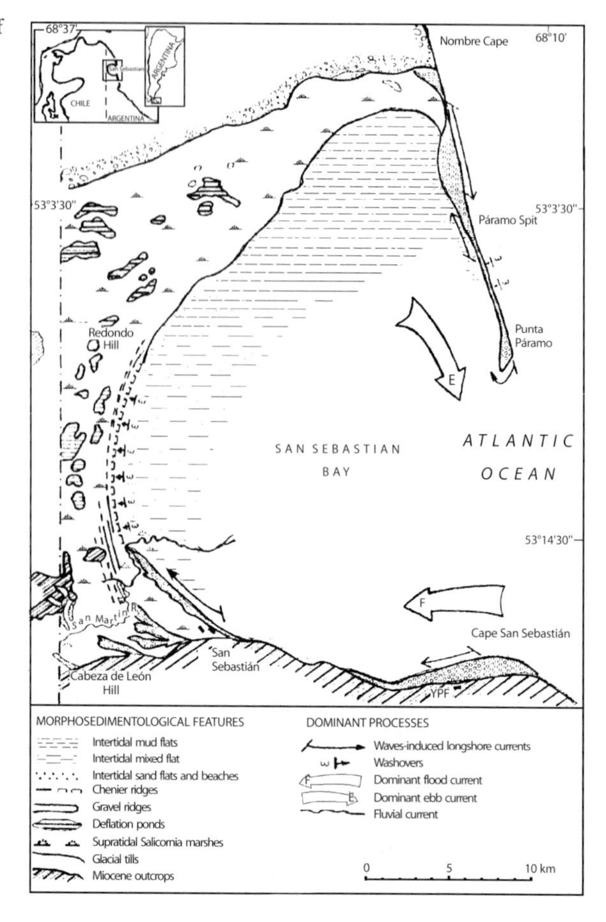

Fig. 6.16. Geological outline of the dominant processes in San Sebastian Bay (modified from Isla et al. 1991)

northern portion of the mouth. This circulation pattern produces a sediment transport from south to north, drifting fine material eroded from the cliffs that border the southern mouth up to the San Martín River.

The bay is the remnant of a glacial tongue that crossed Tierra del Fuego between Inútil and San Sebastian Bays (Codignotto and Maluminan 1981). About 7 000 years B.P. the channel dredged by the glacial tongue has been drowned connecting both bays. This process may have been simultaneous to the formation of the Magellan Strait formed by a similar process. The fact that the Magellan channel was much deeper than the Inútil-San Sebastian channel preserved the former at the subsequent regression.

Tidal flat distribution along the bay is highly related to the gravel spit. From the mouth of the bay northward the width they increase from few tens of metres to finally covering the whole of the northeastern end of the bay up to about half of the spit length. Isla et al. (1991) indicate that the grain size on the flats decrease progressively in relation with the clockwise circulation. The southern flats, mostly exposed to wave attack, are sandy devoid of channels. The mixed flats appear in the stretch from the town of San Sebastián to the mouth of the San Martín River, being characterized by flaser, wavy and lenticular bedding. Finally, the mud flats cover the rest of the environments. Isla et al. (1991) also divide the mud flats in two zones: an upper zone, very flat and uniform, and a lower zone, with meandering tidal channels.

Summary

Argentina coastline has a wide variety of estuaries ranging from the widest in the world (Río de la Plata) to very small ones located in areas of very difficult access. Due to the different climates that characterize the argentine land the estuaries show different discharges, being the Río de la Plata and Negro River the largest ones. Tidal amplitudes also vary significantly, being microtidal in the Río de la Plata to the Quequén Salado, mesotidal in the coast between Bahía Blanca Estuary to Negro River, and macrotidal along the rest of the Patagonia estuaries. Tides and winds are the most important processes that determine the circulation and dynamics of the estuaries. Most estuaries are coastal plain ones, however the Río de la Plata and Colorado Rivers present deltas, only four coastal lagoons are found and few rias at the south of the country.

Acknowledgements

Research dealing to this article were partly supported by Consejo Nacional de Investigaciones Científicas y Técnicas (CONICET) de la República Argentina, Universidad Nacional del Sur, National Geographic Society, and European Economic Community. The authors thank Lic. Walter D. Melo for drawing part of the figures.

References

Aliotta S (1983) Estudio sedimentológico y de deriva litoral entre la desembocadura del Río Negro y playa Bonita (provincia de Río Negro). MSc thesis, Universidad Nacional del Sur, Argentina

Alvarez JA, Alvarez SM, Ríos FF, Ferrante A (1983) General characteristics of the Mar Chiquita Lagoon (Argentina) and its coastal management aspects. Instituto para la Investigación de los Problemas del Mar, Universidad Nacional de Mar del Plata, Argentina, Tech Note

Anger K, Spivak E, Bas C, Ismael D, Luppi T (1994) Hatching rhythms and dispersion of decapod crustacean larvae in a brackish lagoon in Argentina. Helgoländer meeresunters 48:445–466

Balay MA (1956) Determination of mean sea level of Argentine Sea influences of the sea not caused by the tides. Inter Hydrographic Rev 33:31–65

Balay MA (1958) Causas y periodicidad de las grandes crecidas en el Río de la Plata, Servicio de Hidrografía Naval, Tech Note H-611

Balay MA (1961) El Río de la Plata entre la atmósfera y el mar. Servicio de Hidrografía Naval, Tech Note H-621

Barria de Cao MS (1986) Contribución al conocimiento de *Tintinnina* (Protozoa, Cliophora) de la zona de Bahía Blanca, II, Argentina. Boletín Instituto Español de Oceanografía 3:143–150

Barria de Cao MS (1992) Abundance and species composition of *Tintinnina* (Ciliophora) in Bahía Blanca Estuary, Argentina. Estuar Coastal Shelf Sci 34:295–303

Boltovskoy E, Boltovskoy A (1968) Foraminíferos y tecamebas de la parte inferior del Río Quequén Grande. Hidrobiología II:127–164

Boschi EE (1988) El ecosistema estuarial del Río de la Plata (Argentina y Uruguay). An Inst Cienc del Mar y Limnol, Univ Nal Autón México 15:159–182

Castaing PA, Guilcher A (1995) Geomorphology and sedimentology of rias. In: Perillo GME (ed) Geomorphology and sedimentology of estuaries. Elsevier, Amsterdam pp 69–107

Codignotto JO, Weiler NE (1980) Evolución morfodinámica del sector costero comprendido entre Punta Laberinto e isla Olga, provincia de Buenos Aires. Proc Simposio Problemas Geológicos del Litoral Atlántico Bonaerenses, Mar del Plata, pp 35–45

Codignotto JO, Malumian N (1981) Geología de la región al norte del paralelo 54 S de la isla Grande de Tierra del Fuego. Revista Asociación Geológica Argentina 36:44–88
Cuadrado DG, Perillo GME (1996) Características de las corrientes de marea en el estuario del Río Quequén. X Coloquio Argentino de Oceanografía, Bahía Blanca, Argentina, Abstract
Cuadrado DG, Piccolo MC, Perillo GME (1998) Hydrography of the inner shelf offshore Bahía Blanca Estuary. Continental Shelf Res Submitted
Daus FA (1946) Morfología de las llanuras argentinas. In: Geografía de la República Argentina. Sociedad Argentina de Estudios Geográficos, vol III, pp 3–224, Argentina
del Río JL, Colado UR, Gaído ES (1991) Estabilidad y dinámica del delta de reflujo de la boca del Río Negro. Revista de la Asociación Geológica Argentina 46:325–332
Dyer KR (1998) Estuaries: A physical introduction. John Wiley & Sons, London 2nd edn
Elias R (1985) Macrobenthos del estuario de la Bahía Blanca (Argentina), I Mesolitoral. Spheniscus 1:1–33
Evarsa (1995) Estadísticas hidrológicas 1994. Secretaría de Energía, Ministerio de Economía y Obras y Servicios Públicos, Buenos Aires, Argentina, 1:365–366
Fasano JL (1980) Geohidrología de la laguna Mar Chiquita y alrededores, provincia de Buenos Aires. Proc Simposio Problemas Geológicos del Litoral Atlántico Bonaerenses, Mar del Plata, pp 59–71
Fasano JL, Hernández MA, Isla FI, Schnack EJ (1982) Aspectos evolutivos y ambientales de la laguna Mar Chiquita (provincia de Buenos Aires, Argentina) Oceanologica Acta, pp 285–292
Framiñan MB, Brown OB (1996) Study of the Río de la Plata turbidity front, Part I: Spatial and temporal distribution. Cont Shelf Res 16:1259–1282
Framiñan MB, Brown OB (1998) Sea surface temperature anomalies off the Río de la Plata Estuary: Coastal upwelling?. EOS Trans AGU 79, Ocean Sciences Suppl, (abstract)
Freije RH, Asteasuain AO, Schmidt AS, Zavatti JR (1981) Relación de la salinidad y la temperatura del agua de mar con las condiciones hidrometeorológicas en la porción interna del estuario de Bahía Blanca. Instituto Argentino de Oceanografía Tech Rep 57, Argentina
Gagliardini DA, Karszenbaum H, Legeckis R, Klemas V (1984) Application of Landsat MSS NOAA/TIROS AVHRR, and Nimbus CZCS to study the La Plata River and its interaction with the ocean. Remote Sensing of Environment 15:21–36
Grondona MF (1975) Pendiente del Océano Atlántico. In: Geografía de la República Argentina. Sociedad Argentina de Estudios Geográficos, VII:203–211
Glorioso P, Boschi EE (1982) Las condiciones ambientales de la región estudiada en las campañas del BIP "Capitán Canepa" C03/81 y O14/81. Informe INIDEP, Argentina
Gonzalez MA, Weiler NE (1983) Ciclicidad de niveles marinos holocénicos en Bahia Blanca y en el delta del río Colorado (Provincia de Buenos Aires), en base a edades Carbono-14. Actas de la Oscilación del Nivel Medio del Mar durante el último Hemiciclo Deglacial de la Argentina, Mar del Plata, pp 69-88
Guerrero RA, Acha EM, Framiñan MB, Lasta CA (1997) Physical oceanography of the Río de la Plata Estuary, Argentina. Continental Shelf Res 17:727–742
Hansen DV, Rattray M (1966) New dimensions in estuary classification. Limnol Ocean 11:319–326
Hoffmeyer MS (1983) Zooplankton del area interna del la Bahía Blanca (Bs As, Argentina) I: Composición faunística. Historia Natural 3:73–94
Hoffmeyer MS (1994) Seasonal succession of Copepoda in the Bahía Blanca Estuary. Hydrobiologia 292/293: 303–308
INCYTH (1975) Estudio de navegabilidad del Río Negro. Ministerio de Defensa DIGID, Servicio de Hidrografía Naval, Buenos Aires
Isla FI (1980) Evolución morfológica de la zona de la desembocadura de la laguna Mar Chiquita, provincia de Buenos Aires. Proc Simposio sobre Problemas Geológicos del Litoral Atlántico Bonaerense, Mar del Plata pp 89–108
Isla FI (1984) Análisis de variables que rigen la estabilidad y obstrucción de canales de marea: el caso de Mar Chiquita, provincia de Buenos Aires. Proc 9 Congreso Geológico Argentino, SC Bariloche, pp 218–242
Isla FI, Vilas FE, Bujalesky GG, Ferrero M, Gonzalez Bonorino G, Aeche Miralles A (1991) Gravel drift and wind effects on the macrotidal San Sebastian Bay, Tierra del Fuego, Argentina. Marine Geology 97:211–224
Kühnemanm O (1963) Penetración de *Macrocystis pyrifera* en la ría de Puerto Deseado Bol Soc Arg de Botánica 10:105–112
Lanfredi NW, Schmidt SA, Speroni JO (1979) Cartas de corrientes de mareas (Río de la Plata., Servicio de Hidrografía Naval) Tech Rep IC-IT-79/03
Lanfredi NW, Balestrini CF, Mazio CA, Schmidt SA (1987) Tidal sandbanks in Mar Chiquita coastal lagoon, Argentina. J Coastal Res 3:515–520

Lanfredi NW, D'Onofrio EE, Mazio CA (1988) Variations of the mean sea level in the southwest Atlantic Ocean. Continental Shelf Res 3:1211–1220
Lara RJ, Gómez EA, Pucci AE (1985) Organic matter, sediment particle size and nutrient distributions in a sewage affected shallow channel. Marine Pollution Bulletin 16:360–364
López ME (1978) Relevamiento y diagnóstico de la actual situación turística de la laguna y balneario Mar Chiquita y Santa Clara del Mar. Universidad Nacional del Mar del Plata, Tech Rep
Martos P, Reta R, Perillo GME, Ferrante A, Isla FI, Piccolo MC, Guerrero R (1995) Características hidrográficas de la laguna Mar Chiquita. IX Coloquio de Oceanografía, Bahía Blanca
Mazio CA (1990) Modelo hidrodinámico para el Río de la Plata. Frente Marítimo 7:87–94
Mazio CA (1991) Modelación mareológica del efecto no lineal aplicada al Río de la Plata. Frente Marítimo 9:103–108
Melo WD, Ginsberg SS, Perillo GME (1998) Características geomorfológicas del estuario de Río Gallegos. Actas II Jornadas de Geografía Física, Mendoza (in press)
Mianzan HW (1989) Las medusas Scyphozoa de la Bahía Blanca, Argentina. Bol Inst Oceanogr 37: 29–32
O'Connor WP (1991) A numerical model of tides and storm surges in the Río de la Plata Estuary. Continental Shelf Res 11:1491–1508
Oliver SB, Escofet A, Penchazade P, Orensanz JM (1972) Estudios ecológicos de la región estuarial de Mar Chiquita (Buenos Aires, Argentina) I: Las comunidades bentónicas. Anales Sociedad Científica Argentina 193:237-262
Orfila J, Scapini MC (1987) Análisis preliminar de las características fisicoquímicas del estuario del Río Chubut y su actitud para su uso. 1 Jorn Nac Recursos Hídricos en Zonas Aridas y Semiáridas y su Relación con el Hombre y el Medio Ambiente, Río Gallegos
Orfila J, Scapini MC, Perillo GME, Piccolo MC (1987) Características químicas del estuario del Río Chubut. Proc 2° Congreso Latinoamericano de Ciencias del Mar, La Molina, II:69–84
Orozco Storni MS, Lara RS, Pucci AE (1984) Tidal variations of some physico-chemical parameters in Blanca Bay, Argentina. Estuar Coastal Shelf Sci 19:485–491
Parker G, Marcolini S (1992) Geomorfología del delta del Paraná y su extensión hacia el Río de la Plata. Rev de la Asociación Geológica Argentina 47:243–249
Pallares RE (1968) Copépodos marinos de la ría Deseado (Santa Cruz, Argentina): Contribución sistemática-ecológica. I. Servicio de Hidrografía Naval Tech Rep H 1024
Pastor J, Bonilla J (1966) Estudios para la formulación de un plan de desarrollo físico de la ciudad de Río Gallegos. Municipalidad de Río Gallegos
Perillo GME (1995) Definition and geomorphologic classifications of estuaries. In: Perillo GME (ed) Geomorphology and sedimentology of estuaries. Elsevier, Amsterdam, pp 17–49
Perillo GME, Sequeira ME (1989) Geomorphologic and sediment transport characteristics of the middle reach of the Bahía Blanca Estuary (Argentina). J Geophysical Res Oceans 94:14351–14362
Perillo GME, Piccolo MC (1991) Tidal response in the Bahía Blanca Estuary. J Coastal Res 7:437 449
Perillo GME, Arango JM, Piccolo MC (1987a) Parámetros físicos del estuario de Bahía Blanca, Período 1967–1986. Instituto Argentino de Oceanografía Tech Rept 4
Perillo GME, Piccolo MC, Arango JM, Sequeira ME (1987b) Hidrografía y circulación del estuario de Bahía Blanca (Argentina) en condiciones de baja descarga. Proc 2° Congreso Latinoamericano de Ciencias del Mar, La Molina, II:95–104
Perillo GME, Piccolo MC, Scapini MC, Orfila J (1989) Hydrography and circulation of the Chubut River Estuary (Argentina). Estuaries 12:186–194
Perillo GME, Ripley M, Piccolo MC, Dyer KR (1996) The formation of tidal creeks in a salt marsh: New evidence from the Loyola Bay salt marsh, Río Gallegos Estuary, Argentina. Mangroves and Salt Marshes 1:37–46
Piccolo MC (1987) Estadística climatológica de Ingeniero White. Período 1980–1985. Instituto Argentino de Oceanografía, Tech Rep
Piccolo MC, Perillo GME (1989) Subtidal sea level response to atmospheric forcing in Bahía Blanca Estuary, Argentina. Proc 3rd Intern Conf Southern Hemisphere Meteorology and Oceanography, Buenos Aires
Piccolo MC, Perillo GME (1990) Physical characteristics of the Bahía Blanca Estuary (Argentina) Estuar Coastal Shelf Sci 31:303–317
Piccolo MC, Dávila PM (1993) El campo térmico de las planicies de marea del estuario de Bahía Blanca. Actas JNCMAR '91, Puerto Madryn, pp 11–15
Piccolo MC, Perillo GME (1997a) Sea level characteristics in Puerto Quequén. Geoacta 22:144–154
Piccolo MC, Perillo GME (1997b) Geomorfología e hidrografía de los estuarios. In: Boschi EE (ed) El Mar Argentino y sus Recursos Pesqueros. Instituto Nacional de Investigación y Desarrollo, Mar del Plata vol I, pp 133–162

Piccolo MC, Perillo GME, Arango JM (1987) Hidrografía del estuario de Bahía Blanca, Argentina. Revista Geofísica 26:75–89
Piccolo MC, Perillo GME, Martos P, Reta R (1995) Balance energético en la playa de la Laguna Mar Chiquita, Buenos Aires, Argentina. Congreso Latinoamericano de Ciencias del Mar, Mar del Plata
Pino M (1995) Structural estuaries. In: Perillo GME (ed) Geomorphology and sedimentology of estuaries. Elsevier, Amsterdam pp 227–239
Reta R, Martos P, Ferrante A, Perillo GME, Piccolo MC (1995) Efectos de mareas y viento sobre temperatura y salinidad en la laguna Mar Chiquita, Argentina. VI Congreso Latinoamericano de Ciencias del Mar, Mar del Plata
Riggi AE (1944) Cuencas geo-hidrográficas de Argentina. Revista Museo de la Plata Nueva Serie II:185–212
Sala JM (1975) Recursos hídricos. In: Relatorio VI Congreso Geológico Argentino, Buenos Aires, pp 169–193
Schnack EJ, Fasano JL, Isla FI (1980) Los ambientes ingresivos del holoceno en la región de Mar Chiquita, provincia de Buenos Aires. Simposio sobre Problemas Geológicos del Litoral Atlántico Bonaerense, Mar del Plata, pp 229–242
Serman DD (1985) Características de la marea en Bahía Blanca. Bol Centro Naval 103:51–74
Sequeira ME, Piccolo MC (1985) Predicción de la temperatura del agua durante la bajante de la marea en Ingeniero White. Meteorologica 15:59–76
Sericano JL, Pucci AE (1982) Cu, Cd and Zn in Blanca Bay surface sediments, Argentina. Marine Pollution Bull 13:429–431
Soldano FA (1947) Régimen y aprovechamiento de la red fluvial argentina, Parte II: Ríos de la región árida y la meseta patagónica. Ed Cimera, Buenos Aires
Urien CM (1972) Río de la Plata Estuary environments. Geological Society America Memoir 133:213–234
Villa N (1988) Spatial distribution of heavy metals in sea water and sediments from coastal areas of the southeastern Buenos Aires province, Argentina. In: Seeliger U, Lacerda LD de, Patchineelans SR (eds) Metals in coastal environments of Latin America. Springer-Verlag, New York, pp 30–44
Villa N, Pucci AE (1985) Distribution of iron and manganese in the Blanca Bay, Argentina. Marine Pollution Bull 16:369–371
Villa N, Pucci AE (1987) Seasonal and spatial distributions of Copper, Cadmium and Zinc in the sea water of Blanca Bay. Estuar Coastal Shelf Sci 25:67–80
Wagner JM, Hoffmeyer MS, Tejera LA, Nizovoy AM (1991) Proc JNCDM '91 pp 79–86
Weiler NE (1983) Rasgos morfológicos evolutivos del sector costanero comprendido entre Bahía Verde e isla Gaviota, provincia de Buenos Aires. Revista Asociación Geológica Argentina 38:392–404
Wright R (1968) Miliolidae (Foraminíferos) recientes del estuario del Río Quequén Grande (provincia de Buenos Aires). Hidrobiología II: 225–256
Zubillaga HV, Pucci AE (1986) Cu, Cd, Pb and Zn in tributaries to Blanca Bay, Argentina. Marine Pollution Bull 17:230–232

Hydrography and Sediment Transport Characteristics of the Río de la Plata: A Review

Jorge López Laborde · Gustavo J. Nagy

7.1 Introduction

The Río de la Plata is located on the East coast of South America, draining the second largest river basin of the continent and the single largest in terms of population and economic importance. In a recent paper on definitions and geomorphologic classifications of estuaries, Perillo (1995) classified Río de la Plata as:

1. a primary estuary, and
2. a river influenced tidal river.

This means, that:

1. the basic form has been the result of terrestrial and/or tectonic processes and the sea has not changed significantly the original form,
2. due to the high discharge of the tributary rivers, the valley is not presently drowned by the sea; however, the circulation on the lower portions is highly affected by tidal dynamics, including reversing currents, resulting in characteristic morphological patterns, and
3. although it is affected by tidal action, salt intrusion may be limited to the mouth.

In fact, a simple examination of bathymetry, sedimentology and mean salinity distribution charts of the Río de la Plata system, indicates that it is best defined as "a funnel-shaped coastal plain tidal river with a semienclosed shelf area at the mouth and a river palaeovalley at the northern coast that favours river discharge and sediment transport to the adjacent continental shelf".

The present paper reviews previous work on the hydrography and suspended sediment transport characteristics of the Río de la Plata fluviomarine system. Much of the previous works has a limited distribution and some of it appears as technical reports, thesis and symposium abstracts. The aim is to consider the basic Río de la Plata geomorphological and sedimentological characteristics, as well as salinity and suspended sediment distribution and behaviour, to provide a general description and a conceptual model, as useful tools for environmental management purposes. New data, yet to be published, will also be included.

7.2 Regional Setting

7.2.1 Drainage Basin

The Río de la Plata is located on the East coast of South America (Fig. 7.1) between 34°00'–36°10'S and 55°00'–58°10'W, with a surface area of about 38 800 km², a length of 200 km, and a transverse section expanding to the SE (Fig. 7.2).

It is located at the confluence of two well defined physiographic units: the Uruguayan-Brazilian Shield (granitic rocks) and the Argentinean "Pampa" sedimentary basin with a depth of more than 2 000 m of fine sediments. Consequently, the coast has contrasting characteristics:

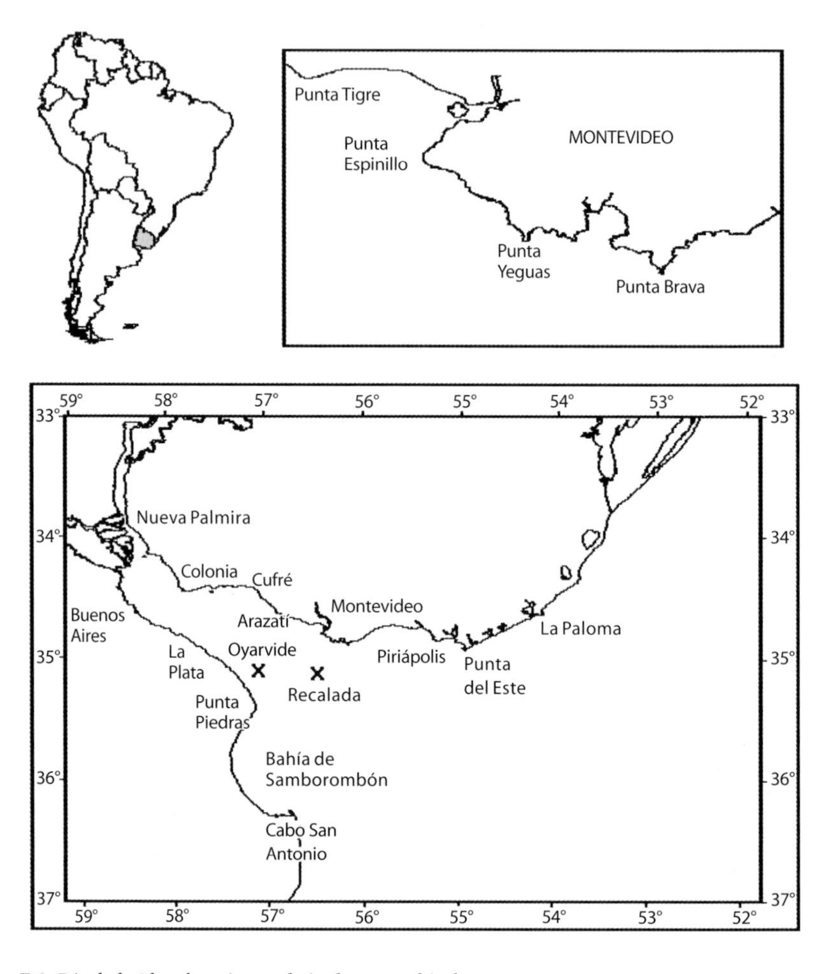

Fig. 7.1. Río de la Plata location and cited geographical names

1. to the north, along the Uruguayan coast, rocky points and sandy beaches predominate (with sand ridges and extensive dune fields, although there are some cliffed areas developed over a variety of geological formations), and
2. to the south, along the Argentinean coast, a low and flat countryside of marshes, lagoons, slikkes, shores, and ancient beach ridges predominates.

The Río de la Plata is the receiving basin of the second largest drainage system of the continent (3 170 000 km²; Tossini 1959), formed by the Uruguay and Paraná-Paraguay Rivers (Table 7.1), with boundary coordinates between 14°05'–37°37'S and 67°00'–43°35'W.

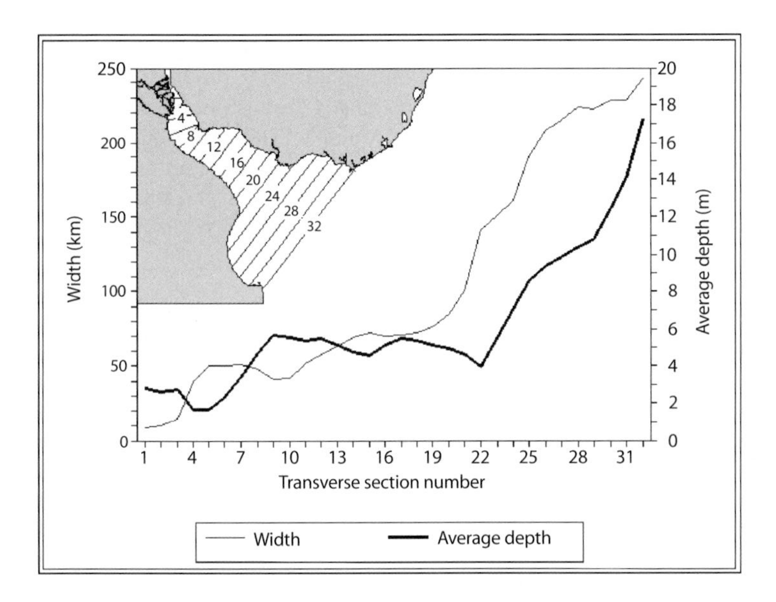

Fig. 7.2. Southeastern width and average depth increase

Table 7.1. Río de la Plata drainage basin characteristics (from Tossini 1959)

	Area (km²)	Annual rainfall (mm)	Rainfall volume (hm³)	Discharge volume (hm³)	River flow rate (m³ s⁻¹)
Upper Paraná	975 375	1 523	1 485 420	371 355	11 780
Paraguay	1 103 000	1 027	1 133 820	142 120	4 506
Lower Paraná	704 815	776	546 800	33 990	1 078
Total Paraná	2 783 190			547 610	17 364
Uruguay	350 250	1 385	485 110	158 720	5 033
Río de la Plata	36 650	962	35 110	6 250	198
Total	3 170 000	1163	3 686 330	712 580	22 505

7.2.2
Tributary River Discharge

Given the vast area of the drainage basin, its relief and precipitation characteristics are highly variable, with elevations ranging from 100–2 000 m and annual average precipitation ranging from 200–3 000 mm. Although snow-rainfall hydrological behaviour can be observed at the upper basin the hydrological characteristics can be considered as purely rainfall related. Table 7.1 shows the principal variables of the Río de la Plata drainage basin and their values.

The analysis of the hydrologic contributions of the main tributaries to the Río de la Plata (Mazio and Martínez 1989) showed significant differences for the period studied (1960–1980). The Uruguay River provides an average flow rate of 4 700 $m^3 s^{-1}$ (Hervidero Gauging Station), with maximum flow rate (6 500 $m^3 s^{-1}$) during the winter months and minimum flow rate (3 500 $m^3 s^{-1}$) during summer. The Paraná River provides an average flow rate between 15 400 $m^3 s^{-1}$ (Rosario Gauging Station) and 17 000 $m^3 s^{-1}$ (Corrientes Gauging Station); however, maximum flow rates occur during the summer months and minimum during the fall. This creates a damping effect on peak flow rates for the combined discharge of the two rivers, producing a combined annual average discharge between 20 000 and 23 000 $m^3 s^{-1}$ with minimum and maximum values of 10 800 and 30 600 $m^3 s^{-1}$.

Recently, many authors have reported an increase in the river discharge; García and Vargas 1994; Genta 1996; Nagy et al. 1996); during the second half of the century (1950s) began a continuous lineal trend expansion of normal and maximum values, well distinguished from the previous behaviour. In fact, Paraná River average flow rate (Rosario Gauging Station) increased from 14 600 $m^3 s^{-1}$ (1884–1975) to 18 400 $m^3 s^{-1}$ (1975–1994); meanwhile Uruguay River average flow rate (Hervidero Gauging Station) increased from 4 400 $m^3 s^{-1}$ (1916–1975) to 5 600 $m^3 s^{-1}$ (1975–1991). Such increments have been associated to many causes:

a deforestation of the upper river basins and, consequently, reducing of the water retention capacity,
b the development of an excessive moisture hydrological period producing greater drain off, and
c "El Niño" phenomenon (Mechoso and Pérez Irribaren 1992; Nagy et al. 1996).

7.2.3
Meteorology

The general atmospheric circulation is controlled by the influence of the quasi-permanent South Atlantic high pressure system and the continuous passage of low pressure systems that come from the south. This general circulation is modified by a low pressure system located in northern Uruguay, which has a NW to SE movement generating NE to SE winds.

Río de la Plata normal weather evolution is controlled by several factors. One of them is the passage of polar air masses. They penetrate into the continent from the Patagonian region with a NE direction. When air masses cross over the Río de la Plata, wind direction changes from North to South, generating strong wind gusts and rainy events. When the air masses reach Brazil their movement is reduced and the air is warmed. Later, warm air masses return to the South originating northerly winds, which may carry new maritime or continental tropical warm air masses. Northern winds stop when a new polar air mass begins to advance from the Patagonian region to the North.

The most frequent storms affecting the Río de la Plata are called "Sudestadas". They are produce by the interaction between frontal systems (with SW-NE trajectories) and the littoral low pressure system, generating strong SE winds. Usually the storms affect the area for several days. The "Sudestada" is responsible for disastrous flooding events along the Argentinean Río de la Plata coastal area.

The coastal area is also affected by southwesterlies winds ("Pampero"). They originate from the passage of polar air masses arriving from the south of the Andean Mountain Range. The Pampero is a strong, dry and cold wind. They originate as dry air moves fast along the Patagonian and Pampa regions arriving at the Río de la Plata from the SW. The "Pampero" present velocities between 50–60 km h^{-1}, and occasionally up to 90 km h^{-1}.

7.3 Hydrographic Characteristics

7.3.1 Morphology

Morphological studies conducted during the "Estudio para la Evaluación de la Contaminación en el Río de la Plata" (Cavallotto 1987; Parker and López Laborde 1988; 1989), based on the 1964–1969 comprehensive survey allowed to recognise areas with particular features, characterized as "morphological units" (Fig. 7.3):

Playa Honda: The subaqueous prolongation of the Paraná River Delta, limited approximately by the 6 m isobath, representing a widespread shallow area crossed by channels.

Sistema Fluvial Norte: This area includes all the channels that extend from the Río de la Plata watersheds to Colonia, characterized by a series of erosive furrows, developed by the erosive action of the Uruguay and Paraná Bravo Rivers. It includes channels, longitudinal benches and subaqueous asymmetrical sandy dunes.

Banco Grande de Ortíz: This unit develops over a large portion of the Río de la Plata between Canal Norte, over the Uruguayan coast, and Gran Hoya del Canal Intermedio. It appears as a large plateau with steep southern slope which flattens to the North and SE, limited by the 6 m isobath.

Gran Hoya del Canal Intermedio: A widespread depression to the South of Banco Grande de Ortíz, formed by three morphological elements known as "Rada Exterior", "Canal Intermedio" and "Sistema de Bancos Chico and Magdalena".

Canal Norte: It extends between the Uruguayan coast and Banco Grande de Ortíz, with a depth around 5 m. It forms a gentle depression related to the Sistema Fluvial Norte channels.

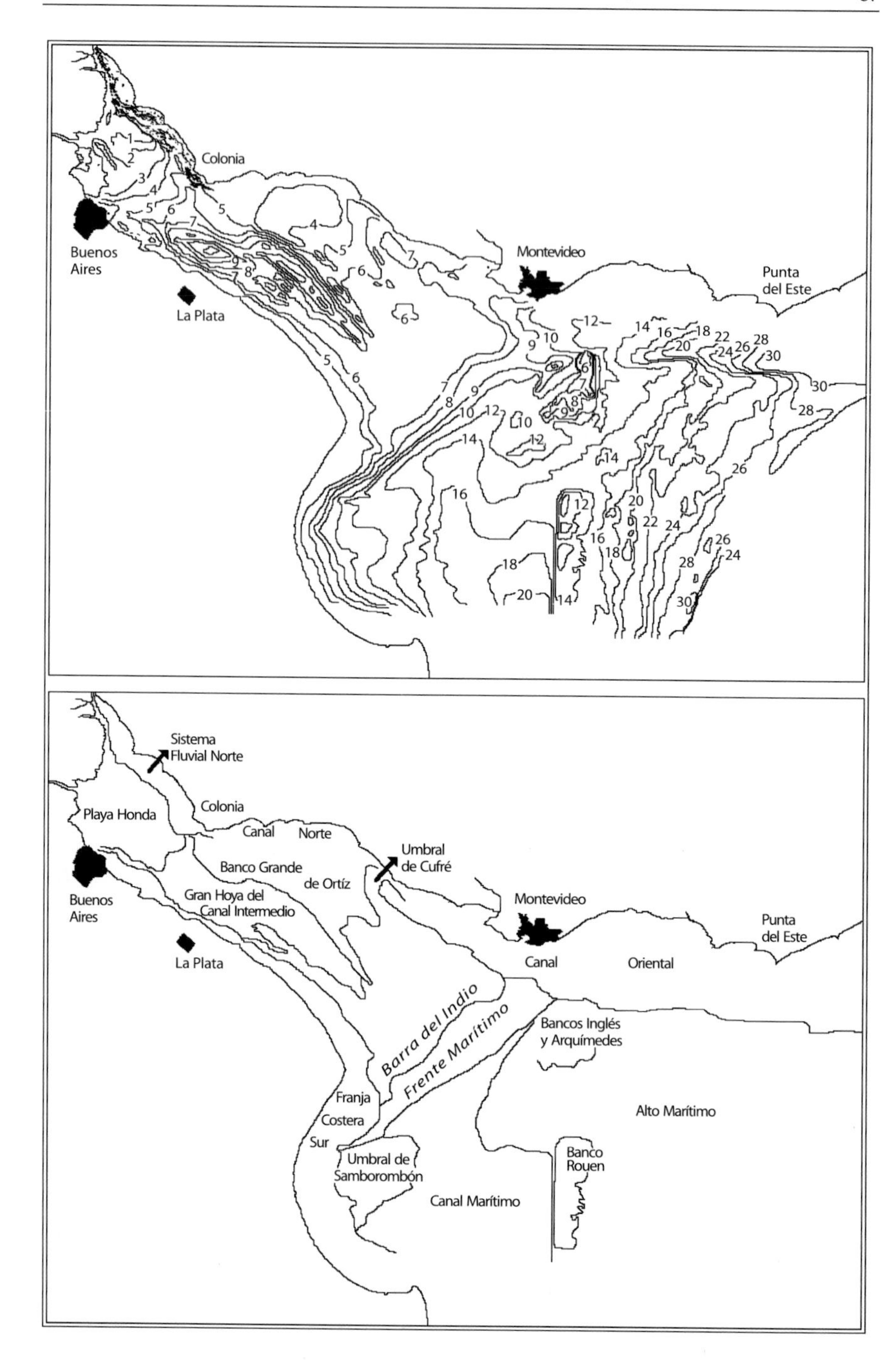

Fig. 7.3. Río de la Plata bathymetry and morphological units (based on Cavallotto 1987, Parker and López Laborde 1988, 1989)

Canal Oriental: A long depression extending, with E-W direction, from the East of the Canal Norte (from which it is separated by the Umbral de Cufré) up to the vicinity of Punta del Este where its direction changes to the NE and it deepens abruptly. Here it is known as "Mud Wells" ("Pozos de Fango").

Barra del Indio: A gentle and wide plain extending to the NE from the Argentinean coast between Punta Indio and Punta Piedras. Morphologically it corresponds to a gentle convex and subhorizontal surface with depths ranging from 6.5–7 m. It is a recent and clayey aggradation form.

Franja Costera Sur: The Argentinean coastal area located between the Luján River and Cabo San Antonio, representing an inclined plane extending from the coast up to the 6–9 m isobath.

Alto Marítimo: The outer Río de la Plata region containing the Inglés, Arquímedes and Rouen banks; the first two represents stable areas.

Umbral de Samborombón: A triangular surface determined by a slope change interposed between the center of Samborombón Bay and the Canal Marítimo.

Canal Marítimo: A widespread and gentle depression, with asymmetrical profile, extending between Barra del Indio, Franja Costera Sur, Umbral de Samborombón and Alto Marítimo.

7.3.2 Surficial Bottom Sediments

The Río de la Plata surficial bottom sediments (Urien 1966; 1967; 1972; Parker et al. 1985; López Laborde 1987a,b; Parker and López Laborde 1989) exhibit a graded distribution with sands in the river watersheds, silts in the middle river and clayey silts in the mouth, where they overlap relict Holocene sands (Urien 1967). The facies distribution is longitudinal, following the main flow directions: parallel to the coast in the upper and middle river, whilst in the outer river they form a parabolic arch (Parker et al. 1985).

According to Parker et al. (1985), in the upper and middle river it is possible to distinguish, along the coasts, two facies associations that are texturally interconnected and formed by sediments with mean diameter decreasing to the SE:

1. along the northern coast the Uruguay and Paraná Guazú Rivers discharge towards Playa Honda there is an association of sands and silty sands, grading to silty sands and silts in the Banco Grande de Ortíz, and then to silts, clayey silts and even silty clays in the outer region, and
2. along the southern coast from the mouth of Luján and Paraná de las Palmas Rivers, finer textured sediments are observed, with greater organic carbon content (Hallcrow 1965), sediments spreads over the coast, forming an unique unit of sandy silts and silts grading to silty clays.

These two main transport pathways remain separated in the middle river, due to the presence of Playa Honda and Gran Hoya del Canal Intermedio, up to Barra del Indio where they form areas with clay content greater than 25% and mean diameter under 25 μm (corresponding to the salt intrusion limit and the turbidity maximum location). The outer limit for sediment dispersion would be the zone or band where recent sediments overlap the relict Holocene sands (Parker et al. 1985; Parker and López Laborde 1989).

Clay mineralogy studies (Urien 1967; Depetris 1968; Depetris and Griffin 1968; Siegel et al. 1968), revealed significant differences between the Uruguayan coast and Samborombón Bay. The differences were originally interpreted as masking of the physical segregation processes of clay minerals by local contributions of the Uruguayan-Brazilian Coastal Shield. A reinterpretation may be appropriate taking into account differential contributions of the main tributaries (Tables 7.2 and 7.3) and the presence of two transport pathways.

Biscaye (1972) analysed rubidium, strontium and strontium isotope composition on whole (carbonate free) surface sediment samples. He revealed significant differences in the main tributaries radiogenic composition (Table 7.4) and fails in the at-

Table 7.2. Clay minerals contribution (percentage) by tributary basin (from Depetris 1968)

River	Montmorillonite	Illite	Kaolinite	Clorite
Paraná	86	95	64	100
Uruguay	14	5	36	–

Table 7.3. Clay fraction mineralogy (percentage; from *a* Siegel et al. 1968 and *b* Urien 1967)

	Montmorillonite		Illite		Kaolinite	
	a	b	a	b	a	b
Uruguayan coast	35	18	31	41	34	31
Samborombón Bay	43	20	28	38	29	27

Table 7.4. Rubidium, strontium and strontium isotope composition of main tributaries basin (adapted from Biscaye 1972)

	% < 20 μ	Rb	Sr	Rb / Sr	Rb^{87}/Sr^{87}	Sr^{87}/Sr^{86}
Uruguay	82	94	105	0.90	2.67	0.7159
Paraná Guazú	67	114	122	0.96	2.77	0.7216
Paraná de las Palmas	33	95	185	0.50	1.44	0.7144

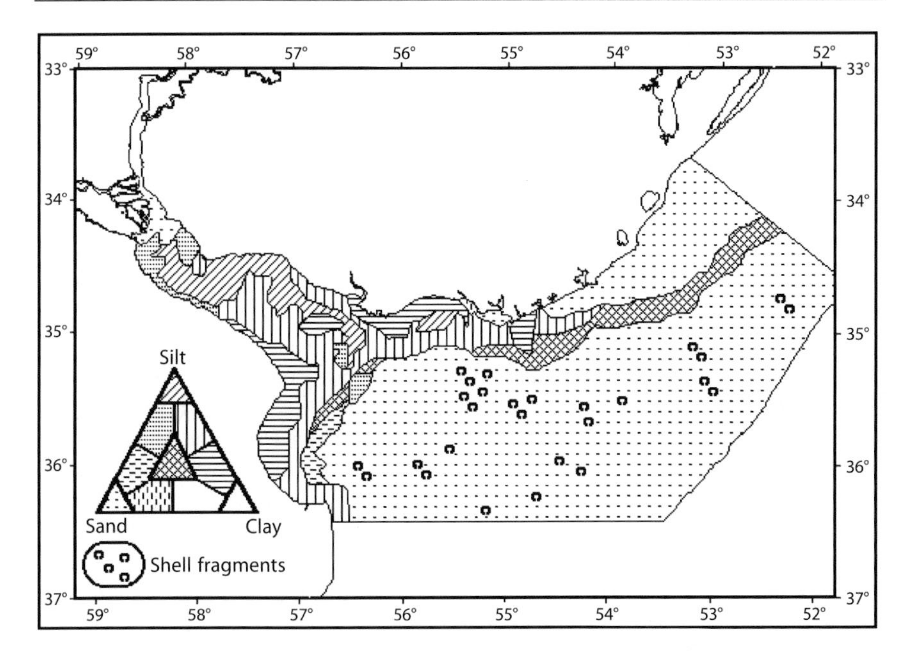

Fig. 7.4. Río de la Plata surface bottom sediment distribution according to Shepard (1954) diagram

tempt to distinguish a local contribution of the Coastal Shield material. Stratigraphical studies (Parker 1985; 1990; de Souza and López Laborde 1988) showed relict sediments outcrops on the bottom surface. Consequently, sediment distribution (Fig. 7.4) must be interpreted in a general context of long-term graded and selective sedimentation beginning in a typical fluvial environment in the upper river, grading to a mixed one (fluvial and tidal action) in the middle river, and ending in a pro-delta on a typical marine environment (Parker et al. 1985). In such a context (López Laborde 1987b), dynamic elements (differential contribution of the main tributaries, two main transport pathways, localisation of sedimentation processes) and historical-stratigraphical elements (relict sediments masking actual processes) are all involved.

7.3.3 Tides and Tidal Currents

Because of the shallow waters of the Río de la Plata, the generation of astronomical tides is not possible within the inner basin, and most water level fluctuations are produced by oceanic processes influenced by weather patterns (Balay 1961). The ocean tide arrives with an approximate wave speed of 200 km h^{-1} and an average amplitude between 1 m along the southern coast and 0.3 m along the northern coast. Tide propagates with an approximate speed of 30 km h^{-1}, taking approximately twelve hours to travel along the river.

According to Mazio and Martínez (1989), the Río de la Plata tides have a semidiurnal regime, with diurnal inequalities. The principal component is M_2, which explains 80%

of the total spectral energy, with a significant participation of the O_1 diurnal component. The size of the basin and the twelve hours that it takes tides to travel up to the whole Río de la Plata, produces a situation where high and low tides can occur simultaneously at different locations.

Balay (1961) stated that lines of equal amplitude shows the tidal differences between the two coasts. These tidal variations are caused by Coriolis force and by the differences in amplitude of the incoming oceanic tides. Since the southern tide has a greater range than the northern one, it has greater velocity, reaches the upper river earlier and creates transverse currents between the southern and northern coasts. The difference in amplitude and separation of tides across the outer section, between Montevideo and San Clemente, are obvious. Further in, along the Colonia–Buenos Aires transverse section, these two tides tend to combine and form a single wave with similar amplitude. The greater strength of the southern wave dominates the combined tide; the tidal amplitude at the Paraná River Delta and Martin Garcia Island is larger than at Colonia. The progression of the tidal maxima up the Río de la Plata and the decay in tidal amplitude away from the Argentinean coast, are consistent with a Kelvin wave response of the Río de la Plata to the tidal forcing on the shelf (O'Connor 1991).

The effect of the weather systems, on the tidal amplitude of the Río de la Plata is very significant and follows two separate fundamental conditions:

1. the incoming waves from the continental shelf, and
2. the influence of local winds and storms (Balay 1961; Mazio and Martínez 1989).

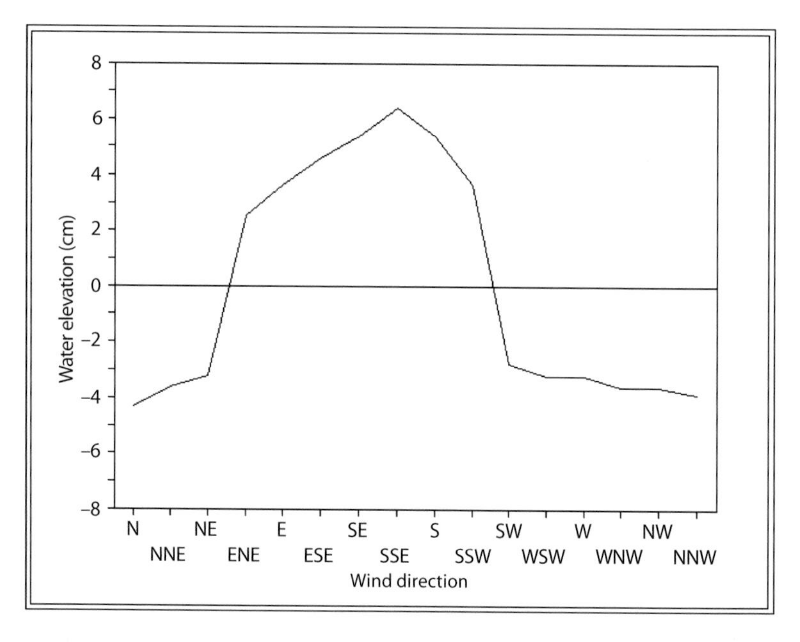

Fig. 7.5. Difference in tide height (cm) by each ms^{-1} of wind variation on the upper Río de la Plata (from Balay 1961)

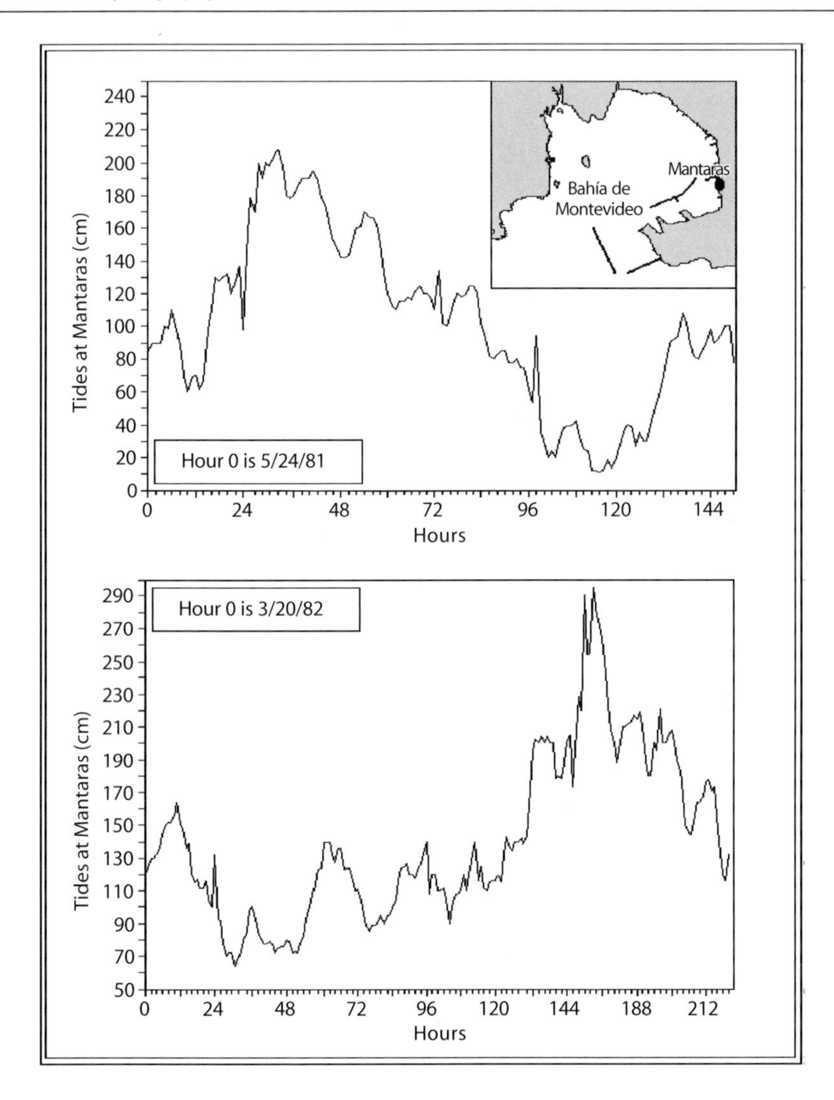

Fig. 7.6. Local wind effects at Mantaras tidal station (34°5336"S to 56°11'54"W, Bahía de Montevideo): a 96-h descent related to 16 up to 20 knots ESE-SSW winds, and a 132-h rise related to 10 up to 20 knots NNE-NNW winds

The effect of the wind on the upper Río de la Plata water surface, as studied by Balay (1961), is shown on Fig. 7.5. Figure 7.6 presents local wind effects at Mantaras tidal station, located at 34°53'36"S to 56°11'54"W (Montevideo Bay) and Fig. 7.7 shows yearly wind direction distribution for some selected weather stations. Mazio and Martínez (1989) provided information about the wind effect at Torre Oyarvide: the water level decreases under the influence of N, NE and W winds and increases with S, SE and SW winds. For E and W winds, the water level can either increase or decrease, depending on the direction of the wind rotation.

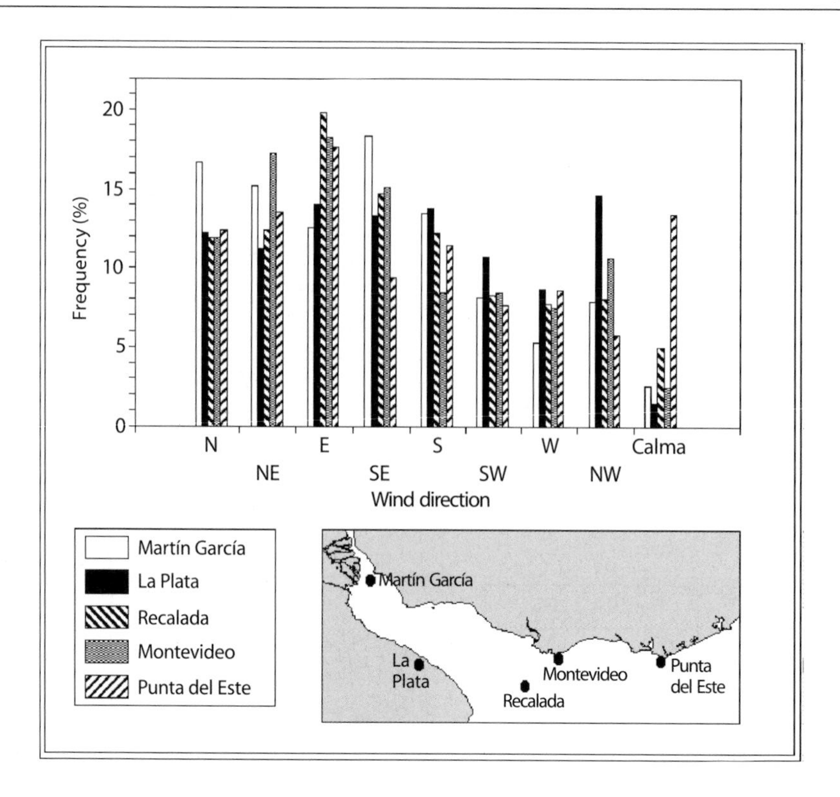

Fig. 7.7. Yearly wind direction distribution at selected weather stations: Martín García (1957–1980), La Plata (1957–1980), Recalada (1961–1980), Montevideo (1977–1986) and Punta del Este (1983–1989)

With respect to the currents encountered in the Río de la Plata (Mazio and Martínez 1989), the net flow follows the main direction of the river discharge. The maximum velocities have been observed approximately two to three hours after the high or low tide at a particular site, with greater velocities recorded along the southern coast. The maximum observed velocities were 59.2 cm s^{-1} along the southern coast (34°47'00"S to 57°49'12"W) with outgoing tide, and 37.6 cm s^{-1} along the northern coast (34°28'42"S to 57°49'50"W) with incoming tide. Bathymetry affects the local currents and incoming and outgoing tidal currents, which are oriented in the direction of the river axes and in the direction of the existing channels.

7.3.4 Salinity

Several studies show the distribution and characteristics of marine and fresh water in the Río de la Plata, as well as how far salinity penetrates and fresh water extends. The first recorded observations may have been made by Charles Darwin. He annotated on the Beagle Expedition diary (July 5th, 1832):

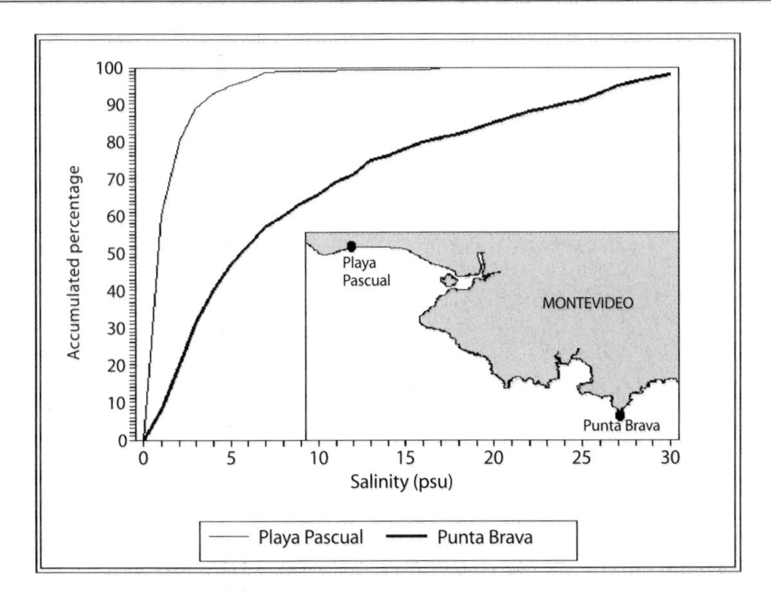

Fig. 7.8. Salinity distribution (1981–1985) at Playa Pascual (34°45'05"S to 56°27'19"W) and at Punta Brava (34°56'07"S to 56°09'36"W) coastal stations (from Jesús 1989)

> "I was very interested by observing how slowly the waters of the sea and the river mixed. The latter, muddy and discolored, from its less specific gravity, floated on the surface of the salt water. This was curiously exhibited in the wake of the vessel, where a line of blue water was seen mingling in little eddies, with the adjoining fluid".

De Buen (1949) refers to an oceanographic cruise aboard R.O.U. "Aspirante" carried out on 1937–1938. In 1953 he reported "a fast and successive salinity increase, with some modest discontinuity" along a profile comprised between 34°55'S to 56°13'W and 35°45'S to 52°41'W.

Ottmann and Urien (1965), from a series of 10 periods of observations carried out over a fortnight, during the years 1963–1964, aboard the lightship "Pontón Recalada" (35°10'S to 56°23'W), inferred the typical features of the area:

1. when W and SE winds are greater than 10 m s^{-1}, the waters are thoroughly mixed and high levels of salinity may be observed, the same on the surface and near the bottom;
2. when high winds are blowing from the N and NW, a maximum layer and fresh water on the surface are found, extending in the direction of the flowing water;
3. with light wind or no wind at all, the variations in the level of salinity follow, approximately, the tide.

Nagy (1983) stated that salinity is the key variable in the Río de la Plata, governing hydrochemistry, suspended sediments and many biological processes. According to Poplawsky (1983), wind effects acts on a spatial and temporal microscale while river runoff acts on a macroscale; salinity will be, to the first degree, a reflection of the river runoff and, to the second degree, a reflection of the salinity evolution history.

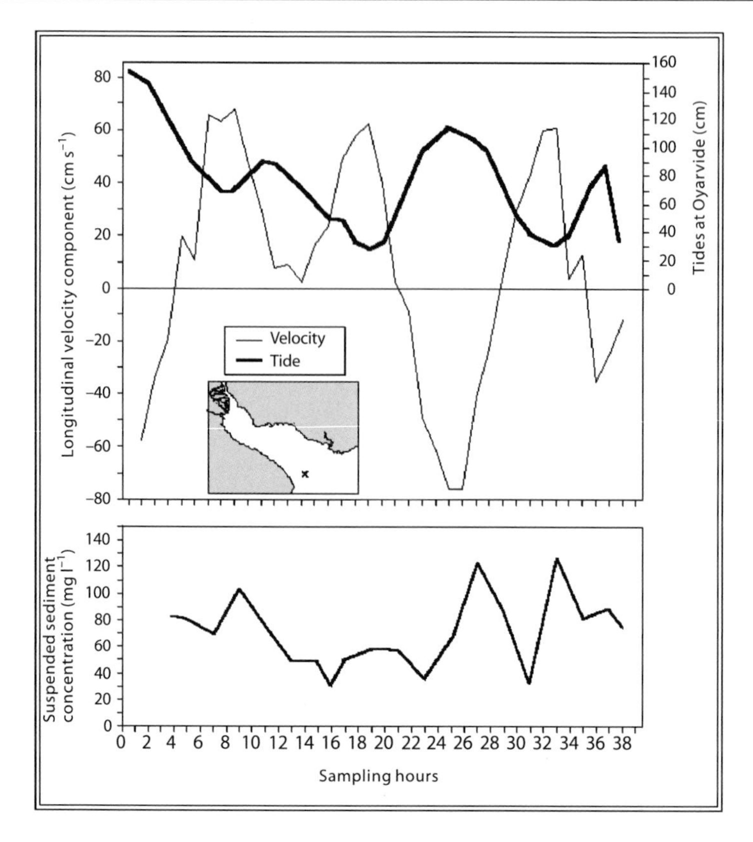

Fig. 7.9. Tides (cm), longitudinal velocity component (cm s^{-1}) and suspended sediments concentration (mg l^{-1}), at one metre above the bottom, during an anchored station near the Argentinean coast (35°14'S to 56°59'W)

Nagy et al. (1986; 1987) and López Laborde and Nagy (1986) applied several statistical techniques (Student-*t* test, proportional similarity coefficients, Student-Neumann-Keuls test, two-way variance analysis and cluster analysis) to salinity and turbidity data. Trying to analyse the discontinuity, homogeneity and spatio-temporal contrast of the data, they determined three main environments: fluvial (well mixed) to stratified (salty-oligohaline), stratified (salty-mesohaline), and marked stratified (mesohaline to polyhaline).

According to Nagy and López Laborde (1987), the salinity of the Río de la Plata is dominated by river discharge, with effective tidal and wind partial mixing up to 10–13 m depth. The most relevant characteristics of the system are:

1. large salinity variation related to time and distance to watersheds, with a range from 0–35 psu,
2. dominant convective circulation, and
3. co-existence and/or alternative stratification patterns.

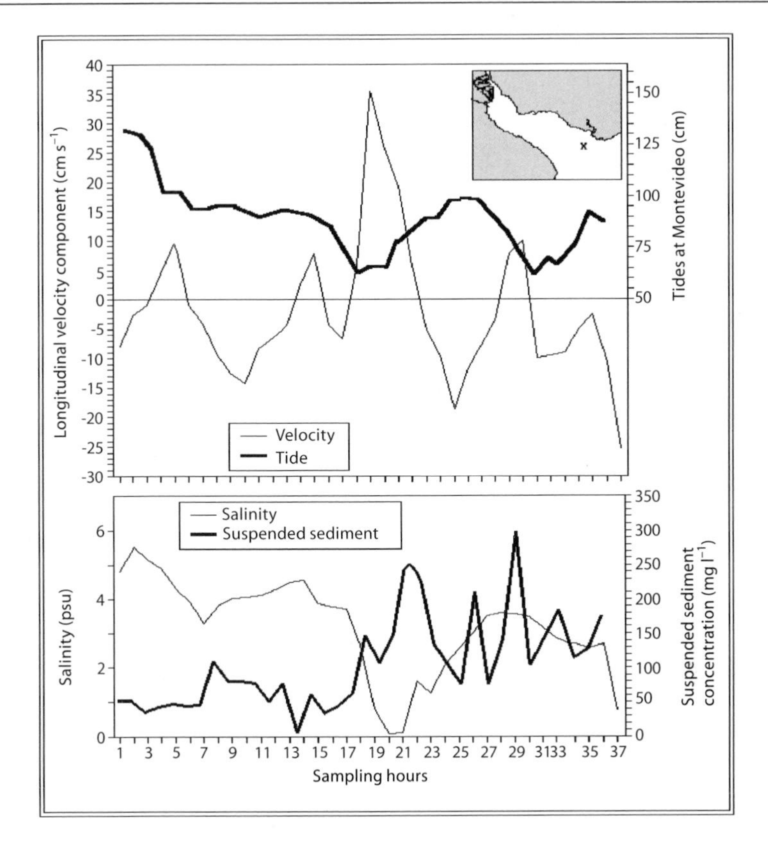

Fig. 7.10. Tides (cm), longitudinal velocity component (cm s^{-1}), salinity (psu) and suspended sediment concentration (mg l^{-1}), at one metre above the bottom, during an anchored station near the Uruguayan coast (35°06'48"S to 56°34'18"W)

Nagy et al. (1987) located the salt intrusion limit between Punta Tigre and Punta Brava transverse sections, shifting to the SE along the Argentinean coast; such patterns may be modified by strong and/or persistent axial winds. Jesús (1989) analysed daily salinity records recorded at Playa Pascual coastal station, located at 34°45'05"S to 56°27'19"W, from 1981 to 1985, obtaining an average monthly salinity ranging from 0.4–1.87 psu with 62% of the data less than 1 psu and 10.87% greater than 3 psu. Similar records from Punta Brava coastal station, located at 34°56'07"S to 56°09'36"W, showed an average monthly salinity ranging from 3.75–15.27 psu with only 9% of the data less than 1 psu and 68.34% greater than 3 psu (Fig. 7.8).

Texeira et al. (1994) analysed daily records obtained at Montevideo coastal stations from 1935–1991. The Montevideo Bay coastal station, located at 34°54'35"S to 56°12'51"W, shows an average salinity of 10.63 psu for the period 1935–1975; while Punta Brava coastal station shows an average salinity of 9.73 psu for the period 1970–1991. López Laborde et al. (1991) analysed data obtained during three tidal cycles at anchored stations located at 35°06'48"S to 56°34'18"W (Uruguayan coast) and

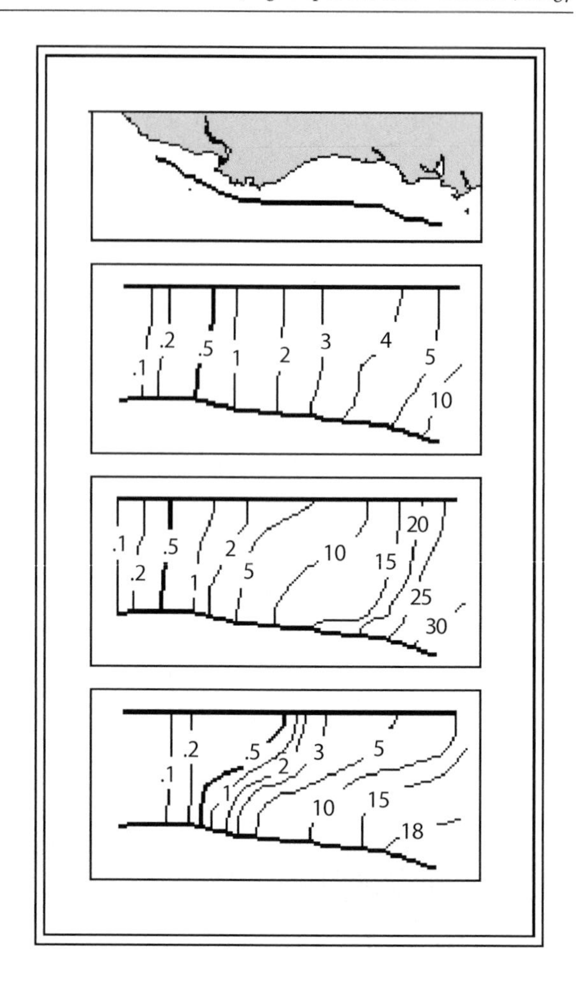

Fig. 7.11. Typical salt structure at the northern coast (from Nagy 1989)

35°14'00"S to 56°59'00"W (Argentinean coast). They reported two different salinity behaviours:

1. at the Argentinean station vertical homogeneity was observed (Fig. 7.9), with freshwater (less than 0.5 psu) always present and net ebb and flood currents;
2. at the Uruguayan station salinity variation was observed (Fig. 7.10), with freshwater at the surface and 0.1–5.5 psu at one metre above the bottom, transverse currents were also present.

This data shows clearly the SE shift along the Argentinean coast of the salt intrusion limit.

López Laborde and Perdomo (1991a) analysed salinity behaviour at the Uruguayan station according to Dyer and New (1986) diagram. During high-tide maximum velocities, at one metre above the bottom or depth averaged, coincided with maximum depth averaged salinity and maximum bottom to surface salinity difference, that was

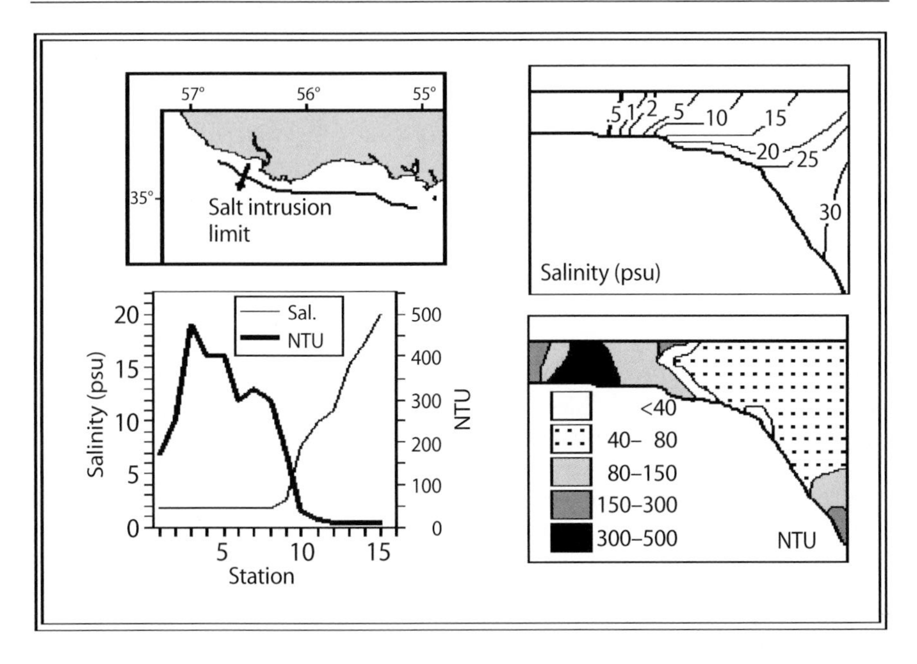

Fig. 7.12. April 1986: an example of turbidity maxima; salinity (psu) and suspended sediments (NTU, nephelometric turbidity units) along the Uruguayan coast

vertical stratification. During ebb-tide mixing was an important phenomenon. During low-tide, bottom induced turbulence provoked intense mixing and bottom sediment resuspension. During flood-tide, this turbulence, together with water advection, destroyed the previous mixing conditions.

López Laborde and Perdomo (1991b) reported an extraordinary transport of marine water up to 34°52'S to 56°48'W (Punta San Gregorio), related to winds blowing, over a period of 34 h, from the SE with speeds of 23–48 km h^{-1}.

Nagy (1989) analysed the vertical salt structure at the northern coast (Canal Oriental). During weak wind conditions, several circulation and stratification patterns were reported as a function of the river discharge (Fig. 7.11). Slight, high or even very high stratification structures dominated; advective processes were more important than diffusive ones until the salinity front is reached (Fig. 7.12). Guerrero et al. (1994; 1995; 1997), analysed salinity front behaviour using historical data (1966–1995), emphasising that surface salinity distribution is controlled by the balance between onshore and offshore winds, the river discharge and Coriolis force, with distinctive fall-winter and spring-summer patterns; whilst bottom salinity is mainly conducted by topography.

López Laborde et al. (1996) analysed stratification and mixing conditions observed at six anchored stations carried on during EcoPlata II Project. Stratificationcirculation diagrams (according to Hansen and Rattray 1966) showed conditions type 1b, 2b and 4, as well as conditions outside of the diagram limits; both, type 4 (salt wedge) and outside of the diagram limits conditions, were explained by the meteorological previous conditions (wind direction and permanence). Mixing evolution diagrams (according to Dyer and New 1986), in agreement with the previous ones, showed:

a conditions with no mixing and maintenance of the previous stratification patterns,
b increasing mixing conditions allowing the break of the previous stratification patterns, and
c decreasing mixing conditions allowing the development of stratification patterns.

7.3.5 Suspended Sediments

There are numerous publications on suspended sediment distribution and behaviour. For instance, Ottmann and Urien (1966) divided the Río de la Plata into three geographical zones, which corresponded to three sedimentological features:

1. an upper zone, to the NW of the Colonia-Buenos Aires transverse section, comprises the Paraná River Delta front and its subaqueous extension (Playa Honda), it is a typical sedimentation area of a fluvial delta;
2. a middle or intermediate zone, which comprises the area between Colonia-Buenos Aires transverse section up to Oyarvide, a fluvial zone where transport of suspended sediments predominates;
3. an outer zone, up to Punta del Este-Cabo San Antonio transverse section, which is the area of greater salinity variation where actual sedimentation processes take place.

These authors pointed out the increasing suspended sediment concentration to the SE and its asymmetrical distribution along the Uruguayan and Argentinean middle Río de la Plata coasts, showing greater values over the southern than over the northern region. This fact was related to the flocculation processes, the lag between suspended sediment and water transport, the opposing river discharge and incoming tide, and the wave and tidal current resuspension processes. They concluded that a "bouchon vaseux" (fluid muds) existed in the outer zone. Table 7.5 presents information on suspended sediment concentrations that has been compiled from various sources.

Studies of composition by Ottmann and Urien (1966) revelled organisms, organic fibers, diatoms, quartz and irregular shaped silt, volcanic ashes, ferruginous clays and heavy minerals (augite, epidote, hornblende, hypersthene and garnet). Ruhstaller and Maihle (1968) stated that Río de la Plata suspended sediments were mainly montmorillonite and illite.

According to Urien (1972) the suspended load delivered to the upper Río de la Plata consists of 75% coarse to medium silt, 15% fine to very fine silt and 10% clay, and has a mean diameter of about 5.8 phí (0.017 mm), with an annual average suspended load of 72.8 million t yr^{-1} for the Paraná River complex and of 7.0 million t yr^{-1} for the Uruguay River.

Also Urien (1972) stated that the actual sedimentary pattern is primary controlled by the estuarine environment. In the upper and middle river, fluvial fresh water conditions exist. There, tides and waves from the East control the fluvial discharge and sediment dispersion. In the outer river the fluvial discharge has influence only along the coast, aided by the bottom topography. The bulk of fluvial discharge flows out more

Table 7.5. Concentration of suspended sediments (mg l^{-1}) in the Río de la Plata

	Source	Area or section	Range or average
Upper Rio de la Plata	Urien (1966)	Paraná de las Palmas	80 – 330
		Río Uruguay	108 – 234
		B. Aires –Punta Gorda	95 – 280
	Ottman and Urien (1966)	B. Aires –Colonia	50 – 200
		B. Aires –La Plata	80 – 500
	Urien (1972)	B. Aires –Colonia	16 – 100
	Pizarro and Orlando (1984)	Río Uruguay	17 – 63
		W Pta. Negra –Quilmes	70.9
	Harris (1992)	Martín García Channels	141 – 825
Middle Río de la Plata	Ottman and Urien (1966)	Whole area	100 – 600
		B. Aires –Recalada	10 – 400 (surface)
			20 – 700 (bottom)
	Orlando and Perdomo (1989)	Whole area	21 – 529
			103.5
	Serra et al. (1992)	Franja Costera Sur	16 – 192
Outer Río de la Plata	Ottman and Urien (1966)	Whole area	10 – 400 (surface)
			15 – 500 (bottom)
	Ayup (1981)	Bahía de Montevideo	8 – 48 (surface)
			33 – 628 (bottom)
	Nagy (1983)	Whole area	2 – 350
	INTECSA (1987)	Bahía de Montevideo	10 – 430
		Approach channel	10 – 446
	López Laborde et al. (1991)	35°14'S –56°59'W (three tidal cycles)	29 – 76 (surface)
			30 – 123 (bottom)
	López Laborde et al. (1991)	35°06'48"S –56°34'18"W (three tidal cycles)	38 – 346 (surface)
			3 – 266 (bottom)
	López Laborde (unpublished data)	Montevideo –Pta. del Este	5 – 129 (surface)
			15 – 206 (bottom)
		Whole area	9 – 166 (surface)
			8 – 223 (bottom)

Exceptional values: 2.5 g l^{-1} (Ottman and Urien 1966); 657.0 mg l^{-1} (López Laborde, unpublished data), 1.2 g l^{-1} (Harris 1992).

readily through the northern channel, whereas the shallowness of Bahía de Samborombón protects it from wave and current action and encourages the deposition of fine sediments.

According to Ayup (1986a,b), outer Río de la Plata does not have an homogeneous circulation pattern, with very important lateral differences associated with suspended sediments behaviour. He concludes that:

1. with a type A circulation pattern (stratified), suspended sediments are carried by freshwater, favouring sediment dispersion, meanwhile processes acting down the halocline (flocculation and aggregation) favours deposition;

2. with a type B circulation pattern (partially mixed), better conditions for suspended sediments transport are developed and the formation of a maximum turbidity is promoted;
3. with a type C circulation pattern (homogeneity), suspended sediments remain trapped in the Río de la Plata.

Ayup (1987) stated that suspended sediment behaviour and morphology along the Río de la Plata northern coast allows active transport to the Uruguayan and southern-Brazilian continental shelf.

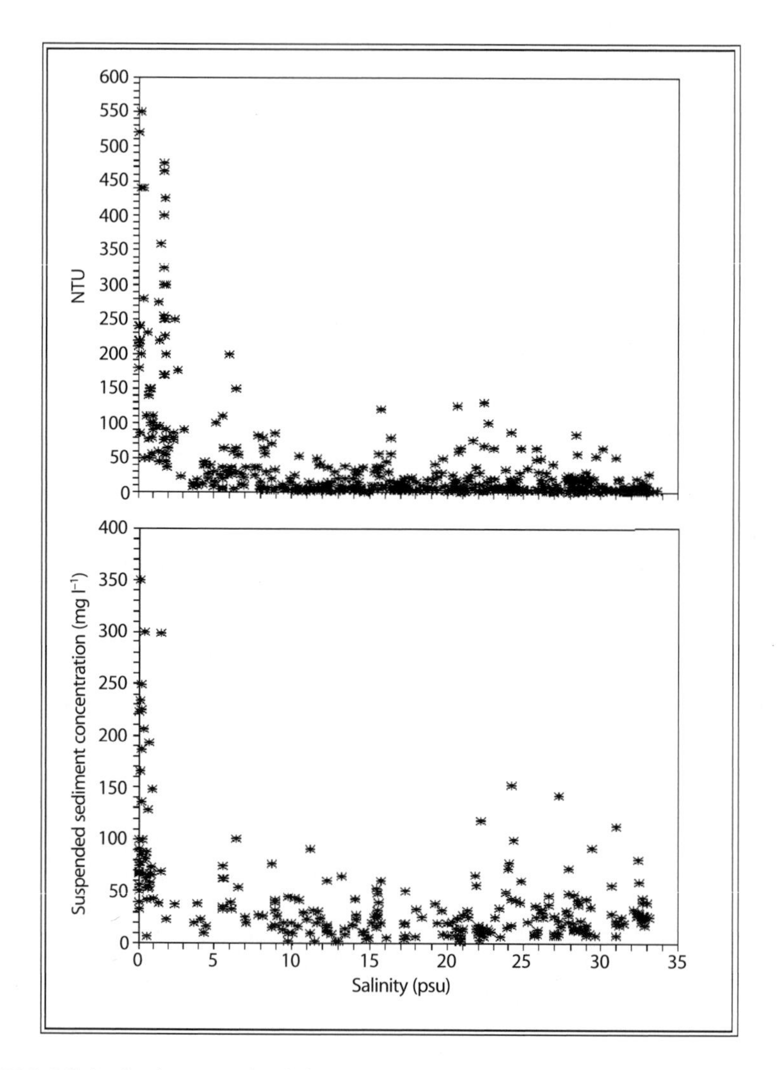

Fig. 7.13. Salinity (psu) vs. NTU (nephelometric turbidity units) and suspended sediment concentration (mg l^{-1}); all water column data is included, bottom resuspension processes are evident

Nagy et al. (1987) located the turbidity maximum between Punta Yeguas and Punta Espinillo transverse sections (see Fig. 7.12 as an example). This location agrees with satellite data (Gagliardini et al. 1984; Karszenbaun et al. 1983; Jackson 1984; Ayup 1987; Nagy 1989), which shows a discontinuous surface distribution of turbidity associated with the frontal area of mixing, where a large portion of the transported solids flocculate (Fig. 7.13). This feature is also reflected in the morphology (Barra del Indio) and in the bottom sediment distribution. Two well defined facies can be found in the area: onlap marine sands and offlap clayey silts.

López Laborde et al. (1991) analysed suspended sediment behaviour during three tidal cycles at an anchored station (Fig. 7.10), concluding that suspended sediment dynamics it is controlled by tidal friction and by salinity induced physico-chemical processes.

Framiñan and Brown (1993) used a four year span of NOAA-AVHRR daily images from September 1986 to August 1990 to determine spatial and temporal distribution of the turbidity front (Fig. 7.14). The results show a high degree of variability of the frontal distribution at the northern coast of the Río de la Plata. In this region, the frontal position varies between 57°00' and 54°12'W. The westernmost location occurs in summer months, which is coincident with minimum river discharge, predominance of easterly winds, and minimum occurrence of southwesterlies. The easternmost location occurs during spring with strong winds from the southwest. At the southern coast the modal position of the front coincides with 5 m isobath, although great variations to this position have been observed during years of large river discharge. During fall and winter, seasons of maximum river discharge, there is a bimodal frontal distribution with maximum values of frontal density at the northern Samborombón Bay and South of Montevideo, and higher variability in the center of the river.

Framiñan and Brown (1993) studied the characteristics of the frontal fluctuations and their relationship with three physical forcing components (river discharge, wind

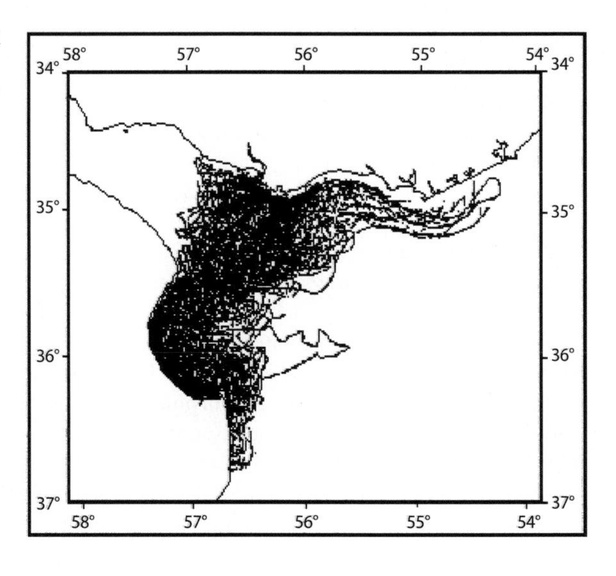

Fig. 7.14. Spatial and temporal distribution of the turbidity front, based on a four year span (September, 1986 to August, 1990) of NOAA-AVHRR daily images (from Framiñan and Brown 1993)

and tides). According to their results, the turbidity frontal area was divided into different regions:

1. a northern area where the principal forcing is the river discharge,
2. a southern area, close to the Argentinean coast, where the tidal and meridional wind effects are more significant, and
3. Samborombón Bay and Punta Rasa areas, where only local wind effects are significant.

7.4 Synthesis

Hydrodynamics of the Río de la Plata is controlled by the discharge of Uruguay River (average flow rate 4700 $m^3 s^{-1}$ at Hervidero Gauging Station) and Paraná-Paraguay Rivers (average flow rate 15400 $m^3 s^{-1}$ at Rosario Gauging Station and 17000 $m^3 s^{-1}$ at Corrientes Gauging Station). Together these rivers produce a combined annual average discharge (1960–1980) between 20000 and 23000 $m^3 s^{-1}$, with minimum and maximum values of 10800 and 30600 $m^3 s^{-1}$ (Mazio and Martínez 1989). Recent data reveals a significant average flow rate increase.

The Uruguay River receives the discharge of Paraná Bravo across from Nueva Palmira, the discharge of Sauce River across from Punta Gorda and the discharge of Paraná Guazú River across from Carmelo. Most of the discharge occurs to the southeast, in the direction of Colonia, by the Sistema Fluvial Norte; however this discharge can shift to the South, in the direction of Playa Honda, along the existing navigation channels (Canal de las Palmas and Pozos de Barca Grande).

In the upper Río de la Plata, bottom morphology appears similar to a braided fluvial pattern. The Paraná River delta front shows a differential lineal growth (10.62 to 30.02 m yr^{-1}). Both have been interpreted as the consequence of the dynamic balance between the subaqueous delta advance and the Uruguay River discharge, causing sediment redistribution (Cavallotto 1987; Parker and López Laborde 1988, 1989).

In the vicinity of Colonia, the Banco Grande de Ortíz produces a concentration of flows along Barra del Farallón, while over Canal Norte, along the Uruguayan coast, a slow and gentle current occurs. Therefore, in the middle Río de la Plata the larger flows occurs against the Argentinean coast, along Gran Hoya del Canal Intermedio.

Two facies associations of bottom sediments have been distinguished in this area (Parker et al. 1985):

1. along the northern coast, sands and silty sands, grade to silty sands and silts in the Banco Grande de Ortíz and to silts, clayey silts and even silty clays in the outer river;
2. along the southern coast, sandy silts and silts grade to silty clays.

Near Oyarvide, due to the influence of Arquímedes and Inglés banks, the main current rotates and divides into two branches: a northern one, along the Uruguayan coast, and a southern one. Bottom morphology (greater depths at Canal Oriental) and Coriolis force are responsible for the greater flow concentration along the northern branch, whereas at Samborombón Bay currents are negligible.

The outer Río de la Plata is an area of complicated current patterns due to tidal amplitude and phase differences along the river mouth transverse section. Winds blowing from the N to W quadrant produce a rapid and simultaneous effect over the whole Río de la Plata adding its effect to the discharge flow and reducing water level. For winds blowing from the S to E quadrant, wind and river runoff effects are opposed and circulation patterns are strongly influenced by the presence of Arquímedes and Inglés banks.

Meteorological forces have great influence on the general Río de la Plata hydrodynamic patterns. In the upper and middle regions, winds are responsible for extreme water levels, while at the outer region wind influence is smaller. The main effect is in producing residual currents and controlling the degree of vertical mixing.

The salt intrusion limit has been located between Punta Tigre and Punta Brava transverse sections, shifting to the southeast along the Argentinean coast. The turbidity maximum has been located between Punta Yeguas and Punta Espinillo transverse sections (Nagy et al. 1987). This observations agrees with satellite data (Gagliardini et al. 1984; Karszenbaun et al. 1983; Jackson 1984; Ayup 1987; Nagy 1989; Framiñan and Brown 1993, 1996) and its also reflected by bottom sedimentology (onlap marine sands and offlap clayey silts) and morphology (Barra del Indio).

Eulerian studies conducted at anchored stations near the upper salt intrusion limit (López Laborde et al. 1991; López Laborde and Perdomo 1991; López Laborde, unpublished data) allowed to observe simple dilution processes of the fluvial load and the development of the turbidity maximum associated with entrapment in the estuarine circulation and with tidal scour (even with salinity greater than 10 psu).

A general conceptual model for salinity and suspended sediment behaviour at the salt intrusion limit could be sketched. During high-tide, maximum velocities at one metre above the bottom or depth averaged coincides with maximum depth averaged salinity and maximum bottom to surface salinity difference, this means vertical stratification. During ebb-tide, mixing is an important phenomenon. During low-tide, bottom induced turbulence may create intense mixing and bottom sediment resuspension. During flood-tide, this turbulence together with water advection tend to destroy the previous mixing conditions.

In fresh water, suspended sediment concentration is clearly related to the axial velocity component. The greater concentrations develops during ebb-tide. When velocity decreases, concentrations fall and continue decreasing even during flood-tide, after current inversion, until axial component velocity (at one metre above the bottom) reaches 15 cm s^{-1}, at this point concentrations begin to increase. When salty water is present, salinity (flocculation and aggregation processes) affects suspended sediment behaviour. The spatial and temporal distribution of the turbidity front shows a high degree of variability, at the northern coast varies between 57°00' and 54°12'W and at the southern coast coincides with 5.0 m isobath (Framiñan and Brown 1996). River discharge and meteorological history are important in determining how far salinity penetrates and fresh water extends, and consequently, in affecting stratification patterns.

Franja Costera Sur and Samborombón Bay, due to their location windward of dominant winds with low wave energy, and due to their relationship with the Gran Hoya del Canal Intermedio, appear as areas with exceptional conditions for sediment deposition by mechanical sedimentation processes. Physico-chemical processes, related to the salt intrusion limit, are mainly responsible for silty clays along the Uruguayan coast and middle Río de la Plata (Barra del Indio), although relict sediments masking modern ones can be observed.

Lusquiños and Valdez (1971), Hubold (1980), and many others, reported the NNE Río de la Plata influence along the continental shelf. Depending on the fluvial discharge,

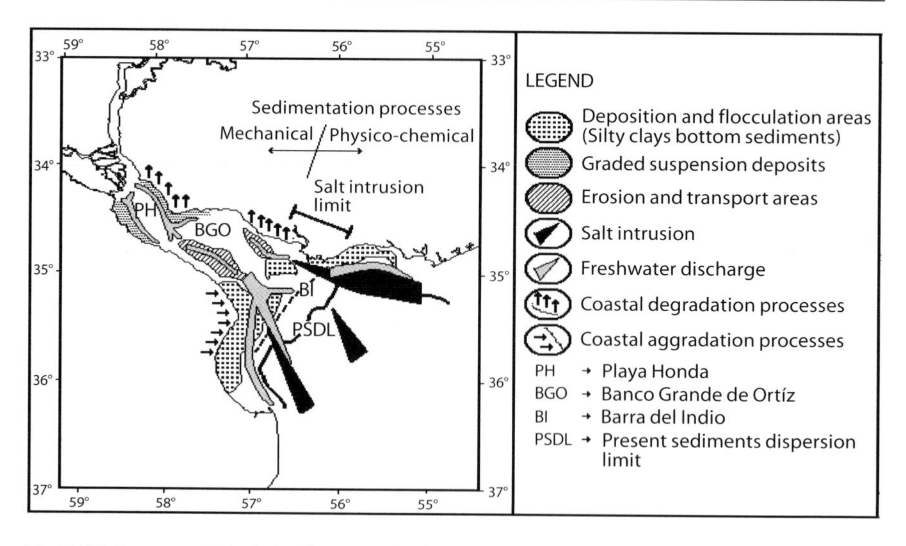

Fig. 7.15. Conceptual Río de la Plata morphodynamic model (from Nagy et al. 1987)

the prevailing stratification type and the associated circulation patterns, an active suspended sediment transport to the Uruguayan and southern-Brazilian continental shelf may be possible (Ayup 1987; Nagy 1989).

The present sediment dispersion limit has been located at the zone or band where recent sediments overlap relict Holocene sands (Parker et al. 1985; Parker and López Laborde 1989). Parker et al. (1985) proposed a schematic model where:

1. the banks (Playa Honda, Banco Grande de Ortíz, Barra del Indio) act directing river discharge, trapping sediments by competence loss and dispersing them by wave action (Barra del Indio has its own characteristics due to its relationship with the salt intrusion limit);
2. the erosive basins (Barra del Indio western side, Gran Hoya del Canal Intermedio) due to tidal action act alternatively as temporary receivers and sediment supply areas;
3. the channels, subject to tidal action, represent the river discharge routes.

Figure 7.15 shows a conceptual Río de la Plata morphodynamic model where bottom morphology interacting with fresh water discharge, marine water intrusion, tidal characteristics and meteorological forces, determines circulation patterns and surface bottom sediment distribution. Hydrochemistry, biological processes and suspended sediments behaviour will be affected by this general picture.

Acknowledgements

The authors desire to express their very special gratefulness to Dr. Gerardo M.E. Perillo (Instituto Argentino de Oceanografía) for the constant encouragement and to Dr. John T. Wells (Institute of Marine Sciences, University of North Carolina at Chapel Hill) for the careful correction of the original manuscript and the very valuable suggestions.

References

Ayup RN (1981) Contribución al conocimiento del material en suspensión de la Bahía de Montevideo. SOHMA Publ 81-01

Ayup RN (1986a) Aspectos da dinámica sedimentar do Río de la Plata Exterior. Msc thesis, UFRGS

Ayup RN (1986b) Comportamento dos sedimentos em suspensao no Río de la Plata exterior e proximidades. Pesquisas 18:39-68

Ayup RN (1987) Intercámbio sedimentar entre o Río de la Plata Exterior e a plataforma continental adjacente. Pesquisas 19:106-206

Balay MA (1961) El Río de la Plata: Entre la atmósfera y el mar. Bol SHN H-641

Biscaye PE (1972) Strontium isotope composition and sediment transport in the Río de la Plata Estuary. Geol Soc Am Mem 133:349-357

Buen F de (1949) El Mar de Solís y su fauna de peces. Publ Cient SOYP 1:1-43

Buen F de (1950) La oceanografía frente a las costas del Uruguay. Anales Museo Historia Natural VI:3-7

Cavallotto JL (1987) Dispersión, transporte, erosión y acumulación de sedimentos en el Río de la Plata. Informe final de beca de iniciación Comisión de Investigaciones Científicas

Darwin C (1906) The voyage of the Beagle. JM Dent and Sons Ltd, London

Depetris PJ (1968) Mineralogía de algunos sedimentos fluviales de la cuenca del Río de la Plata. Rev Asoc Geol Arg 23:317-325

Depetris PJ, Griffin JJ (1968) Suspended load in the Río de la Plata drainage basin. Sedimentology 11:53-60

Dyer KR, New AL (1986) Intermittency in estuarine mixing. In:Wolfe DA (ed) Estuarine variability. Academic Press Inc, pp 321-339

Framiñan MB, Brown OB (1993) Análisis del campo de temperatura de superficie en el Río de la Plata y la plataforma continental adyacente. Jornadas Nacionales de Ciencias del Mar Puerto Madryn Abstract

Framiñan MB, Brown OB (1996) Study of the Río de la Plata turbidity front Part I: Spatial and temporal distribution. Cont Shelf Res 16:1259-1282

Gagliardini DA, Karszenbaun H, Legeckis R, Klemas V (1984) Application of Landsat MSS, NOAA-TIROS AVHRR and Nimbus CZCS to study La Plata River and its interaction with the ocean. Remote Sensing Environ 15:21-36

García NO, Vargas W (1994) Análisis de la variabilidad climática en la cuenca del Río de la Plata a través de sus caudales. VIII Congr Brasilero de Meteorología – II Congr Latinoamericano e Iberoamericano de Meteorología, pp 213-217

Genta JL (1996) Análisis de tendencia de caudales en el Sureste de Sudamérica, y de precipitaciones en la cuenca del Río Negro. Taller sobre Vulnerabilidad y Adaptación al Cambio Climático en América Latina y el Caribe Montevideo

Guerrero R, Acha E, Framiñan M, Lasta C (1994) El frente de salinidad en el Río de la Plata. Simposio Comisión Técnico Mixta del Frente Marítimo Mar del Plata

Guerrero R, Acha E, Framiñan M, Lasta C (1995) Patrones estacionales de la salinidad en el estuario del Río de la Plata. VI Congr Latinoamericano de Ciencias del Mar, Mar del Plata

Guerrero R, Acha E, Framiñan M, Lasta C (1997) Physical oceanography of the Río de la Plata Estuary – Argentina. Cont Shelf Res 17:727-742

Hallcrow W (1965) Estudio y proyecto del Canal de Vinculación entre el Puerto de Buenos Aires y el Río Paraná de las Palmas. Sir W Hallcrow & Partners, Ingenieros Consultores

Hansen DV, Rattray M (1966) New dimensions in estuary classification. Limnol Oceanogr 11:319-326

Harris FH (1992) Feasibility study for channel deepening of the Martín García Channel and the expansion of the port installations at Nueva Palmira. Frederic R Harris, Hidrosud SRL, Ingenieros Consultores

Hubold G (1980) Hydrography and plankton off southern Brazil and Río de la Plata. Atlántica 4:1-22

INTECSA (1987) Plan de desarrollo a largo plazo para el puerto de Montevideo, incluyendo un estudio de las condiciones hidráulicas de la Bahía de Montevideo y del Canal de Acceso. Internacional de Ingeniería y Estudios Técnicos SA

Jackson JM (1984) Contributions to the geology and hydrology of southeastern Uruguay based on visual satellite remote sensing interpretation. Münchener Geographische Abhandlungen

Jesús CB (1989) Iniciación al conocimiento de los procesos de transporte y sedimentación en el área de mezcla fluvio-marina del Río de la Plata. Tésis Lic Universidad de la República

Karszenbaun H, Gagliardini DA, Klemas V, Dominguez F, Legeckis R (1983) The applicability of TIROS-NOAA advanced very high resolution radiometer data to studies of large estuaries. 17th Int Symp on Remote Sensing of the Environment, Ann Arbor, Michigan

López Laborde J (1987a) Distribución de sedimentos superficiales de fondo en el Río de la Plata exterior y plataforma adyacente. Invest Oceanolog 1:19-30

López Laborde J (1987b) Caracterización de los sedimentos superficiales de fondo del Río de la Plata exterior y plataforma adyacente. Anales Cient Univ Nac Agraria La Molina II:33–47

López Laborde J, Nagy GJ (1986) Asociaciones espaciales por salinidad y turbiedad óptica en el Río de la Plata Exterior. In: Sistemas Costeros Templados de America Latina. Inf UNESCO Ciencias del Mar 47:40–41

López Laborde J, Perdomo AC (1991a) Condiciones de mezcla y transporte en el límite de intrusión salina del Río de la Plata. Sem Intern sobre Preservación Ambiental y Desarrollo Costero en America Larina y el Caribe (unpublished)

López Laborde J, Perdomo AC (1991b) Efectos metereológicos en el Río de la Plata: Agosto/1990. Jorn de Invest Cient en materia de Contaminación de las Aguas, Montevideo, Uruguay

López Laborde J, Perdomo AC, Bazán JM (1991) Comportamiento del material en suspensión del Río de la Plata en condiciones de flujo y reflujo. Jorn de Invest Cient en materia de Contaminación de las Aguas, Montevideo, Uruguay

López Laborde J, Nagy GJ, Martínez CM (1996) Observations about Río de la Plata stratification, Montevideo vecinities, during Ecoplata II Project. Conferencia Internacional Ecoplata '96 "Hacia el desarrollo sostenible de la zona costera del Río de la Plata" Montevideo, Uruguay

Lusquiños A, Valdéz AJ (1971) Aportes al conocimiento de las masas de agua del Atlántico Suroccidental. Bol SHN H-659

Mazio C, Martínez CM (1989) Aspectos físicos. In: Estudio para la Evaluación de la Contaminación en el Río de la Plata. SHN-SOHMA-CARP Buenos Aires, pp 73–206

Mechoso CR, Pérez Irribaren G (1992) Streamflow in Southern South América and the Southern Oscilation. J Climate Res 5:1535–1539

Monestier H, Miguez JC (1993) Evolución de caudales del Río Uruguay. Rev de Ingeniería 15:5–14

Nagy GJ (1983) Caracterización de los procesos hidroquímicos del Río de la Plata. Tésis Lic Universidad de la República

Nagy GJ (1989) Bilan des connaissance sur l'hydrologie et l'hydrodynamique sedimentaire du Río de la Plata. Apports de la teledetection et consequences sur l'environment biologique. Raport du DEA Univ de Bordeaux I

Nagy GJ, López Laborde J (1987) Sinopsis del sistema fluvio-marino del Río de la Plata. 2do Congr Latinoam Ciencias del Mar Res Abstract

Nagy GJ, Anastasía LH, López Laborde J (1986) Zonación ambiental del Río de la Plata exterior. I: Salinidad y turbiedad óptica. In: Sistemas Costeros Templados de America Latina. Inf UNESCO Ciencias del Mar 47:39

Nagy GJ, López Laborde J, Anastasía LH (1987) Caracterización de ambientes del Río de la Plata Exterior (salinidad y turbiedad óptica). Invest Oceanolog 1:31–56

Nagy GJ, Gómez M, Acuña A, Martínez CM, Severova V, López Laborde J, Perdomo AC (1996) Vulnerabilidad de la costa Norte del Río de la Plata a la variabilidad del caudal del Río a corto y largo plazo. Taller sobre Vulnerabilidad y Adaptación al Cambio Climático en América Latina y el Caribe Montevideo

O'Connor WPO (1991) A numerical model of tides and storm surges in the Río de la Plata. Cont Shelf Res 11:1491–1508

Orlando A, Perdomo AC (1989) Aspectos químicos. In: Estudio para la Evaluación de la Contaminación en el Río de la Plata. SHN-SOHMA-CARP Buenos Aires, pp 207–412

Ottmann F, Urien CM (1965) La melange des eaux douces et marines dans le Río de la Plata. Cahiers Oceanographiques 17:213–234

Ottmann F, Urien CM (1966) Sur quelques problemes sedimentologiques dans le Río de la Plata. Rev Geogr Physique et Geol Dinamique 8:209–224

Parker G (1985) El subsuelo del Río de la Plata (Recopilación de perforaciones). Servicio de Hidrografía Naval Informe Técnico N° 36/85

Parker G (1990) Estratigrafía del Río de la Plata. Rev Asoc Geol Arg 45:193–204

Parker G, López Laborde J (1988) Morfología y variaciones morfológicas del lecho del Río de la Plata. SHN-SOHMA Informe Técnico 4

Parker G, López Laborde J (1989) Aspectos geológicos. In: Estudio para la Evaluación de la Contaminación en el Río de la Plata. SHN-SOHMA-CARP Buenos Aires pp 1–72

Parker G, Marcolini S, Cavallotto JL, López MC, de León A, Maza MT, Ayup RN, López Laborde J (1985) Distribución de sedimentos en la superficie del fondo del Río de la Plata. SHN-SOHMA Inf. Técnico 3

Perillo GME (1995) Definitions and geomorphologic classifications of estuaries. In: Perillo GME (ed) Geomorphology and sedimentology of estuaries. Elsevier, Amsterdam, pp 17–47

Pizarro MP, Orlando AM (1984) Distribución de fósforo, nitrógeno y silíceo disueltos en el Río de la Plata. SHN Publ H-625

Poplawsky R (1983) Introducción al estudio de la variabilidad temporal de la salinidad en la costa uruguaya. Tésis Lic Universidad de la República

Ruhstaller RE, Mailhe AR (1968) Contribución al conocimiento del material inorgánico en suspensión del Río de la Plata. Comunicaciones Mus Arg de Cien Nat 2:31–37
Serra AE, Toschi NO, Anvaria O, Altieri A, Orlando AM, Bazán JM (1992) Calidad de las aguas, Franja Costera Sur. AGOSBA-OSN-SIHN
Shepard FP (1954) Nomenclature based on sand-silt-clay ratios. J Sediment Petrol 24:151–158
Siegel FR, Pierce JW, Urien CM, Stone LC (1968) Clay mineralogy in the estuary of Río de la Plata, South America. 23th Int Geol Congr 8:32–37
Souza S de , López Laborde J (1988) Consideraciones preliminares sobre la estratigrafía del Río de la Plata en las proximidades de Montevideo. VI Panel de Geología del Litoral y 1ra Reunión de Geología del Uruguay
Texeira J, Robatto, P, Falcón M (1994) Salinidad y temperatura costera en las proximidades de Montevideo (período 1935–1991). SOHMA Informe Técnico
Tossini L (1959) Sistema hidrográfico y cuenca del Río de la Plata Contribución al estudio de su régimen hidrológico. An Soc Cient Arg 167:41–64
Urien CM (1966) Distribución de los sedimentos en el Río de la Plata Superior. Bol SHN 3:197–203
Urien CM (1967) Los sedimentos modernos del Río de la Plata Exterior. Bol SHN 4:113–213
Urien CM (1972) Río de la Plata Estuary environments. Geol Soc Am Mem 133:213–234

Chapter 8

Physical Characteristics and Processes of the Río de la Plata Estuary

Mariana B. Framiñan · María P. Etala · Eduardo M. Acha · Raúl A. Guerrero
Carlos A. Lasta · Otis B. Brown

8.1 Introduction

The Río de la Plata Estuary, located at 35°00'S on the Atlantic coast, drains the second largest basin in South America. The drainage area covers over 3.1×10^6 km^2 and includes parts of Argentina, Bolivia, Brazil, Paraguay and Uruguay. It is the most developed basin in South America: the population of the hinterland is estimated to be 30 million inhabitants. Two of the major ports in the region, the cities of Buenos Aires, Argentina, and Montevideo, Uruguay, lie along its shores. The Río de la Plata also provides access to ports upstream along the Paraná, Paraguay, and Uruguay Rivers. Biologically, the estuary is also very important (Cousseau 1985; Boschi 1988). The outer region, where the fresh waters rich in nutrients interact with the coastal waters, is the spawning and nursery area of many coastal species. Because of its regional importance, the Río de la Plata Estuary has been the subject of many research programs which try to understand different aspects of the environment and ecosystem.

The aim of this chapter is to review the physical characteristics of the estuary, especially those of the outer estuary. Discussion will focus mainly on the analysis of results from recent studies presented by the authors. These studies are based on in situ information, satellite imagery and modelling. They are part of ongoing research programs carried out as a joint effort by institutions from Argentina and USA.

In the following section, a brief description is presented of the Río de la Plata environment, its physiographic setting, climatology, and hydrology. Section 8.3 analyses the tidal regime in the estuary. In Section 8.4 results of the study of the salinity field are presented; the stratification, the effects of river and wind forcing, and the implications of the salinity distributions on circulation, are discussed. The temperature field and its variability are analysed in Section 8.5, based on in situ and remote sensing information. The last section presents a study of the turbidity front, a characteristic feature of the Río de la Plata Estuary.

8.2 Río de la Plata Environment

8.2.1 Physiographic Setting

The Río de la Plata is located on the east coast of South America, between 34°00'S and 36°20'S, and 55°00'W and 58°30'W. It is a shallow, large-scale, coastal plain estuary that covers an approximate area of 35 000 km^2 (Fig. 8.1). It is 320 km long, and its width

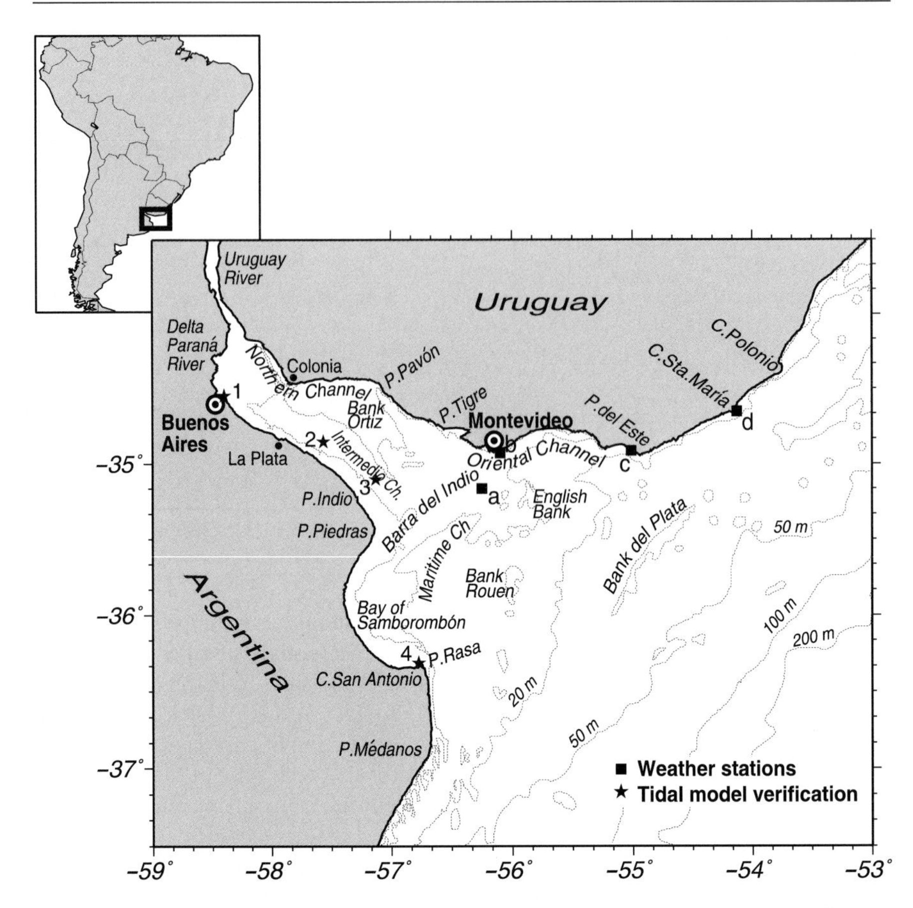

Fig. 8.1. The Río de la Plata Estuary study area. Contours are given at: 5, 10, 20, 50, 100 and 200 m respectively. Location of stations used in tidal model verification: *1* Buenos Aires, *2* Magdalena Channel, *3* Oyarvide, and *4* San Antonio lighthouse; Weather stations: *a* Pontón Prácticos Recalada, *b* Punta Brava, *c* Laguna del Sauce, and *d* La Paloma

varies from 38 km in the upper region to 230 km at the mouth between Punta Rasa and Punta del Este. The northern coastline is 416 km long, while the southern is 393 km.

The estuary can be split into two main regions, upper and lower, based on its morphology and dynamics (Comisión Administradora del Río de la Plata, CARP 1989). The two regions are divided by the Barra del Indio, a shallow area across the river in the line Punta Piedras-Montevideo. The upper region has a fluvial regime, with a two-dimensional flow and it is almost entirely occupied by fresh water. In the lower estuary, fresh water interacts with the shelf water and three-dimensional flow characteristics are present. A turbidity maximum, as is clearly observed in visible satellite imagery (Gagliardini et al. 1984; Framiñan and Brown 1996), is located in the transition zone between the two regions.

A complete description of the geomorphology and sedimentology of the estuary can be found in Cavallotto (1987, 1988), Parker et al. (1986a,b); Ottman and Urien

(1965, 1966); Urien (1966, 1967, 1972); López Laborde (1987a,b); Depretis and Griffin (1968). A brief description of the main morphological features is presented here.

The upper region is characterized by shallow areas, 1–4 m deep (Playa Honda and Great Ortiz Bank) separated from the coasts by channels 5–8 m deep (North Channel System, west part of the Oriental Channel and the Intermediate Channel). The Barra del Indio is a wide and gentle bar extending from the Argentinean coast between Punta Indio and Punta Piedras to the northeast. It has a slightly convex surface with a depth of 6.5–7 m. It borders the Maritime Channel to the east, a wide depression which separates the Barra and the Samborombón Bay from the Alto Marítimo (Cavallotto 1987; CARP 1989).

The Maritime Channel, the Samborombón Bay, the Alto Marítimo and the east part of the Oriental Channel are the main characteristics of the lower estuary. The Maritime Channel has a depth of 12–14 m increasing to the south where it reaches 20 m. The Oriental Channel runs along the Uruguayan coast and is the deepest channel in the estuary: it is 25 m deep off Punta del Este. Alto Maritimo is a plain formed by the Bank Arquímedes, English Bank and the Bank Rouen (Cavallotto 1987; CARP 1989). Bank Arquímedes and the English Bank are stable areas, 6–8 m deep, and with approximate surface areas of 45–350 km^2, respectively. They act to divide the river flow, separating the east part of Oriental Channel and the Maritime Channel (Urien 1967; CARP 1989). The Bank Rouen is 30 km south of them, has a north-south orientation, covers an approximate area of 700 km^2, and is 10–12 m deep. The other major feature in the outer region is the Samborombón Bay, a shallow area with depth ranging from 2–10 m, which extends from the north at Punta Piedras to the south at Punta Rasa.

The suspended sediment concentration in the Río de la Plata water varies from 100–300 mg l^{-1} (Urien 1967; CARP 1989). The estimated average annual suspended sediment load is 79.8×106 t yr^{-1} (Urien 1972).

8.2.2 Climatology

The atmospheric general circulation in the region is controlled by the influence of the quasi-permanent South Pacific and South Atlantic high pressure systems. Southwestward circulation associated with the South Atlantic high advects warm, moist air from subtropical regions. Cold anticyclones over southern Argentina periodically (particularly in winter) drive cold maritime air masses from the Southwestern Atlantic over the littoral area. The Río de la Plata is located within one of the most cyclogenetic areas in the Southern Hemisphere (Taljaard 1967). The highest frequency of occurrence is in winter. A recent, comprehensive review of possible mechanisms that produce this cyclogenesis is given by Seluchi (1995). An important feature in this region is the presence of upper air troughs, usually associated with frontal systems approaching the region with southwest-northeast trajectories, that interact with subtropical air masses over northeastern Argentina, northern Uruguay and southern Brazil. This interaction produces cyclogenesis resulting in strong winds over 30 m s^{-1} from the southeast in extreme events, and storms affecting the area for several days. This phenomenon is called "sudestada" and is responsible for disastrous flooding in the littoral of the Río de la Plata. Statistics of the wind field (Servicio Meteorológico Nacional (SMN) 1982), based on 10 years of data from the weather station Pontón Prácticos Recalada, located in the outer estuary (Fig. 8.1), show that the winds generally range from southwestward to northward with a predominance of northwestward. There is a marked seasonal variability with predominance of westward winds in summer and northwestward winds in winter. The highest speeds correspond to the northeastward and northwestward winds during winter.

8.2.3 River Input

The Río de la Plata drains the second largest basin in South America, which, with an area of 3.1×10^6 km^2, extends through Argentina, Bolivia, Brazil, Paraguay and Uruguay (Depretis and Griffin 1968). The two major tributaries are the Paraná and the Uruguay Rivers, with annual average discharges of 16 000 m^3 s^{-1} (15 970 m^3 s^{-1}) and 6 000 m^3 s^{-1} (5 817 m^3 s^{-1}) respectively. These values, provided by the Instituto Nacional de Ciencia y Técnicas Hídricas (INCYTH), Argentina, are based on an 111 years river discharge record of the hydrological station Rosario (Paraná River) and 16 years record of station Salto Grande Abajo (Uruguay River). The Paraná River flows into the estuary forming a delta; the main channels are the Paraná de las Palmas and the Paraná-Guazú. The Uruguay River regime has smaller transport values than the Paraná, but extremely high variability. Maximum and minimum transport correspond to fall and spring for the Paraná, and fall and summer for the Uruguay River, respectively. The mean transport of minor tributaries is several orders of magnitude smaller; for example, the annual average local discharge into the Samborombón Bay, from the Salado River and the channel system, is estimated as 26 m^3 s^{-1} (Consejo Federal de Inversiones 1969). Therefore, the discharge of the Río de la Plata can be evaluated as the result of the transport of the two major rivers. An average value of 22 000 m^3 s^{-1} is estimated, with maximum values during fall. Previous studies suggested a mean value of 20 000 m^3 s^{-1} (Urien 1967; CARP 1989), and Tossini (1959) estimated an average of 23 000 m^3 s^{-1} at the Río de la Plata mouth.

8.3 Tides and Tidal Currents

8.3.1 Tides in the Estuary

The tidal waves associated with the South Atlantic amphidromes reach the continental shelf break while propagating northward. Bottom topography and coastline geometry modify their propagation over the continental shelf, so that they enter the estuary, in general, from the southeast. Tidal charts for the five main constituents, M_2, S_2, N_2, K_1 and O_1 (CARP 1989) are shown in Fig. 8.2. The extremely shallow water shortens tidal wavelengths after they enter the estuary. Due to this effect and to the considerable length of Río de la Plata, semidiurnal constituents have the very unusual feature of a nearly complete wavelength within the estuary at all times, as can be seen in Fig. 2a–c. Tidal amplitudes are generally not amplified towards the upper part; the estuary is long and converges only at its innermost part, where it is extremely shallow and bottom friction plays a fundamental role in controlling the wave amplitude.

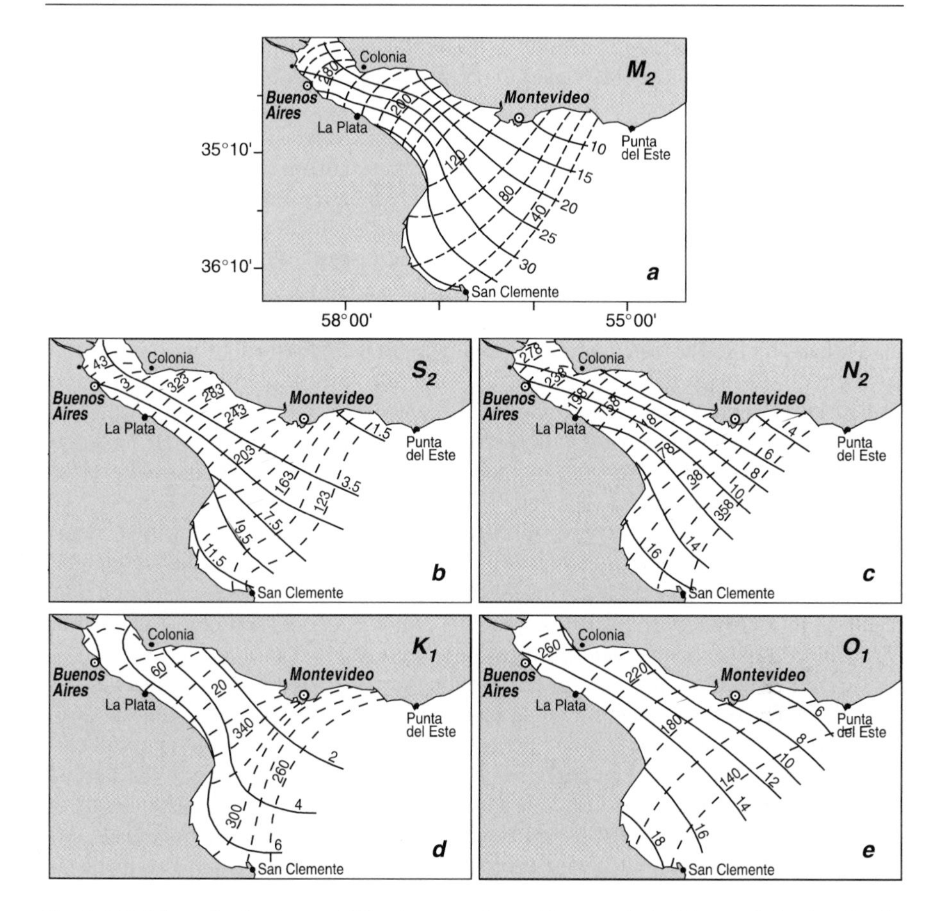

Fig. 8.2. Tidal amplitudes (cm, *solid lines*) and cophases (degrees, *dashed*) in Río de la Plata for the **a** M_2, **b** S_2, **c** N_2, **d** K_1 and **e** O_1 components (redrawn from CARP 1989; Fig. II)

Some constituents may present amphidromes very close to the mouth of the estuary, as was indicated by the results of Glorioso and Flather (1995) and in Schwiderski (1981) for K_1. Some evidence of this characteristic was also found by Etala (1995). This feature complicates tidal charting in the area making the joint use of tidal data and models effective.

Several hydrodynamical models have either been developed or adapted to simulate the tides and storm surges in Río de la Plata. Alvarez (1973) presented the first model for the estuary, in an attempt towards routine storm surge forecasting; the method could not be used in daily practice due to technical limitations at that time. Molinari (1986) modelled the tides and the currents produced by the incoming water from the main sources, Paraná (Paraná de la Palmas and Paraná-Guazú) and Uruguay.

This work provided a basis for further studies, e.g., hydrodynamical conditions influencing sedimentation on channels (Marazzi and Menéndez 1991). Mazio (1987) considered the response of an unidimensional channel of variable width and depth to model tidal elevations and currents in the estuary; a description and validation of the model can be found in CARP (1989). O'Connor (1991) adapted a model, that had been extensively used elsewhere, to the estuary and part of the continental shelf, and typified situations based on historical information on storm surges. A similar model is presented by Guarga et al. (1992), in a work mainly focused on engineering applications.

As the mass of water in the estuary is small, direct astronomical forcing is considered negligible. For that reason all the models mentioned above are forced by some previous knowledge of the tides at the open ocean. Also all of them are based on the so-called shallow water equations, a vertically integrated approach to Navier-Stokes equations, making use of the fact that tide and surge wave lengths are much greater than the depth. Results from a model of this type (Etala 1996a) will be used in the following sections to illustrate some aspects of the tide and surge-induced water levels and currents in the estuary. The constituents considered in this model are M_2, S_2, N_2 and O_1. Although the agreement with their counterparts, for instance CARP (1989), is good, it is considered desirable to include more constituents in the model to represent the tide more accurately.

The tidal regime in the estuary is mixed, dominantly semidiurnal. At the upper zone, where the wave amplitude to depth ratio becomes sufficiently large, the wave starts deforming by nonlinear effects during low tide. Ebb tide is then slower than flood tide, resulting in an asymmetric wave shape. This effect is already noticeable in Buenos Aires, and it increases upstream in the estuary, as noted by Balay (1961).

Río de la Plata tides effect on the tidal regimes of some of the main tributaries has been studied by several authors. For example, in the Paraná de las Palmas, Junod (1996) finds that the transition area, where the tidal effects are not observable, is usually located between Campana and San Pedro. Its position varies according to the Paraná River stage, the surge in the Río de la Plata, and the tidal amplitude. The effects of extraordinary floods in the Río de la Plata, provided the Paraná River stage is favourable, can be detected up to Rosario (approximately 250 km upstream of Paraná de las Palmas mouth).

8.3.2 Tidal Currents

Water levels are easier to measure than currents and consequently these data are more widespread. Hydrodynamical models allow the specification of currents using knowledge of water level. The intention is not to substitute measured with modelled current values, but a properly verified model can be a useful tool to provide information that otherwise would not be obtained for logistical reasons. Model limitations are associated with model resolution, bottom topographic specification, limited number of constituents, uncertainty in open boundary tides, and other assumptions made in the governing equations, for example in relation to parameterizations of subgrid processes, such as bottom stress and horizontal diffusion.

The model run presented here (Etala 1996a) does not include flow from fresh water sources. Consequently, a better representation of the currents is expected in the outer estuary rather than in the inner part, where the effect of this discharge on the current pattern might be more pronounced. For the verification that follows, constant values of discharge current have been added to model results, following the proce-

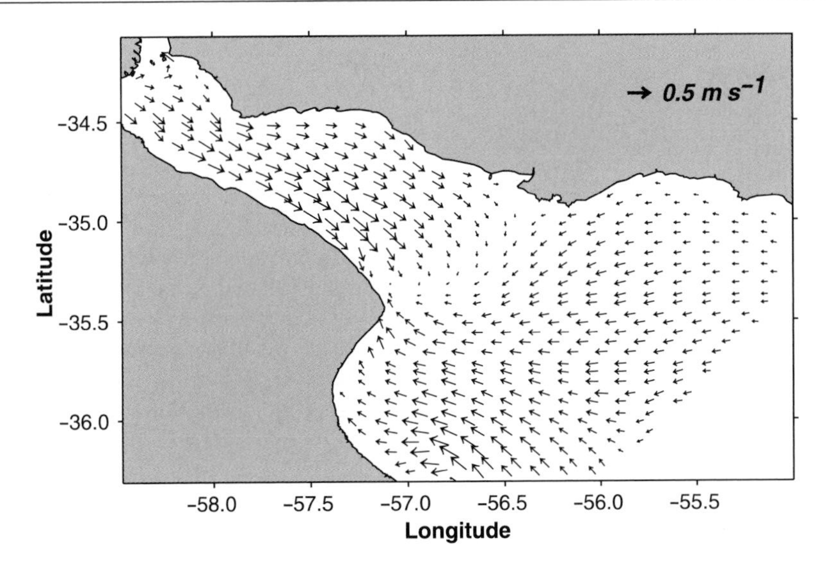

Fig. 8.3. Tidal currents (m s^{-1}) provided by the numerical model for January 21, 1996 0:00 GMT, within a period of maximum semidiurnal tidal range

dure used to calculate the currents presented in tidal tables (Servicio de Hidrografía Naval, SHN 1996). Maximum tidal currents and the time of the slack produced by the model are compared to harmonic prediction values (based on observations), during apogee and perigee spring tides in January 1996. The location for verification is the De la Magdalena Channel (Fig. 8.1). The maximum modelled current speed for a rising tide is 0.42 m s^{-1} while maximum tabulated value is 0.46 m s^{-1}. Instantaneous differences may be greater, and flood currents tend to be exaggerated. The most probable difference to be found is slightly greater than 0.05 m s^{-1}. During ebb tide, currents tend to be underestimated, and the most probable error is similar to that for rising tide, but negative. The maximum modelled current in this case is 0.51 m s^{-1} and the maximum tabulated value is 0.57 m s^{-1}. Current directions have a mean error of 3°, with a standard deviation of 3°. The time of the slack tide is, on average, moved forward 15 min, with a standard deviation of 40 min. Due to the strong dependence of currents on bottom topography, this comparison is not expected to be strictly valid elsewhere.

With the previous verification in mind, the complex map of tidal currents in the river, produced by the effects of channels and banks and of the different states of the tide throughout the estuary at a given time, can be, to some extent, illustrated by the results of the numerical model. An example of instantaneous tidal current (Fig. 8.3), corresponds to the perigee close to spring tides of January 1996, when maximum semidiurnal amplitudes are expected. According to Balay (1961), the maximum tidal current speeds in the outer estuary are found in the vicinity of San Antonio lighthouse (Fig. 8.1), where a maximum value of 0.62 m s^{-1} occurs. This feature, as well as the minimum at the Uruguayan coast close to Punta del Este, is in agreement with the current fields provided by the numerical model. Another maximum is seen in the upper estuary, steered by the Intermediate Channel.

8.3.3 Storm Surges

The estuary is frequently affected by an extraordinary rise in the water level due to atmospheric forcing. Northwestward winds are the major mechanism responsible for this rise, by piling up of water. Winds blow onshore over the continental shelf, producing storm surges which increase their magnitude when they enter the very shallow water close to the Argentine coastline north of 38°S and at the mouth of the estuary. The worst flooding occurs during the atmospheric situation that local tradition calls "sudestada". Strictly, this situation involves a migrating anticyclone moving over the sea off Patagonia and an extratropical cyclogenesis that develops and evolves as it moves to the northeast of Río de la Plata (Seluchi 1995). Under this synoptic conditions, the "sudestada" may last several days. The estuary's low southwestern coastline, highly populated, suffers severe flooding during the most intense of these events. Moreover, the "sudestadas" are accompanied by localised and sustained heavy rain that contributes to urban flooding. The "sudestadas" is not always related to cyclogenesis, as was noted by Ciappesoni and Salio (1996), but it is always associated with a high pressure system located south of the estuary. These authors identified twenty one cases during the period 1990–1994; only six of them included cyclogenesis to the north.

Other mechanisms have been considered to produce water level rise in the estuary, although their consequences are not so extreme. Winds blowing equatorward can generate surge waves on the shelf which propagate into the estuary from south to north. Such events have been identified by Alvarez and Balay (1970). Romero (1994) studied several cases, found correlations and established the propagation time along the shelf for forecasting purposes. They found that in most cases these waves are associated with strong extratropical cyclones moving off the coast and isobars parallel to the

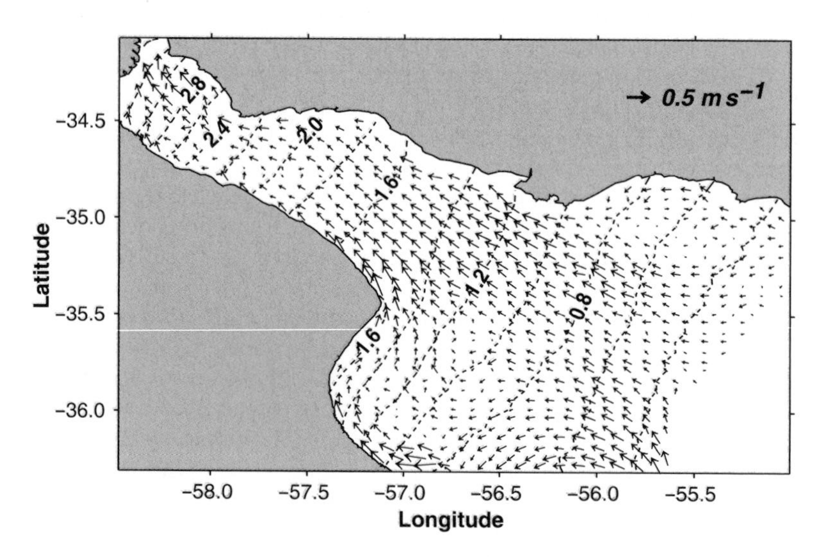

Fig. 8.4. Surge elevations (m) and currents (m s^{-1}) for August 30, 1993 18:00 GMT, 5 h before the peak of the surge in Buenos Aires

coast. Within the estuary, the surge shows minimal or no correlation with local winds. Sometimes, an intense high pressure system establishes over Patagonia and the center of Argentina intensifying the equatorward winds, with a fetch that covers almost the entire Argentine continental shelf and a duration close to one day. This situation produces strong alongshore currents on the shelf. In steady state, the elevated water level in Río de la Plata is geostrophically adjusted to the currents. The periodicity of extraordinarily high sea levels events in the Río de la Plata was analysed by Balay (1959). He found a significant relationship between the maximum annual heights over a period of 37 years, and the mean longitude of the lunar perigee and the ascending node of the moon. According to his results, these astronomical factors produce a favourable environment in which the local meteorological forcings described above in this section, would lead to the largest floods in the estuary. On these basis, he proposed a method for prediction of the maximum annual heights, which was tested on a 21 year series.

A severe surge occurred on 30 August 1993 is illustrated by the numerical model products, i.e., elevations and depth average currents. The model takes account of the tides and meteorological forcings in one run and of the tides only, in another run. The difference between both results represents the storm surge, including surge-tide interactions. The surge currents and elevations so produced, during the rising stage prior to the sea level peak in Buenos Aires, are presented in Fig. 8.4. Northwestward winds over 15 m s^{-1} were reported by the meteorological station Pontón Prácticos Recalada for 24 h, with 20.5 m s^{-1} observed for at least 12 h. Surge currents with speeds greater than 0.5 m s^{-1} can be seen at the beginning of the upper estuary, between Punta Piedras and Montevideo. The surge elevations and currents at the mouth of the estuary are provided to the Río de la Plata model by a similar, coarser resolution continental shelf

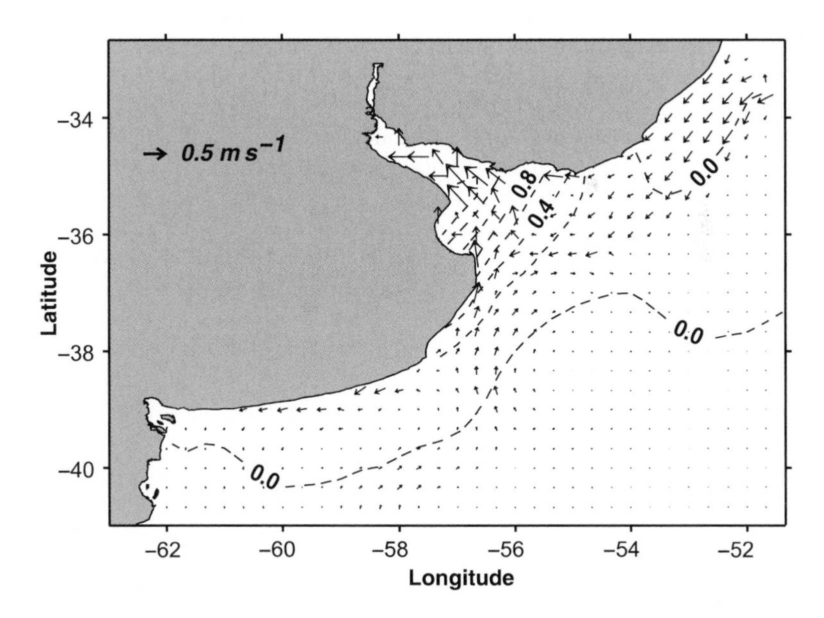

Fig. 8.5. Output of the continental shelf tidal model. Surge elevations (m) and currents (m s^{-1}) for August 30, 1993 18:00 GMT, 5 h before the peak of the surge in Buenos Aires

model (Fig. 8.5). The location of the maximum surge currents is coincident with those of the fine mesh model and their values are similar. Another maximum is observed entering the estuary from the south. This alongshore current is generated as a response to the higher surge levels towards the coast of Punta Rasa (Fig. 8.5) and is a typical feature in these situations. The water level at Buenos Aires at the time of the peak was accurately reproduced by the model, although its decrease was delayed probably due to the poor temporal and spatial resolution of the available wind fields (Etala 1996a). The analysis of a similar, but less intense, event can be found in Etala (1996b), where an improved version of the model is tested in a pre-operational stage.

8.4
Salinity and Water Stratification

8.4.1
Background

Studies in the Río de la Plata are technically and financially difficult because of its great geographic extent. A large number of observations are needed to obtain a synoptic description of the estuary and to resolve the spatial and temporal variability. Many investigations of its physics, chemistry, geology and/or biological properties have been conducted in the area. Most of them include salinity measurements, which were used by several authors to describe the salinity fields of the upper and lower estuary.

The upper part of the Río de la Plata has salinity less than 0.4, and the outer region presents high spatial variability, with salinity values from 0–33 (CARP 1989). Descriptions of the salinity field in the upper estuary can be found in CARP (1989), Bazán and Janiot (1991), Bazán and Arraga (1993); here, the focus will be on the salinity fields of the outer estuary. On the southern coast, based on in situ observations, studies (Urien 1967; CARP 1989) have located approximately the landward limit of salt intrusion in the area of "El Codillo". On the northern coast, Nagy et al. (1987) located this limit between Punta Tigre and Punta Brava. Remote sensing studies give a more complete description of its spatial and temporal distribution, based on the distribution of the associated turbidity maximum at the surface (Framiñan and Brown 1996), as reviewed earlier. Previous studies of the vertical structure suggest a highly variable regime. Ottman and Urien (1965) analysed the relation between the evolution of the salinity profile and the tides and wind. They compared hourly surface and bottom salinity data measured at a fixed station in the outer region (at Pontón Prácticos Recalada), during periods of 15 days for each season in three consecutive years, with tidal height and wind speed and direction. Their conclusion was that, at times of eastward winds, a fresh-water layer was found at the surface, and that in light wind or calm conditions, the variation in salinity follows the tide. The sampling was limited to two levels which restricts analysis of the vertical structure. CARP (1989) presented the analysis of vertical profiles obtained from bottle-samples at several stations in the outer estuary taken between 1981 and 1985. High spatial variability of the vertical structure was found, as well as strong stratification, during events of high river flow.

Based on the distribution of salinity, the parameter controlling the density in the estuary, several attempts have been made to understand the river circulation. Ottman and Urien (1965) suggested that the shallow banks in the outer estuary split the river flow into two branches. One branch flows along the northern coast up to Punta del Este and the other turns to the south, into Samborombón Bay and reaches Cabo San Antonio. They presented a distribution of surface and bottom salinity showing a clear tendency for southward flow. Urien (1967, 1972), showed vertical salinity sections for February, March and October along the northern coast and for March and May in the Samborombón Bay. Based on these sections, he suggested that the saline water movement is greater on the northern coast through the deep northern channels. In the Samborombón Bay, he said, the saline water movement is quite restricted, due to the shallowness of the area. Brandhorst and Castello (1971) and Brandhorst et al. (1971) presented seasonal maps of temperature, salinity and other chemical properties on the continental shelf. From the analysis of the distribution of properties on the shelf adjacent to the estuary, they concluded that the influence of the Río de la Plata is more important to the north, and proposed the southern discharge as a "periodic" event. But they did not explain the characteristics and the frequency of observation of these southern extensions. The northward influence of the river plume was inferred by Hubold (1980), Carreto et al.(1986) and Lusquiños and Figueroa (1982) working with in situ data from hydrographic stations on the shelf. Nagy et al. (1987) also proposed a northward flow based on observations at coastal stations. A southward discharge pattern was reported by Brandhorst et al. (1971) and Carreto et al. (1982). The mean surface distribution presented by CARP (1989) shows higher salinity values along the northern coast, although they noted very low surface salinity in this area during periods of high river flow.

Based on this brief review, it is evident that previous studies provide information to support the occurrence of both possible paths proposed initially by Ottman and Urien (1965). The northward flow was the case more frequently observed and the hypothesis most likely accepted. It corresponds to the general picture for the dispersion of a buoyant plume injected on an eastern coast in the Southern Hemisphere. The low salinities along the southern coast have been explained in terms of a more complex tidal regime and mixing processes associated with the shallowness of the Samborombón Bay (Urien 1972; CARP 1989). Both situations have been observed and documented. Nevertheless, no clear explanation of the mechanism that generates the bimodal discharge pattern was proposed, and the frequency of occurrence was not established.

Many rivers around the world exhibit a seasonal pattern for the river plume as the result of seasonal variability of the forcing. The Columbia River plume off Oregon is a well known example. During winter the river plume is a low temperature, low salinity feature against the coastline to the north. During summer, however, the plume extends over an area of about 105 km^2 and is deflected southward by prevailing summer winds and coastal currents (Barnes et al. 1972; Conomos et al. 1972; Bowman 1988; Landry et al. 1989). Another example is the Mississippi-Atchafalaya River in the Texas-Louisiana continental shelf. Essentially, there is downcoast (westward to southward) nearshore flow, except during the summer months. In July and August the average wind has an upcoast component and the nearshore flow is reversed (Cochrane and Kelly 1986; Li et al. 1997). During these months the influence of the Mississippi-Atchafalaya River freshwater discharge is observed in the salinity field to the east. In the case of the Río de la Plata, the shallowness of the estuary and the adjacent shelf might increase the role of the wind in the force balance.

Over the past six years, several cruises have been conducted in the lower estuary and the adjacent shelf as part of an ongoing inter-disciplinary project to study the coastal ecosystem. During these cruises, continuous profiles of conductivity and temperature (CTD) were obtained, together with biological and chemical samples. Comparison of the resultant salinity distribution with the wind and river flow conditions suggests the importance of these two forces for seasonal distributions. The seasonal cycle of the forcing, and the response of the salinity and temperature distributions to it, was investigated by Guerrero et al. (1997). Here, results from cruises conducted during 1994, 1995 and 1996, are added to the original database. Following the seasonal scheme proposed by Guerrero et al. (1997), new surface and bottom salinity and temperature fields are obtained. The new information improves the definition of salinity field features especially in the outer estuary. The analysis of the horizontal salinity fields (and temperature fields in the next section) obtained with this expanded database, a review of the results obtained by Guerrero et al. (1997) for the vertical salinity sections, and the analysis of the vertical stratification are presented in the following sections.

8.4.2
Seasonal Distribution

The effects of two forces, wind and river discharge, was considered when studying the seasonal variability of the salinity field. Guerrero et al. (1997) considered river and longitudinal wind forcing; here the study is expanded to include the effect of alongshore winds in the analysis. The coast in the Río de la Plata region has approximately a general orientation of 35° from the north. The wind field was divided into onshore (southwestward to northward winds) and offshore (northeastward to southward) relative to a perpendicular to the river axis. In the alongshore direction, the winds were divided into downwelling-favourable (northwestward to eastward) and upwelling-favourable (southeastward to westward). Monthly frequencies and mean speeds for each group were computed based on 10 year monthly statistics of mean wind speed and direction (expressed as frequency of occurrence from an 8-point wind rose) at weather station Pontón Prácticos Recalada (SMN 1982). The monthly distribution of the ratio frequency of onshore to frequency of offshore winds, and the ratio frequency of upwelling-favourable to frequency of downwelling-favourable winds are shown (Fig. 8.6a), and the average speed for each group is presented in Fig. 8.6b. Monthly mean transports of the Río de la Plata (Fig. 8.6c) were estimated from discharge data of the two major tributaries, Paraná and Uruguay Rivers. The period of September 1982 to August 1983 was excluded in the seasonal analysis of the river discharge cycle, because it presents highly anomalous values. Data from cruises during this period were also set aside (for a brief discussion see Guerrero et al. (1997)). Based on the monthly distribution of the wind and river discharge, two seasons were defined to study the salinity distribution. The period from April to August, fall/winter, characterized by large river flow, minimum frequency of onshore winds, maximum frequency of downwelling-favourable winds and the highest average speed of offshore and downwelling-favourable winds; and October to February, spring/summer, with minimum discharge and predominance of onshore and upwelling-favourable winds. March and September are considered periods of transition.

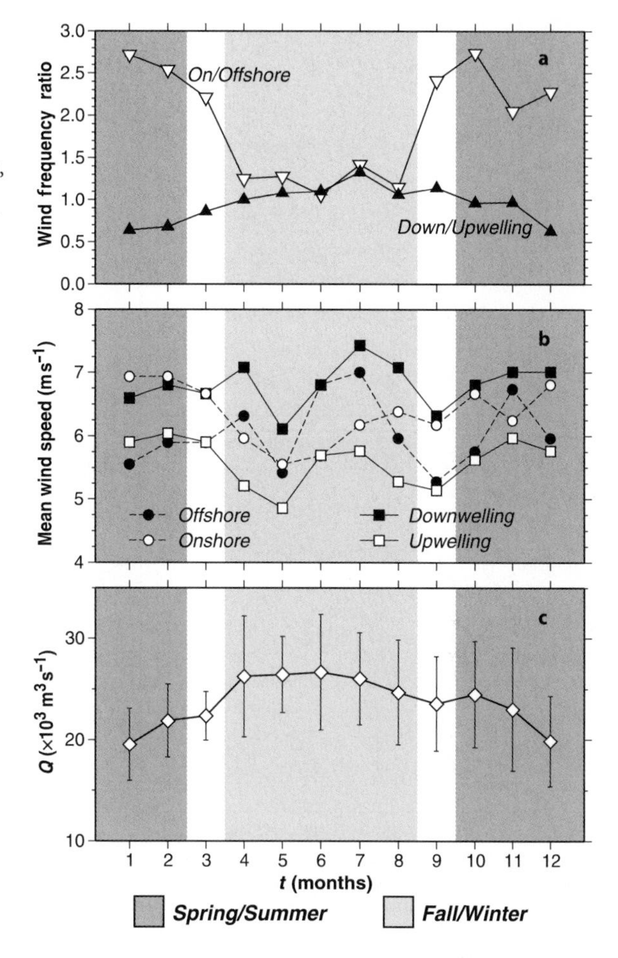

Fig. 8.6. Seasonal wind and river discharge forcing cycle. **a** Onshore to offshore, and downwelling to upwelling-favourable wind frequency ratio, **b** mean speed: onshore (m, *dashed*), offshore (l, *dashed*), upwelling-favourable (o, *solid line*) and downwelling-favourable (n, *solid line*) wind, **c** mean river discharge (error bars are ± 1 S.D.)

The original database used by Guerrero et al. (1997) includes a total of 1 600 hydrographic stations from 44 cruises conducted in the lower estuary and the shelf over the last 30 years. A large portion of the database (34%) corresponds to CTD stations acquired from 1991 to May 1994. A detailed explanation of the origin of the information, units, and the quality control procedures, can be found in Guerrero et al. (1997). The analysis presented here adds information from seven cruises performed from June 1994 to July 1996. Of the 442 CTD stations added, 80% correspond to spring/summer and 20% to fall/winter. As the present study is focused on the river plume and the inner shelf, data from stations deeper than 100 m were not included in the analysis. The reason is to avoid the influence of the highly variable (spatial and temporal) regime characteristic of the outer shelf and the shelfbreak.

The salinity fields were gridded at 12-min separations in latitude and longitude. Contours were produced using the objective analysis method of the Generic Mapping Tools (GMT) software package (Wessel and Smith 1991).

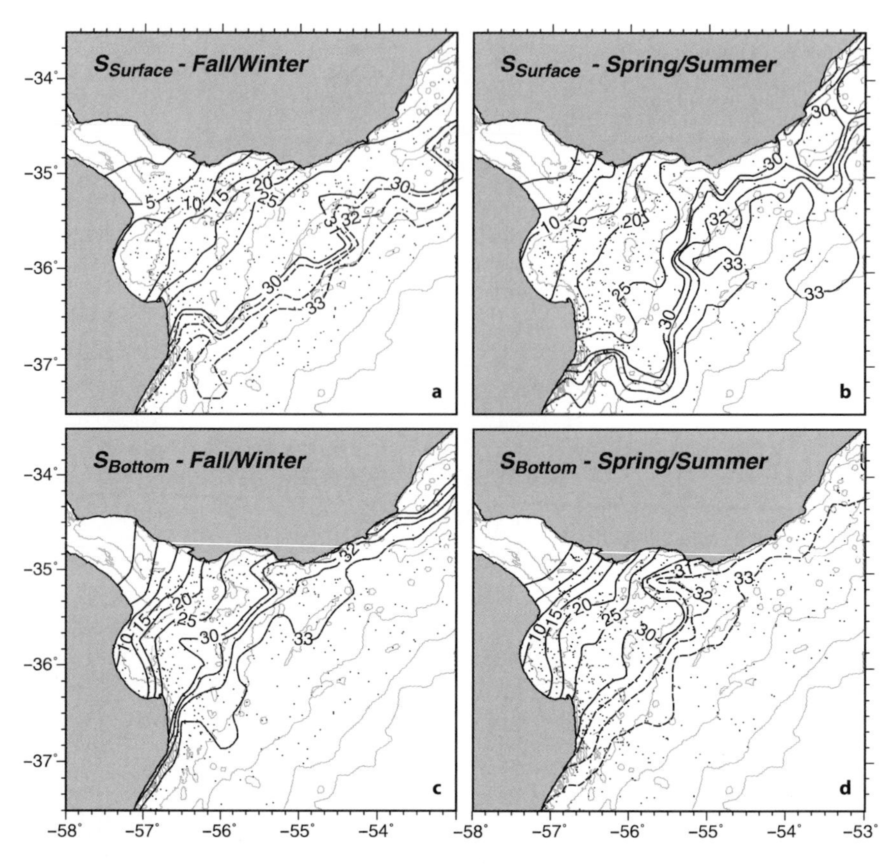

Fig. 8.7. Seasonal salinity fields. Contour interval is 5 su (*solid line*) for salinity less than 30 su and 1 su (*dashed*) for salinity greater than 30 su. *Dots* indicate location of hydrographic stations. **a** and **c** Surface and bottom distribution for fall/winter (April–August); **b** and **d** Surface and bottom distribution for spring/summer (October–February)

The surface salinity distribution for both seasons (Figs 8.7a,b) shows largest gradients between the isohalines of 10 and 25 salinity units (su; salinity is reported as salinity units, because historical data were not converted to Practical Salinity Scale). The horizontal gradient decreases offshore of the 25 su isohaline. The location of this isohaline is used to analyse the extent of the freshwater influence. The fall/winter surface distribution shows a northeastward flow along the Uruguayan coast, with salinity between 25 and 30 su. The Samborombón Bay has minimum horizontal variation, with salinity values between 10 and 15 su. In spring/summer the 25 su isohaline contour is located at an inner position at the northern coast, and offshore in the central and southern part of the estuary, compared with the fall/winter distribution. Higher salinity values occur on the shelf north of the Río de la Plata during spring/summer, and the coast near Punta del Este has salinity close to that of the shelf water. Lower salinity values occur south of Cabo San Antonio along the Argentinean coast during spring/summer; the Samborombón Bay during this period is homogeneous.

The bottom salinity has a smaller seasonal variation than the surface. The general pattern is similar in both seasons, with isohalines mainly following the bottom topography. Fresher water is associated to shoals and more saline water is found in the deeper channels. Similar effects of bottom topography on the transverse salinity distribution has been studied in Delaware and Chesapeake Bay (Wong 1994; Valle-Levinson and Lwiza 1995; Valle-Levinson and O'Donnel 1997). The importance of the shallow banks is suggested by the locus of the 25 su isohaline, especially during fall/winter; the influence of the deepest Oriental Channel along the Uruguayan coast is clearly seen in the position of the 30–32 su isohalines. Another difference between the two seasons is that during fall/winter a stronger gradient along the bottom occurs at the northern end of Samborombón Bay, where the Maritime Channel borders the central shallow area of the bay. In spring/summer, the 25 and 30 su isohalines are located at greater depths in the bay and the channel. This difference could be associated with an increase in the freshwater flow through the southern part of the estuary during spring/summer compared to fall/winter.

Estuarine-shelf interaction processes are the responses to tidal, buoyancy and atmospheric forces, and the relative significance of these forces varies in time and space (Beardsley and Boicourt 1981; Wiseman 1986; Simpson 1997). Local and remote forcing mechanisms are important, and the response of the estuary and the buoyant plume occurs over a wide range of time-scales. Although the lack of long time-series of currents and physical properties in the Río de la Plata limits a complete analysis of the exchange processes, the characteristics of the salinity fields (Fig. 8.7) allows a qualitative explanation of the seasonal response to winds and continental runoff.

In fall/winter, the lower salinity values observed on the shelf northeast of the estuary can be explained as the result of an anticyclonic turn of the buoyancy flow after leaving the estuary. The position of the 25 su isohaline parallel to the northern coast suggests the existence of a coastal current. During this season, the river discharge is maximum, the offshore wind frequency almost equals the occurrence of onshore winds, downwelling favourable winds have their maximum mean speed and are more frequent than upwelling-favourable winds. Under this condition, buoyancy dominates and a river-forced plume (Garvine 1987; Chao and Boicourt 1986; Chao 1988a) flowing to the left of the estuary is observed. Modelling and observational studies show that downwelling-favourable winds narrow the plume and reinforce the coastal jet, while upwelling-favourable winds spread the buoyancy flow offshore and tend to arrest the alongshore current (Chao 1988b; Münchow and Garvine 1993; Kourafalou et al. 1996; Fong et al. 1997). These effects of the alongshore winds may explain some of the characteristics observed, such as the better definition of the coastal current in the fall/winter. During spring/summer, river discharge is minimum, and onshore and upwelling-favourable winds are predominant. No coastal current is observed (Fig. 8.7b), and the offshore position of 25 isohaline in the central and southern part of the estuary might indicate increased offshore flow at the surface, both characteristics that can be explained by the action of upwelling-favourable wind. In the longitudinal direction, onshore (offshore) winds has been shown to reduce (increase) the estuarine circulation (Hansen and Rattray 1965; van der Kreeke and Robaczewska 1989; Geyer 1997). Most of the studies assume lateral homogeneity, in rectangular channels or small estuaries. The effect of longitudinal winds on the cross-section characteristics of circulation and properties was analysed in modelling studies, for example, by Chao (1988b) and Wong (1994). With the information available, the effect of the longitudinal winds on the circulation can not be properly evaluated, but, due to the large cross-section of the Río de la Plata, transverse circulation generated by longitudinal winds might be expected. Onshore winds might produce southward advection of water in the estuary, as suggested by Guerrero et al. (1997); this mechanism, superposed onto the effect of upwelling-favourable winds, can explain the presence of the fresh water extension along the Argentinean coast during spring/summer.

From the analysis of the salinity distribution, a seasonal pattern exists when the seasons are based on the forcing cycles. Although a dynamical analysis of the circulation and forcing is needed to fully understand the complex processes of the estuarine-shelf interaction, the qualitative force balance described above gives a first order explanation of the seasonal variability observed.

8.4.3
Vertical Distribution and Stratification

Data from three cruises were used by Guerrero et al. (1997) to characterize the vertical structure. Three transects with a common origin (35°00'S, 57°00'W) from each

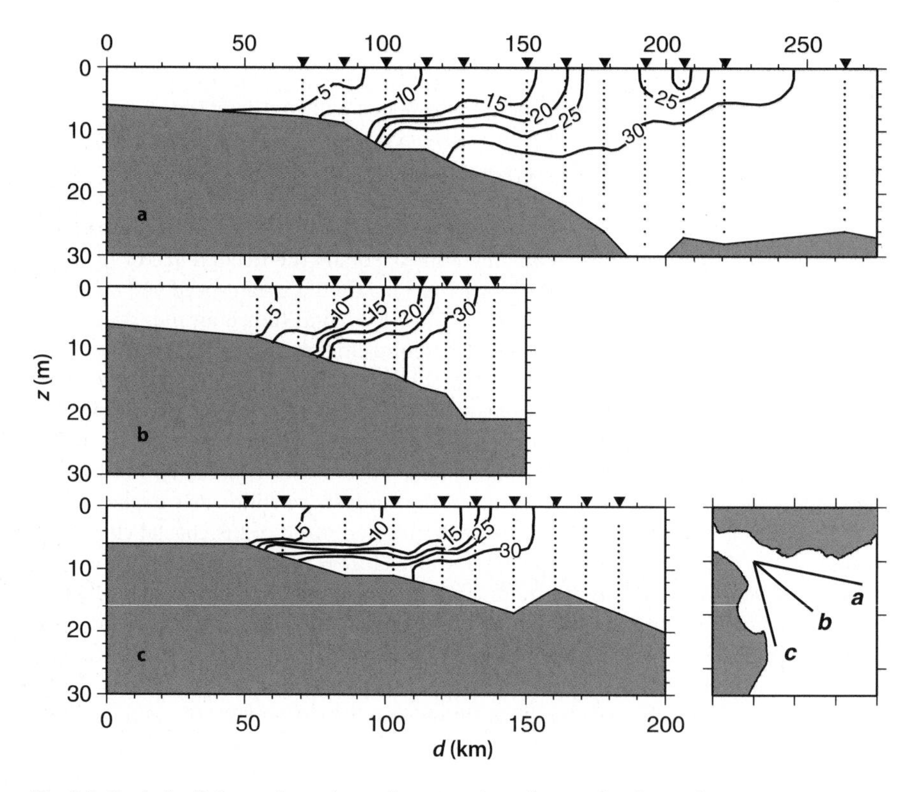

Fig. 8.8. Vertical salinity section – August '92. **a** Northern, **b** central and **c** southern transects respectively. ▼ indicate station location; *dots* show CTD observations (redrawn from Guerrero et al. 1997)

cruise were selected; the analysis of these transects is reviewed here. The northern transect follows the Oriental Channel along the Uruguayan coast; the central approximately corresponds to the river axis and the southern transect describes the vertical structure along the Argentine coast. Data from cruises during August 1992 (fall/winter) and November 1995 (spring/summer) are presented in Figs 8 and 9, respectively. The condition during these cruises was of calm to moderate wind. The salinity sections for a cruise under moderate to strong wind condition during April 1993 (fall/winter) are shown in Fig. 8.10.

During the August 1992 cruise (Fig. 8.8) the wind was offshore with an average speed of 2.7 m s^{-1}. Under this light wind condition, the vertical stratification was strong. A maximum vertical gradient of salinity of 23.87 m^{-1} (salinity from CTD data was computed using the Practical Salinity Scale) was measured between 6 and 7 m depth (total depth 8 m) in the northern area of Samborombón Bay at the limit with the Maritime Channel. The depth of the halocline remains almost constant, and a salt wedge exists. The extension offshore of the fresh water is greatest at the northern section; the 30 isohaline reaches the surface at 240 km from the origin. The same isohaline reaches the surface at 130 and 150 km in the central and southern sections, respectively. The sections corresponding to the November 1995 cruise (Fig. 8.9) were acquired

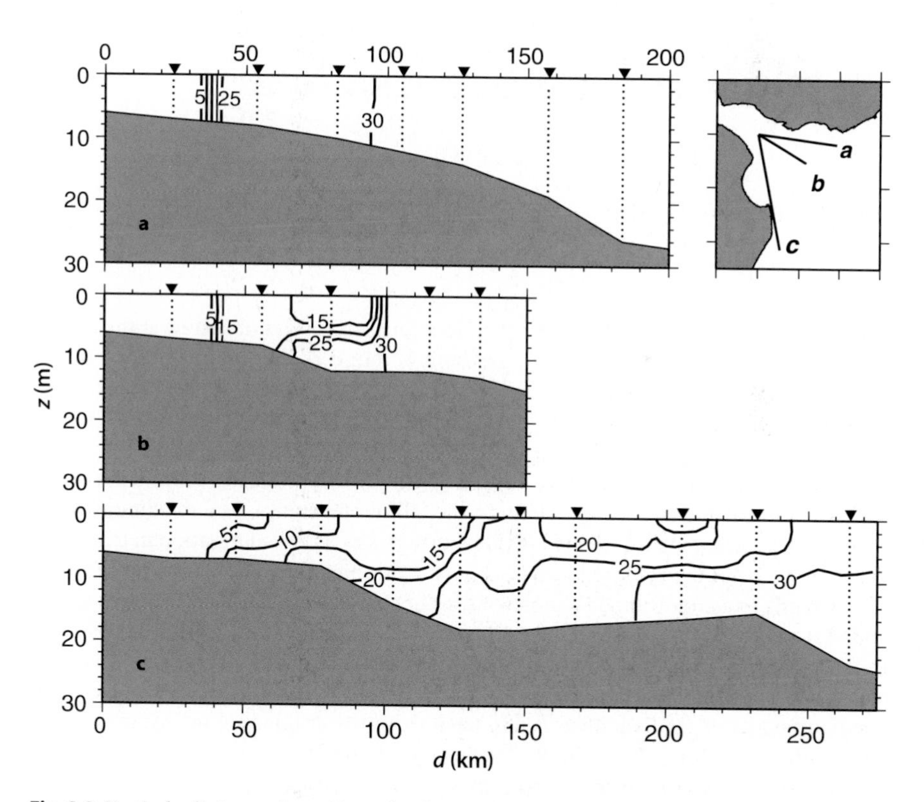

Fig. 8.9. Vertical salinity section – November '95. **a** Northern, **b** central and **c** southern transects respectively. ▼ indicate station location; dots show CTD observations (redrawn from Guerrero et al. 1997)

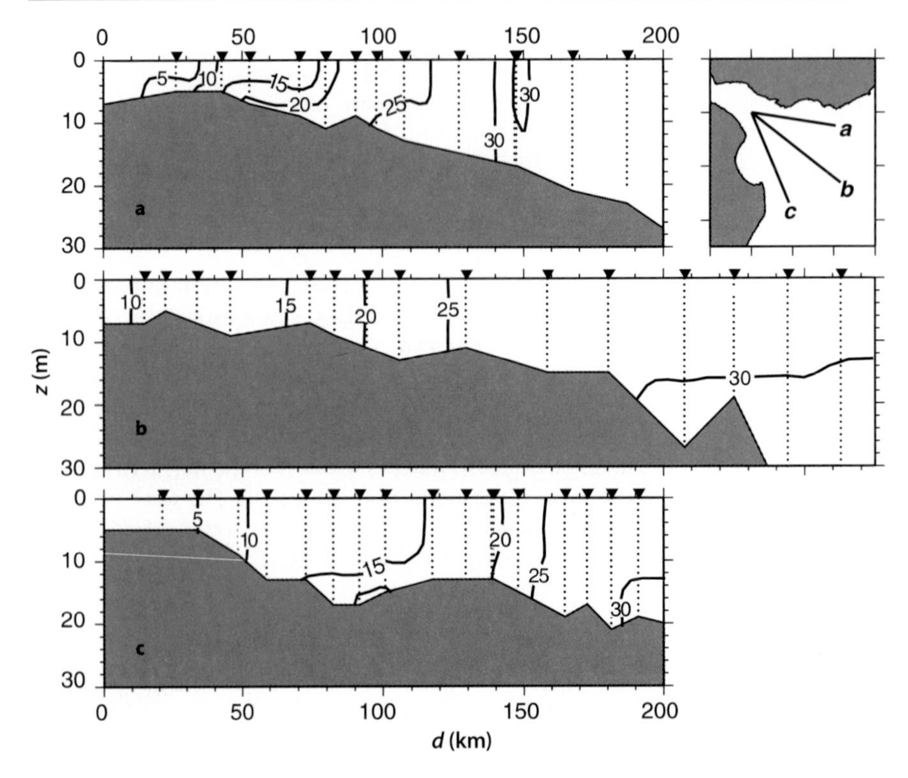

Fig. 8.10. Vertical salinity section – April '93. **a** Northern, **b** central and **c** southern transects respectively. ▼ indicate station location; dots show CTD observations. (redrawn from Guerrero et al. 1997)

under average onshore winds of 7.4 m s^{-1}. No stratification was observed at the northern section; and a strong horizontal gradient is found at the upstream part. In the central section, waters are stratified between two zones of strong horizontal gradient at the surface located at 40 and 100 km from the origin. The maximum vertical gradient is 8.3 m^{-1}. The southern transect exhibits the southward extension of the river flow; the 30 isohaline reaches the surface beyond 275 km from the origin, south from Punta Médanos. Stratification is weaker than in the fall/winter cruise. The maximum vertical salinity gradient measured during this cruise was 14.47 m^{-1} at a station in 6 m water depth (not part of the transects), southwest from Montevideo. The cruise in April 1993 corresponds to a condition of strong winds. The winds were onshore, from southeast and east-southeast, with speeds of 10–14 m s^{-1} for 60 h. No stratification is observed in the estuary, except for the inner part of the northern section (Fig. 8.10a), and an isolated deep layer close to the bottom in the center of the southern transect. The maximum vertical gradient observed in the northern section is 2.6 m^{-1}. At the central and southern sections, the halocline has been destroyed.

Some examples of the vertical profiles acquired during the November 1995 cruise are shown (Fig. 8.11) to display the spatial variability across the estuary, with some areas strongly stratified and others completely mixed. Due to the highly variable range

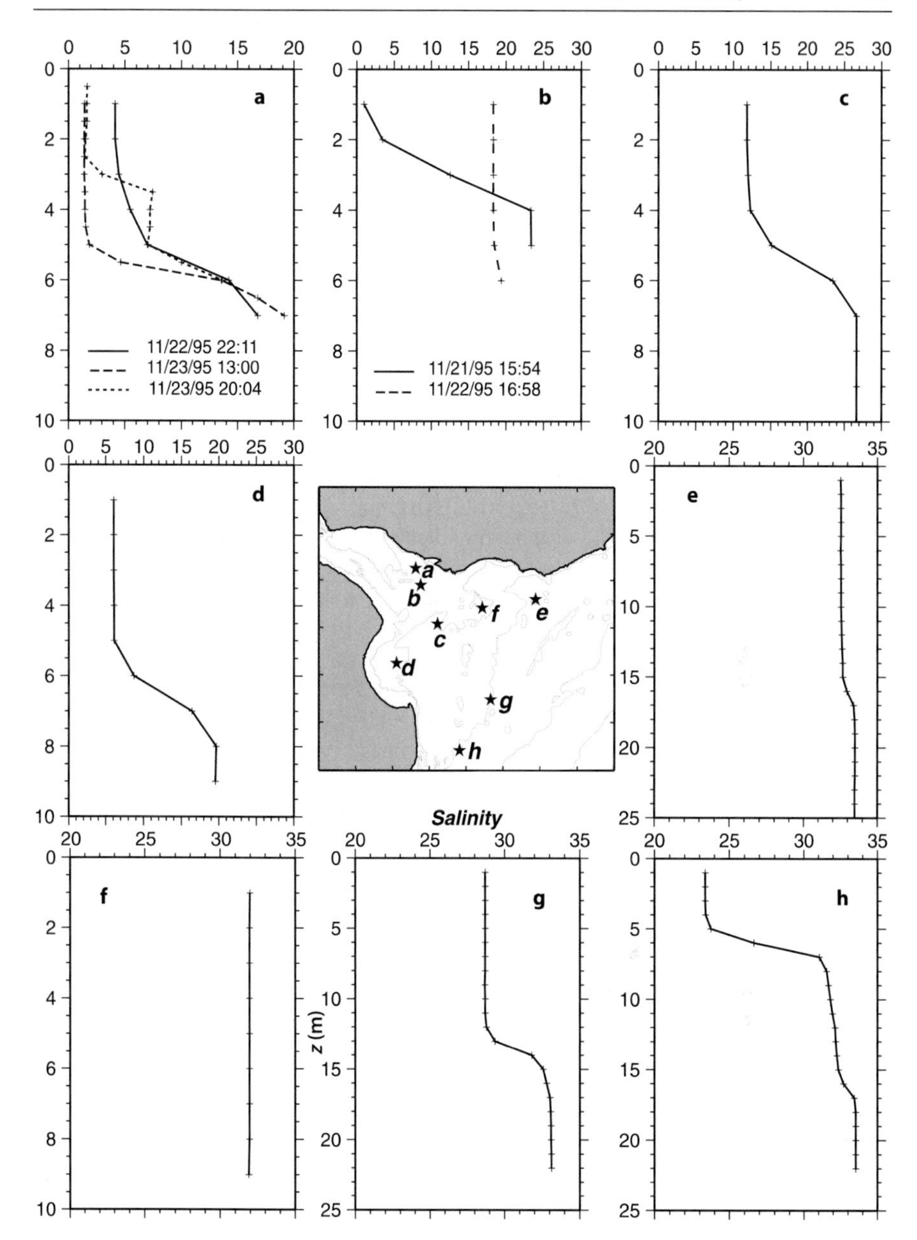

Fig. 8.11. Vertical salinity profiles from selected CTD stations of the November '95 cruise

of salinity, it was not possible to use one scale for all profiles; hence offshore stations have an expanded salinity scale. Repeated profiles at two locations are presented as an example of local variability (Fig. 11a,b); they were taken under different wind conditions in the inner part of the salt intrusion. In both stations, intensification of the halocline occurs when wind decreases. The three profiles of the innermost station (Fig. 8.11a) were acquired under similar tidal condition, during flood tide, according to the predicted values at Montevideo. The first profile was obtained with west-northwestward wind of 6.2 m s^{-1}, the second with calm 3 m s^{-1} north-northwestward wind, and the last with westward wind of 6.7 m s^{-1}. The maximum vertical gradients of the three profiles were: 7.3, 8.9 and 4.4 m^{-1}, respectively. Another example, from a station located southeast from the previous analysed (Fig. 8.11b) shows similar results. These two profiles were taken during ebb tide. The first profile was acquired during calm condition and has a difference of salinity of 22.4 units from surface to bottom. The next day at the same location and similar tidal condition, but with north-northwest-ward winds of 8 m s^{-1}, the water column was almost homogeneous with a small vertical gradient (1.02 m^{-1}) near the bottom. At the center of the estuary (Fig. 8.11c) and at the Samborombón Bay (Fig. 8.11d) there is a typical two-layer stratification, with a salinity difference of 15 between surface and bottom layer. Over the shallow banks, the water column can be homogeneous even with light wind conditions (Fig. 8.11f). At the eastward part of the Oriental Channel (Fig. 8.11e) an almost homogeneous column occurs, with a very weak stratification (0.44 m^{-1}) at 16–17 m. Two stations on the inner shelf in the southern part (Figs 8.11g,h) still have strong stratification with maximum gradients of 2.46 and 4.43 m^{-1}. The upper layer at the southernmost station (Fig. 8.11h) is fresher than in the station located to the north (Fig. 8.11g), showing good agreement with the southern extension of the river plume observed in the mean seasonal distribution (Fig. 8.10b). The upper layer of fresher water (salinity less than 25) at the southernmost station (Fig. 8.11h) is 7 m deep, occupying 30% of the water column. The thickness of the upper layer shows the importance of this buoyant plume.

The previous analysis gives a qualitative description of the vertical structure in the estuary, and the effect of the wind on stratification. Analysis of vertical profiles together with wind conditions, as well as field-work experience, suggest that atmospheric forcing plays a major role in the mixing process in the estuary. The authors are aware that, to fully understand local, short-term variations in the vertical structure, detailed information of tidal currents should be included in the analysis. But, the combined effects of tidal and storm-driven currents must be considered, as was discussed in section II, so these two forces are not independent in terms of mixing processes.

8.5 Temperature

A description of the temperature field in the upper estuary have been presented in CARP (1989). In the present study, as was done for salinity, the focus is on the outer estuary. Guerrero et al. (1997) presented a seasonal analysis of temperature distribution in the lower estuary based on field observations. Part of these results are reviewed here, along with the analysis of the horizontal fields obtained using the expanded database. A description of surface temperature fields obtained from remote sensing data is also presented. Remote sensing information has been especially useful for detecting features that are difficult to observe using in situ techniques because of their limited duration and spatial coverage and resolution.

8.5.1 Seasonal Distribution

Following Guerrero et al. (1997), two seasons, summer and winter, were defined based on the time series of surface and bottom temperature observations from the estuary (Fig. 8.12). Four months centred at the maximum and minimum values defined each season. December to March correspond to the warm period, while June to September characterize the cold season. The seasonal cycle observed in temperature differs from the one in salinity due to their different forcing, e.g. winds and river discharge for salinity and air-sea heat flux for temperature. Data from three cruises were added to the original database used by Guerrero et al. (1997). Of a total of 155 stations added, 57% of them correspond to winter and 43% to summer. Contour maps of surface temperature, 30 m depth and bottom temperature distribution are presented (Fig. 8.13). The depth of 30 m corresponds approximately to the base of the mixed layer at the shelf (SHN 1973), and was selected to show the relation of the estuarine and the shelf temperature field, without the effect of the bottom slope. The temperature fields were gridded at 15-min separations in latitude and longitude, and the objective analysis method of the Generic Mapping Tools (GMT) software package was used to generate the contours.

The surface temperature during summer is homogeneous in the lower estuary and the shelf north of the river mouth with values of 21–22° C. To the south, the boundary between river and shelf water seems to follow the 20° C, where a larger horizontal gradient is observed (Fig. 8.13b and Plate 8.1b). The 30 m distribution presents a pattern similar to that at the surface: no strong difference in temperature occurs between the estuary and the water of the upper layer on the inner shelf. The bottom distribution during summer has a strong gradient associated with the intersection of the isotherms with the sloping bottom. This maximum gradient zone is located approximately at a depth of 50 m. No advection of cool shelf water into the estuary is observed in the mean temperature field. A weak temperature stratification of less than 1° C occurs in the estuary, except in Samborombón Bay where the water column is homogeneous.

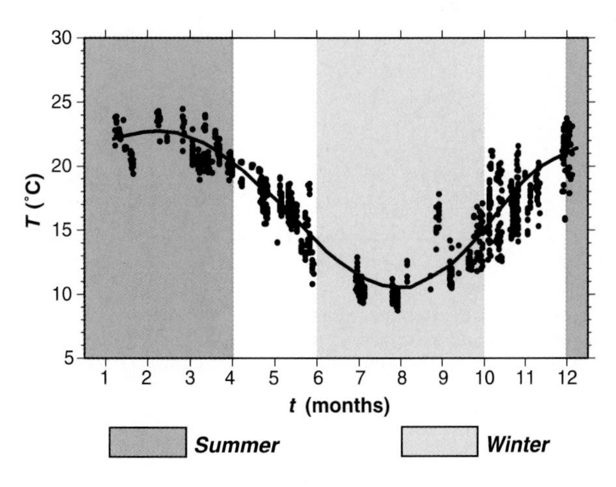

Fig. 8.12. Seasonal temperature cycle (redrawn from Guerrero et al. 1997)

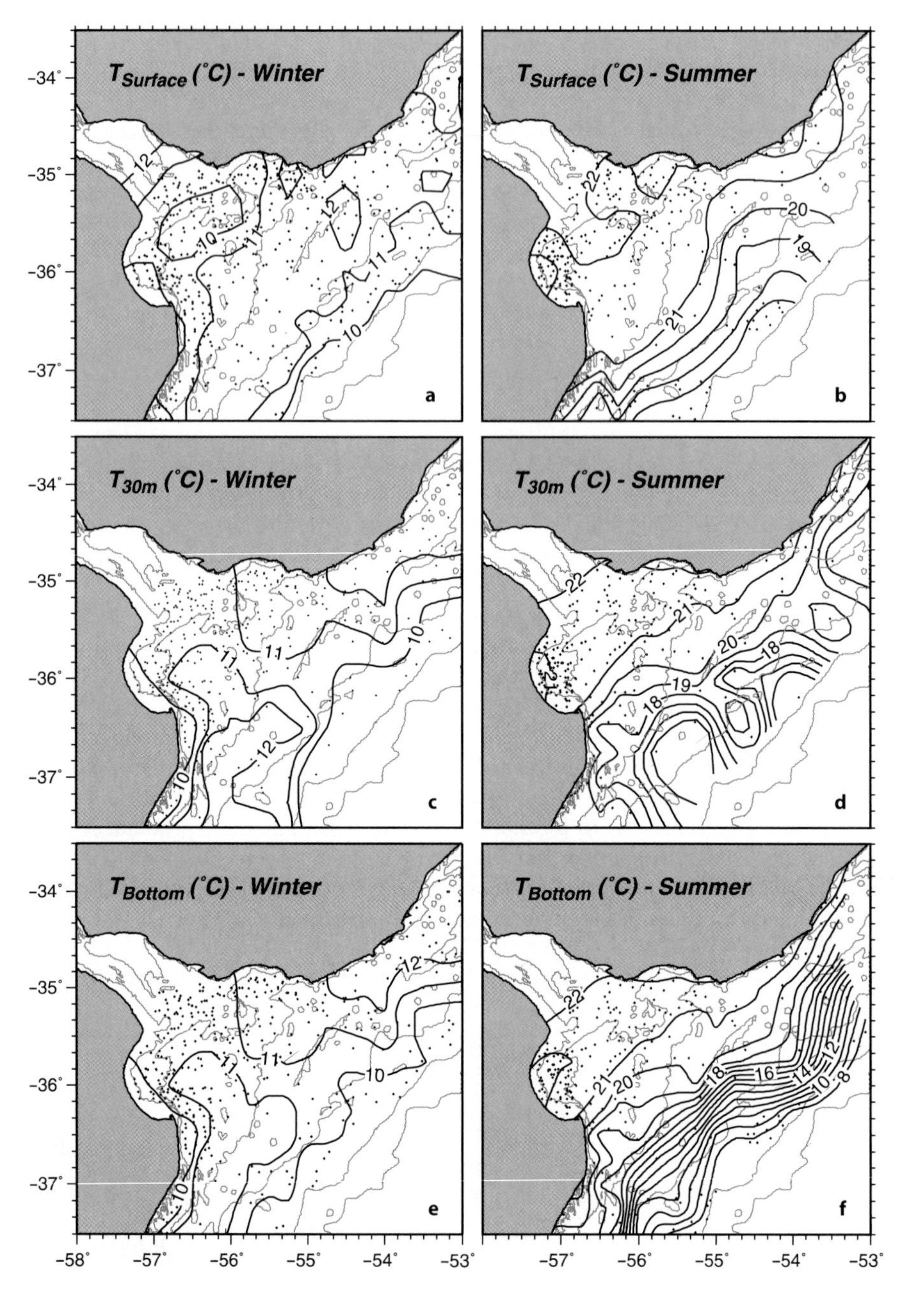

Fig. 8.13. Seasonal temperature fields. Contour interval is 1° C. *Dots* indicate location of hydrographic stations. **a**, **c** and **e** Surface, 30 m and bottom distribution for winter (June–September); **b**, **d** and **f** Surface, 30 m and bottom distribution for summer (December–March)

During winter, the temperature field in the estuary and shelf is quite homogeneous, with values of 10–12° C. Cooler surface water is located in the center of the outer estuary with temperature of 9–10° C, 1° C cooler than the surroundings. This surface cooling results in a negative difference of 1° C between surface and bottom. The adjacent shelf is not stratified.

In spite of the strong vertical stratification generated by salinity differences, the temperature in the estuary remains almost vertically homogeneous during the warm and cool seasons. The mean horizontal fields show that there is not a strong difference in temperature between the estuarine and the inner shelf water, except during summer in the southern extension of the river plume.

8.5.2 Sea Surface Temperature

Remote sensing information, such as the images obtained by the Advanced Very High Resolution Radiometer (AVHRR) of the NOAA satellites, is especially useful for the study of the Río de la Plata Estuary. NOAA-AVHRR imagery gives simultaneous coverage of the whole region, something that is not possible with in situ information obtained by shipboard sampling due to the large areal extent of the river. Sea surface temperature (*SST*) is computed using a multichannel algorithm for the infrared channels. The images used in this study are individual, daily *SST* images with maximum horizontal resolution (1.1 km).

Two images are presented as examples of the cool and warm season (Plate 8.1). Homogeneity in the estuary and the inner shelf, with cooler water near the coast and in the Samborombón Bay occurs during winter (Plate 8.1a). During summer, there is a strong gradient between the river plume and the shelf water in the southern part (Plate 8.1b). This frontal zone follows approximately the geometry of the 50 m isobath, and its occurrence was first reported by Framiñan and Brown (1993) based on a 4 year span of satellite images. The frontal zone begins to appear in the *SST* images in early September, and it is observed until late March. The temperature gradient at the surface obtained from the images associated with this transition zone has values up to 0.25° C km^{-1}. The front defines two zones on the continental shelf; the limit is associated with the 20° C isotherm, as suggested in the mean temperature distribution derived from in situ data. Another feature that can be observed in the *SST* images (Plates 1b and 2b) is a temperature minimum associated with a cool extension of shelf water along the Argentinean coast entering the Samborombón Bay at Cape San Antonio. This northward flow can be observed in all seasons, but is more noticeable during summer due to the greater temperature gradient. In summer, the flow has a temperature difference with the surrounding of 3–4° C, and it extends 20–25 km offshore (Framiñan and Brown 1993). There are manifestations of this coastal water intrusion in in situ data, as well as remote sensing information; Lasta et al. (1996) discussed its importance in the estuarine ecosystem.

A temperature minimum on the northern coast of the outer estuary (Plate 8.2) is observed during summer months in *SST* images, especially from mid December to early February (Framiñan and Brown 1998). The difference of temperature between the center and the surrounding water is of approximately 4–5° C. The minimum starts developing very close to the coast in the proximity of Punta del Este. It can reach large horizontal extent on the inner-shelf offshore the estuary as in December 1989 (Plate 8.2a). This temperature minimum has the surface characteristics of coastal upwelling. However, the in situ observations available are not appropriate to describe the water column during these events, so the characterization of this phenomenon is incomplete. Between Punta del Este and Cabo Polonio the coast has approximately an orientation of 60°; southward to northwestward winds are upwelling favourable. The wind field statistics from the weather station Pontón Prácticos Recalada, located in the estuary (Fig. 8.1), for the period 1981–1990 (SMN 1992) shows that in January and December such winds are predominant, with 71% occurrence, and 69% in February. During these three months, westward winds are predominant (26%, 22% and 27% of the total distribution for January, February and December, respectively) followed by the southwestward winds. Wind time series recorded at several weather stations in the area during December 1989 to January 1990 are presented (Plate 8.2). Weather stations Punta Brava (Montevideo), Laguna del Sauce (located east of Punta del Este) and La Paloma (C. Santa María) are located on the Uruguayan coast (Fig. 8.1). The records consist of four observations per day; direction is from an 8-point wind rose at P.P. Recalada station and a 16-point rose at the other stations. The reference system is rotated 30° counterclockwise and vectors point in the downwind direction. Upwelling favourable winds on the P. del Este-C. Polonio coast are represented on this reference system to the left of page. In weather station P. Recalada 78% of the observations are southward to northwestward winds; 85% at Punta Brava, 58% at Laguna del Sauce and 78% on the northern coast at La Paloma.

The analysis shows that during December and January the wind field is favourable for driving coastal upwelling between P. del Este and C. Polonio. But, the shallowness of the area and the homogeneity observed in the water column in the estuary and inner shelf during summer (Fig. 8.13) suggest a more complex picture. Strong onshore advection of cool water from deeper layers on the shelf is necessary to generate the temperature gradient observed at the surface in satellite images. Further studies are necessary to fully understand the phenomenon. Bottom topography and shelf circulation might play an important role in the process.

8.6
Turbidity Front

In the transition zone between the upper and lower estuary, the processes associated with the interaction of the fresh river water, the saline shelf water, and tidal-stirring generate a turbidity front. The turbidity maximum is clearly defined in the NOAA-AVHRR visible channel as a strong gradient in reflectance and a sharp change in water color (Plate 8.3a). Framiñan and Brown (1996) used these characteristics to determine the position of the front; results of their study are reviewed here.

Framiñan and Brown (1996) used a four-year span of NOAA-AVHRR daily images, from September 1986 to August 1990. Channels 1 and 2 (visible and near-infrared) were used to digitise the fronts, and channel 4 and sea surface temperature were used as complementary information for cloud detection. From a total of 2 578 images, 1 274 daytime images allowed determination of 333 positions of the front (Plate 8.3b). This information served to estimate the distribution of the *frontal density*, a probalistic measure of frontal occurrence in a given area (for a full description of the data and methodology see Framiñan and Brown (1996).

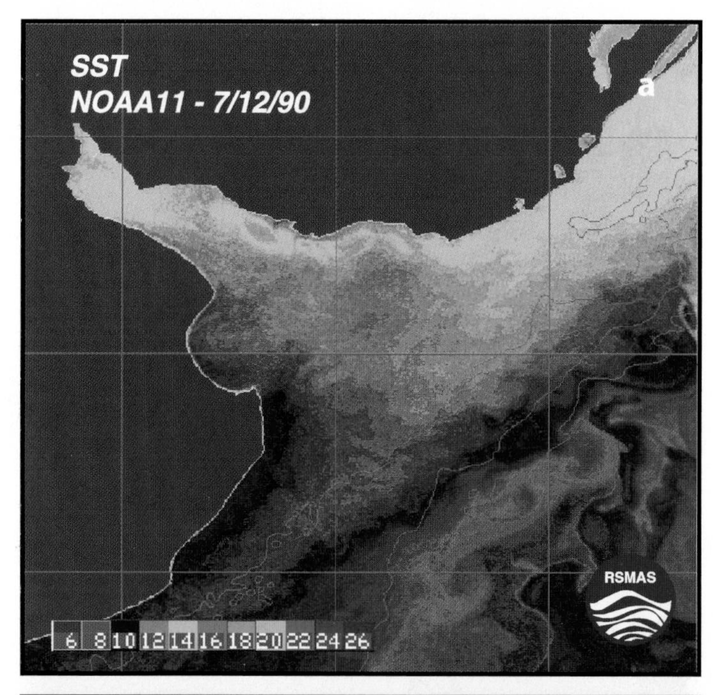

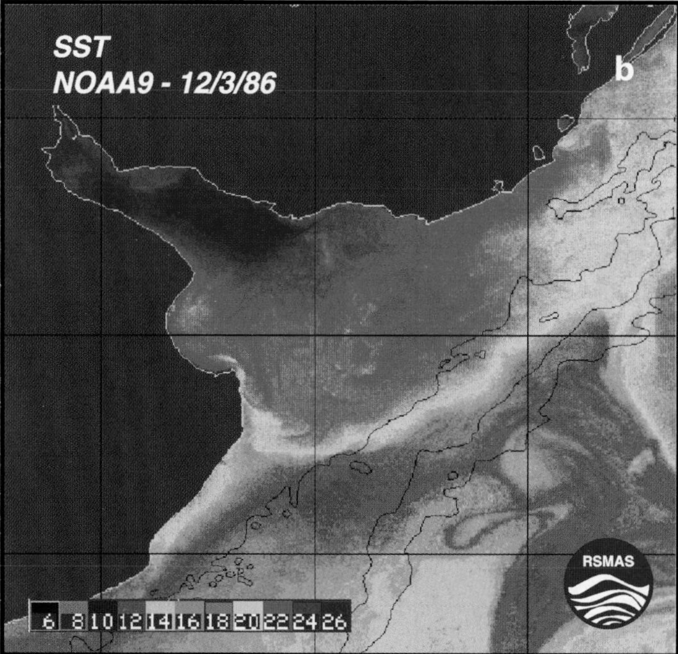

Plate 8.1. Sea surface temperature (*SST*) compute with multichannel algorithm for the infrared channels of the NOAA-AVHRR image. **a** Winter example **b** Summer example

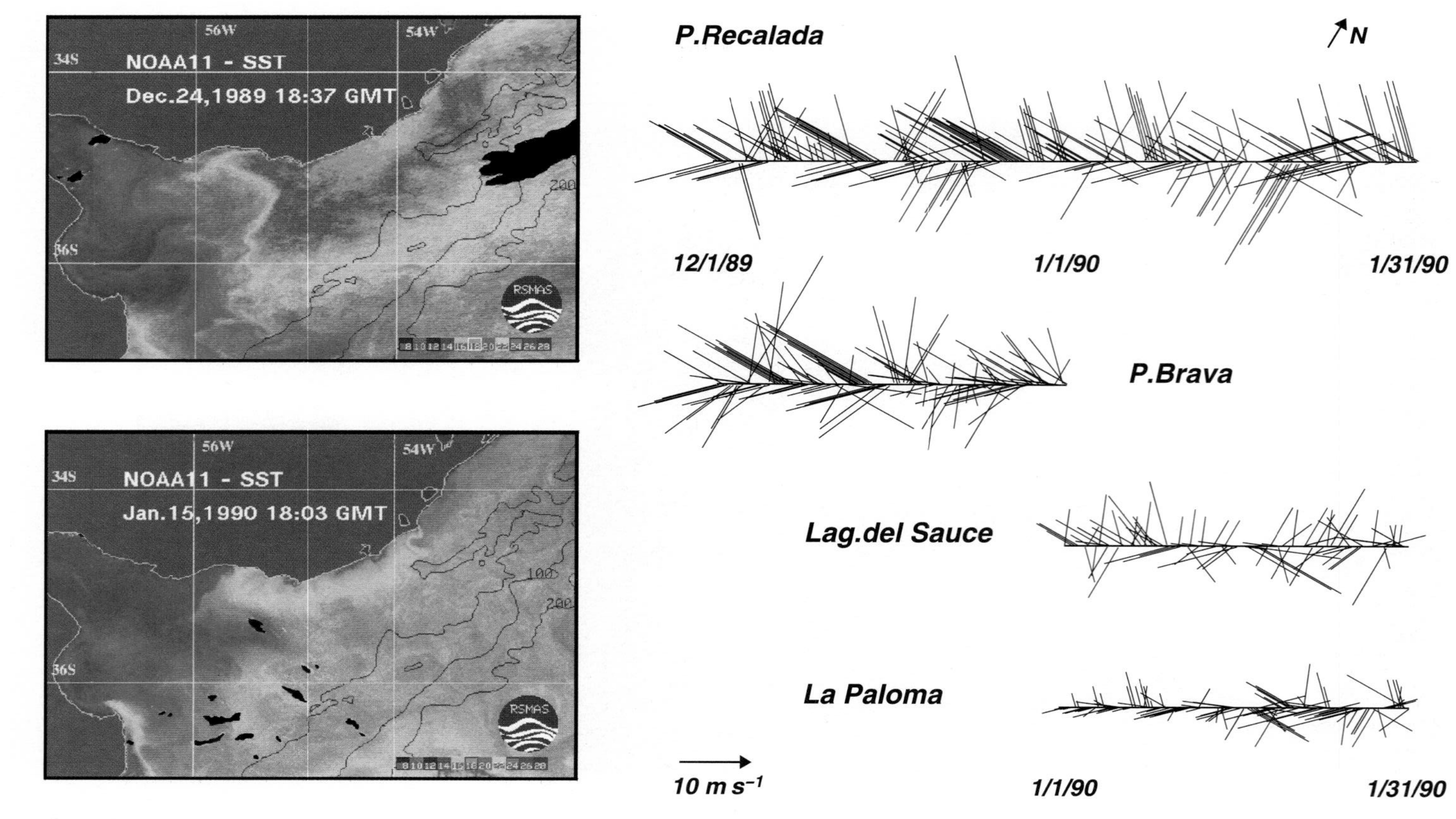

Plate 8.2. Sea surface temperature (*SST*) and wind field during upwelling events

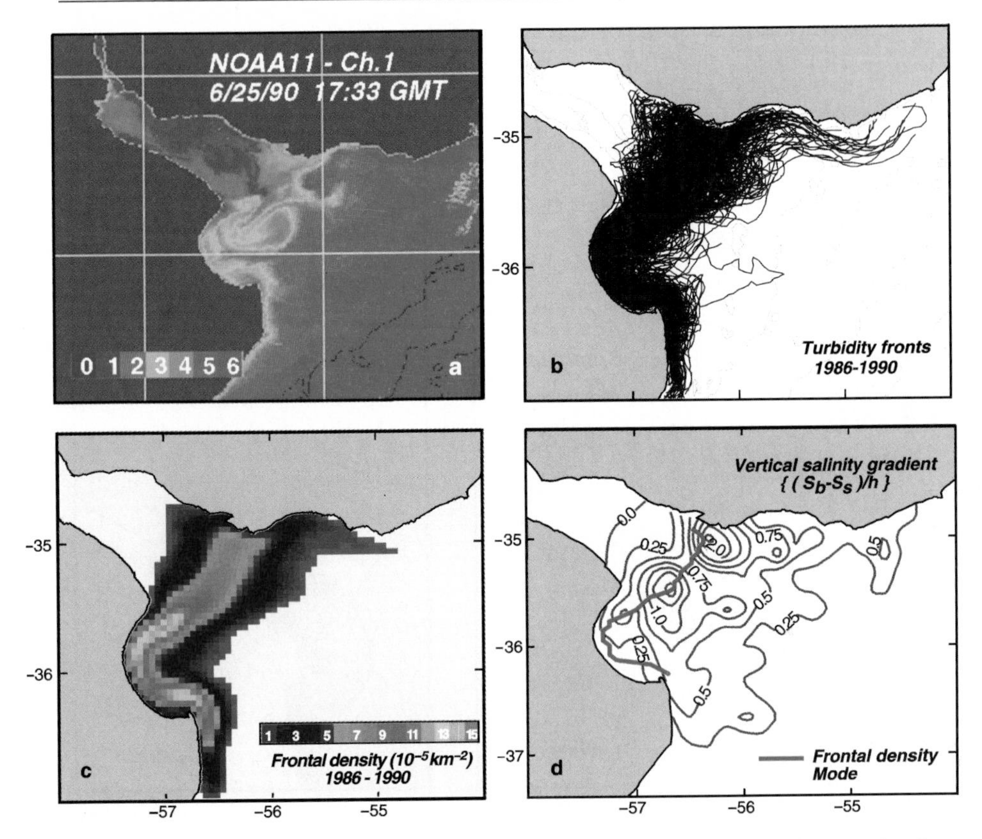

Plate 8.3. Turbidity front study. **a** NOAA-AVHRR Channel 1 image for June 25, 1990; **b** composite of the turbidity fronts obtained from the satellite images; **c** mean distribution of frontal density; **d** vertical salinity gradient distribution estimated from 29 yr of in situ observation. The contour interval is 0.25 m^{-1}. Superposed (*heavy red line*) is the mode of the frontal density distribution obtained from the satellite imagery (*b*, *c* and *d* redrawn from Framiñan and Brown 1996)

The structure of the turbidity front is temporally variable, taking complex shapes. However, the image time series permits the identification of some typical patterns: linear, convex, V-shape, Z-shape, wavy, eddy, hammer-head, and finger-like extensions. Eddies are rare features, but can be observed in several scenes; they are anticyclonic and can have exceptional development as in 25 June 1990 (Plate 8.3a).

8.6.1 Mean Distribution

The mean distribution of the turbidity front (Plate 8.3c) has a high degree of variability at the northern coast of the estuary. In this region, the frontal position varies between 57°00' and 54° 12'W, a distance of approximately 200 km. At the southern coast the modal position of the front (red line, Plate 8.3d) coincides with the 5 m isobath, although great variations in this position occur during years of large river discharge.

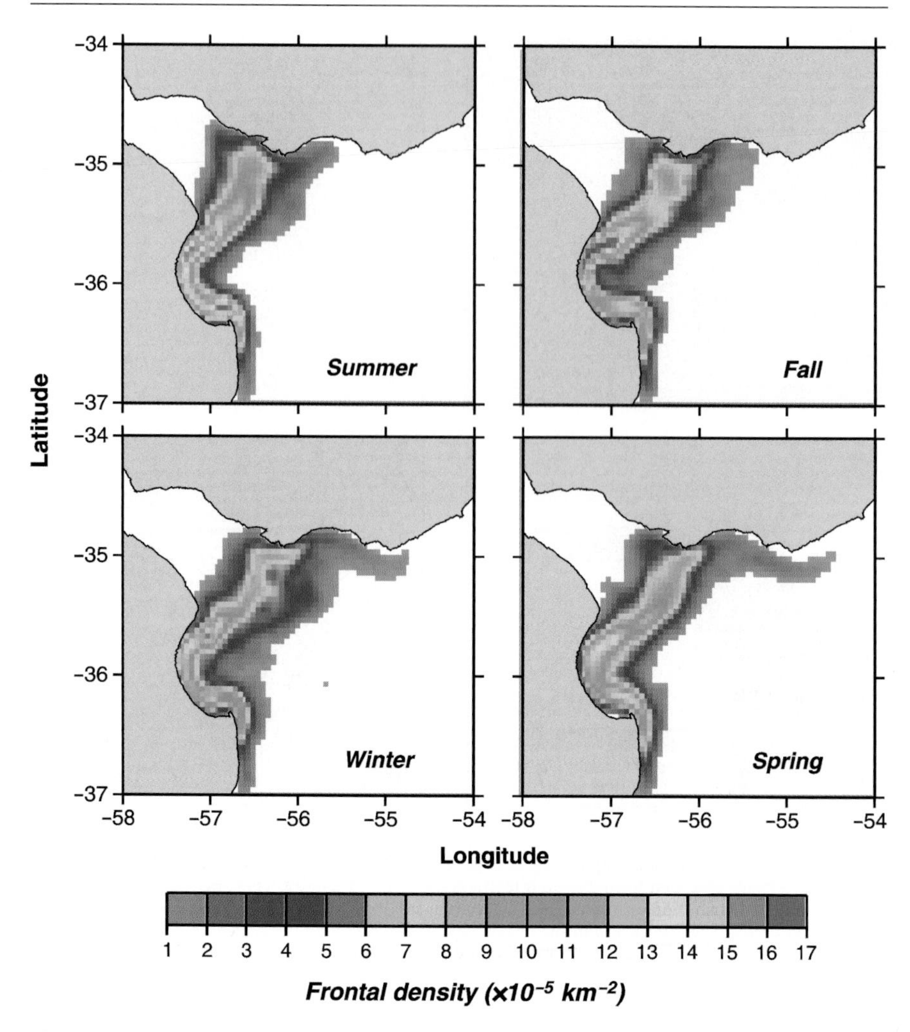

Plate 8.4. Seasonal distribution of turbidity front (redrawn from Framiñan and Brown 1996)

The distribution has three areas with maximum values of frontal density. Values of 14×10^{-5} km^{-2} are found in the southern and northern area of the Samborombón Bay, centred in 36°13'S, 56°50'W and 35°40'S, 57°00'W, respectively, close to the Argentine coast. This maximum decreases across the estuary toward the Uruguayan coast, where the maximum density value is less than 7×10^{-5} km^{-2}, south of Montevideo (35°00'S, 56°20'W). These are the areas where it is most probable to find the turbidity front located in the outer region of the Río de la Plata. The mean distribution is strongly related to the bottom topography. At the Uruguayan coast, the greater depth apparently allows a higher degree of mobility, resulting in higher variability of the frontal density distribution. At the southern coast, the shallow areas appear to constrain the frontal movement, locking the front along the isobaths.

The relationships between turbidity maximum and estuarine circulation, physico-chemical processes, tidal-stirring, sediment dynamics and stability of the water column have been extensively reviewed in estuarine literature (e.g., Postma 1967; Schubel 1968; Officer 1976; Allen et al. 1976; Schubel and Carter 1976, 1983; Dyer 1986, 1995; Geyer 1993). Differences in the location of the turbidity maximum are associated with variation in tidal regime, river flow, and sedimentological processes (Allen et al. 1980; Gelfenbaum 1983; Uncles et al. 1985; Uncles and Stephens 1989). However, over long-term time scales, the turbidity maximum occurs near the landward limit of the salt water intrusion (Dyer 1986, 1995; Grabemann and Krause 1989). A relationship between the frontal location and stability of the water column is then expected. Based on archived hydrographic data, Framiñan and Brown (1996) com-puted the stratification parameter $\{\Delta S / h\}$, where ΔS is the bottom-surface salinity difference and h is the total depth (Plate 8.3d). A good visual correlation exists between the areas of maximum gradient and the areas with maximum frontal density, i.e., the locations of the salinity intrusion and the satellite-derived loci are in good agreement.

8.6.2 Seasonal Distribution

Framiñan and Brown (1996) also developed monthly, seasonal, and annual analyses of the frontal density. The seasonal distribution (Plate 8.4) shows that the turbidity front has its westernmost location during the summer (seasonal composite data sets are constructed such that: summer is January, February, March; fall is April, May, June; winter is July, August, September; and spring is October, November, December), when the river discharge is minimum and the predominant winds are southwestward and westward. During spring, the distribution reaches its easternmost location; larger values of frontal density are associated with deeper areas in the central region and in Samborombón Bay. In fall and winter, seasons with maximum river discharge, the distribution is bimodal with maximum values of density near the Uruguayan and Argentinean coast and higher variability in the center of the estuary. A Student's-t test of significance finds that the summer mean location (Fig. 8.14b) is

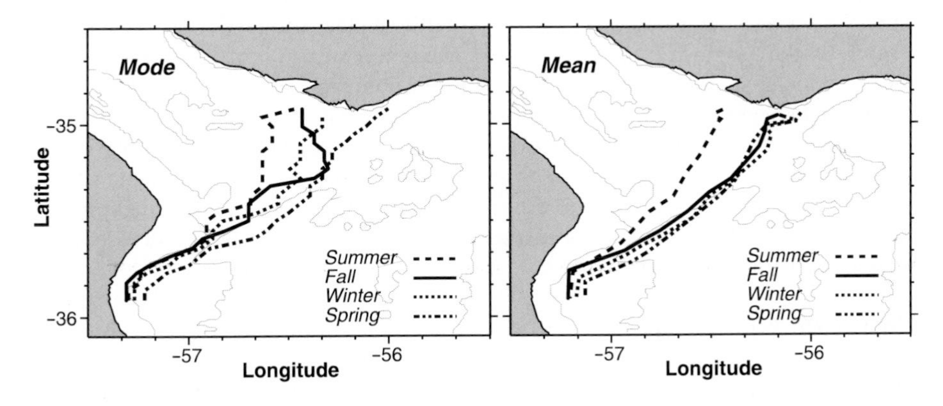

Fig. 8.14. Seasonal variability of the turbidity frontal mode and mean loci

significantly different from the others at a 99% confidence interval all across the estuary. All the loci of the mean are significantly different from each other close to the Uruguayan coast, at a level of 90%. The mode loci (Fig. 8.14a) demonstrate the seasonal variability, with the eastward movement of the turbidity fronts at the northern coast from summer to spring, and the eastward position of the frontal distribution during spring months.

Summary

A description of the physical characteristics and processes of the Río de la Plata Estuary is presented based on in situ data, remote sensing information, and modelling. The large geographical extension and a complex bottom topography, associated with the existence of shallow banks and channels, play fundamental roles in how the river water mixes with the shelf water. Because of the geometry of the estuary, the balance between atmospheric, tidal and buoyancy forcing varies not only in time, but also in space. As a result, high temporal-spatial variability is a distinctive attribute of the estuarine processes in the Río de la Plata.

A brief description of the physiographic setting, climatology and hydrology of the estuary was presented. Previous studies of tides and tidal currents were reviewed, and results obtained using a hydrodynamical model were analysed. The effect of atmospheric forcing on the water level in the estuary was discussed and the model outputs for a case of storm surges was presented. The seasonal distribution of salinity was described, based on field observations and the analysis of the seasonal cycles of the atmospheric and river discharge forces. A bimodal pattern is observed in the orientation of the river plume: northeastward during fall/winter and southward, along the Argentinean coast, during spring/summer. The analysis of vertical salinity sections shows that a salt-wedge is a quasi-permanent feature at the estuary, however, the stratification may be destroyed by moderate to strong winds. A strong influence of bottom topography in the estuarine circulation can be inferred from the bottom salinity distribution.

The mean temperature fields in the estuary and the inner shelf are quite homogeneous. The seasonality in temperature is controlled by air-sea heat flux exchange. Some specifics properties of the surface temperature field, such as fronts and cold water intrusions, were described using NOAA-AVHRR satellite imagery. A characteristic feature of the Río de la Plata is a turbidity maximum located in the transition zone between the upper and lower estuary. Results of a study of the mean spatial distribution and the seasonality of this turbidity front, based on satellite images, indicate that the turbidity front has a higher degree of variability at the northern coast. At the southern coast the modal position of the front is related to the 5 m isobath. During summer, the distribution has it innermost location at the northern coast, and during spring, the frontal distribution is located eastward, and larger values of frontal density are associated with deeper areas.

Acknowledgements

The authors are very grateful to Dr. C.N.K. Mooers for his encouragement and support during the preparation of this manuscript, and his valuable comments and re-

view. M.P. Etala would like to thank Ing. E. D'Onofrio for his useful comments on a first draft of the description of tides, and N. Cattáneo for those interesting discussions on local cyclogenesis. We would like to acknowledge the assistance of Ms J. Splain and RSMAS Remote Sensing Laboratory, Univ. of Miami, for satellite data preparation. The in situ information was graciously provided by: The Servicio de Hidrografía Naval (hydrographic and tide data), Instituto Nacional de Investigación y Desarrollo Pesquero (hydrographic data), Servicio Meteorológico de la Armada Argentina (wind data), Servicio Meteorológico Nacional, Argentina (wind data), Servicio Oceanográfico e Hidrográfico, Uruguay (wind data) and the Instituto Nacional de Ciencia y Técnicas Hídricas, Argentina (river discharge data). M. Framiñan and O. Brown's research was supported by NASA Grant NAS531361 and ONR Grant N000149510166.
[*Received 4 March 1997; accepted 6 October 1997*]

References

Allen GP, Sauzay G, Castaing P (1976) Transport and deposition of suspended sediment in the Gironde Estuary, France. In: Wiley ML (ed) Estuarine processes. Academic Press, New York, vol II, pp 63–81
Allen GP, Salomon JC, Bassoullet P, Du Penhoat Y, De Grandpré C (1980) Effects of tides on mixing and suspended sediment transport in macrotidal estuaries. Sediment Geol 26:69–90
Alvarez JA (1973) Predicción numérica de la onda de tormenta en el Río de la Plata. PhD thesis Univ Nac de Buenos Aires
Alvarez JA, Balay MA (1970) Ondas de tormenta en el Atlántico Sur. Bol SHN 7:15–29
Balay MA (1959) Causes and periodicity of large floods in Río de la Plata. Int Hydrographic Rev 36:123–151
Balay MA (1961) El Río de la Plata: entre la atmósfera y el mar. Bol SHN H–641
Barnes CA, Duxbury AC, Morse BA (1972) Circulation and selected properties of the Columbia River effluent at sea. In: Pruter AT, Alverson DL (eds) The Columbia River Estuary and adjacent ocean waters. Univ of Washington Press, Seattle pp 41–80
Bazán JM, Janiot LJ (1991) Zona de máxima turbidez y su relación con otros parámetros del Río de la Plata. Servicio Hidrografía Naval Inf Técnico 65/91 Buenos Aires
Bazán JM, Arraga E (1993) El Río de la Plata, un sistema fluvio-marítimo frágil? Acercamiento a una definición de la calidad de sus aguas. Conf de Limnología, La Plata 71–82
Beardsley RC, Boicourt WC (1981) On estuarine and continental shelf circulation in the Middle Atlantic Bight. In: Warren BA, Wunsch C (eds) Evolution of physical oceanography. MIT Press Cambridge pp 198–233
Boschi EE (1988) El ecosistema estuarial del Río de la Plata (Argentina y Uruguay). An Inst Cienc del Mar y Limnol, Univ Nal Autón México, 15:159–182
Bowman MJ (1988) Estuarine fronts. In: Kjerfve B (ed) Hydrodynamics of estuaries: estuarine physics. CRC Press Boca Raton, vol I, pp 85–132
Brandhorst W, Castello JP (1971) Evaluación de los recursos de anchoíta (*Engraulis anchoita*) frente a la Argentina y Uruguay. I. Las condiciones oceanográficas, sinopsis del conocimiento actual sobre la anchoíta y el plan para su evaluación. Proyecto de Desarrollo Pesquero FAO Tech Rep 29 Mar del Plata
Brandhorst W, Castello JP, Perez Habiaga R, Roa BH (1971) Evaluación de los recursos de anchoíta (*Engraulis anchoita*) frente a la Argentina y Uruguay. IV. Abundancia relativa entre las latitudes 34°30'–44°10'S en relación a las condiciones ambientales en agosto–septiembre de 1970. Proyecto de Desarrollo Pesquero FAO Tech Rep 36 Mar del Plata
Carreto JI, Negri RM, Benavidez HR (1982) Fitoplancton, pigmentos y nutrientes. Resultados Campañas III y VI del B/I Shinkai Maru 1978. In: Lesculescu V (ed) Campañas de investigación pesquera realizadas en el Mar Argentino por los B/I "Shinkai Maru" y "Walter Herwig" y el B/P "Marburg" 1978 y 1979. Contrib INIDEP 383 pp 181–201
Carreto JI, Negri RM, Benavidez HR (1986) Algunas características del florecimiento del fitoplancton en el frente del Río de la Plata. 1: Los sistemas nutritivos. Rev Inv y Des Pesq 5:7–29
Cavallotto JL (1987) Morfología y dinámica sedimentaria en el Río de la Plata. Informe de Beca CIC Buenos Aires
Cavallotto JL (1988) Descripción e interpretación morfológica del Río de la Plata. Simposio Internacional sobre el Holoceno en America del Sur, Comm on the Quaternary of S America Argentina

Chao S-Y (1988a) River-forced estuarine plume. J Phys Oceanogr 18:72–88
Chao S-Y (1988b) Wind-driven motion of estuarine plumes. J Phys Oceanogr 18:1144–1166
Chao S-Y, Boicourt WC (1986) Onset of estuarine plumes. J Phys Oceanogr 16:3849–3860
Comisión Administradora del Río de la Plata (CARP) (1989) Estudio para la evaluación de la contaminación en el Río de la Plata, CARP Montevideo, Buenos Aires
Ciappesoni HH, Salio P (1996) Pronóstico de sudestadas en el Río de la Plata a corto plazo. VII Congr Latinoam e Ibérico de Meteorología, Buenos Aires, Abstract
Consejo Federal de Inversiones (1969) Los recursos hidráulicos de Argentina, análisis y programación tentativa de su desarrollo, Comisión Económica para América Latina, vol 2
Conomos TJ, Gross MG, Barnes CA, Richards FA (1972) River-ocean nutrient relations in summer. In: Pruter AT, Alverson DL (eds) The Columbia River Estuary and adjacent ocean waters. Univ of Washington Press, Seattle, pp 151–175
Cousseau MB (1985) Los peces del Río de la Plata y su frente marítimo. In: Yañez-Arancibia A (ed) Fish community ecology in estuaries and coastal lagoons: Towards an ecosystem integration. UNAM Press México pp 515–534
Cochrane JD, Kelly FJ (1986) Low-frequency circulation on the Texas-Louisiana continental shelf. J Geophys Res 91:10645–10659
Depetris PJ, Griffin JJ (1968) Suspended load in the Río de la Plata drainage basin. Sedimentology 11:53–60
Dyer KR (1986) Coastal and estuarine sediment dynamics. J Wiley and Sons, New York
Dyer KR (1995) Sediment transport processes in estuaries. In: Perillo GME (ed) Geomorphology and sedimentology of estuaries. Elsevier, Amsterdam, pp 423–449
Etala MP (1995) Un modelo para onda de tormenta en el Río de la Plata y plataforma continental, VI Congr Latinoam Cs Mar (COLACMAR) Mar del Plata, Abstract
Etala MP (1996a) Modelado de onda de tormenta en el Río de la Plata. Rev Inst Arg Navegación 5:13–25
Etala MP (1996b) Primeras experiencias en la determinación en tiempo real del efecto atmosférico sobre el nivel del agua. Congr Latinoam e Ibérico de Meteorología, Buenos Aires, Abstract
Fong DA, Geyer WR, Signell RP (1997) The wind-forced response on a buoyant coastal current: observations of the western Gulf of Maine plume. J Marine Systems 12:69–81
Framiñan MB, Brown OB (1993) Descripción del campo de temperaturas de superficie en el Río de la Plata exterior y la plataforma continental adyacente. Workshop on Comparative Studies of Oceanic, Coastal and Estuarine Processes in the Temperate Zones, Inter-American Institute Montevideo, Abstract
Framiñan MB, Brown OB (1996) Study of the Río de la Plata turbidity front Part I: Spatial and temporal distribution. Cont Shelf Res 16:1259–1282
Framiñan MB, Brown OB (1998) Sea surface temperature anomalies off the Río de la Plata Estuary: Coastal upwelling?. EOS Trans AGU 79 Ocean Sciences Suppl 128
Gagliardini DA, Karszenbaun H, Legeckis R, Klemas V (1984) Application of Landsat MSS, NOAA-TIROS AVHRR and Nimbus CZCS to study La Plata River and its interaction with the ocean. Remote Sensing Environ 15:21–36
Garvine RW (1987) Estuarine plumes and fronts in shelf waters: a layer model. J Phys Oceanogr 17: 1877–1896
Gelfenbaum G (1983) Suspended-sediment response to semidiurnal and fortnightly tidal variations in a mesotidal estuary. Columbia River, USA, Mar Geol 52:39–57
Geyer WR (1993) The importance of suppression of turbulence by stratification on the estuarine turbidity maximum. Estuaries 16:113–125
Geyer WR (1997) Influence of wind on dynamics and flushing of shallow estuaries. Estuarine Coastal Shelf Sci 44:713–722
Glorioso PD, Flather RA (1995) A barotropic model of the currents off SE South America. J Geophys Res 100:13427–13440
Grabeman I, Krause G (1989) Transport processes of suspended matter derived from time series in a tidal estuary. J Geophys Res 94:14373–14379
Guarga R, Vinzón S, Rodríguez H, Piedra Cueva I, Kaplan E (1992) Corrientes y sedimentos en el Río de la Plata. *CARP*
Guerrero R, Acha E, Framiñan M, Lasta C (1997) Physical oceanography of the Río de la Plata Estuary – Argentina. Cont Shelf Res 17:727–742
Hansen DV, Rattray Jr M (1965) Gravitational circulation in straits and estuaries. J Marine Res 23:104–122
Hubold G (1980) Hydrography and plankton off southern Brazil and Río de la Plata, August–November 1977. Atlântica 4:1–22
Junod J (1996) Aspectos mareológicos de la Hidrovía. Rev del Inst Arg de Navegación 5:45–52
Kourafalou VH, Oey L-H, Wang JD, Lee TN (1996) The fate of river discharge on the continental shelf 1. Modeling the river plume and the inner shelf coastal current. J Geophys Res 101:3415–3434

Lasta C, Gagliardini DA, Milovich J, Acha M (1996) Seasonal variation observed in surface water temperature of Samborombón Bay, Argentina, using NOAA-AVHRR and field data. J Coastal Res 12:18–25
Landry MR, Postel JR, Peterson WK, Newman J (1989) Broad-scale distributional patterns of hydrographic variables on the Washington/Oregon Shelf. In: Landry MR, Hickey BM (eds) Coastal oceanography of Washington and Oregon, Elsevier Amsterdam, pp 1–40
Li Y, Nowlin Jr WD, Reid RO (1997) Mean hydrographic fields and their interannual variability over the Texas-Louisiana continental shelf in spring, summer, and fall. J Geophys Res 102:1027–1049
López Laborde J (1987a) Distribución de sedimentos superficiales de fondo en el Río de la Plata exterior y plataforma adyacente. Invest Oceanolog 1:19–30
López Laborde J (1987b) Caracterización de los sedimentos superficiales de fondo del Río de la Plata exterior y plataforma adyacente. Anales Cient Univ Nac Agraria La Molina II:33–47
Lusquiños AJ, Figueroa H (1982) Influencia del Río de la Plata en el mar epicontinental. Servicio Hidrografía Naval Tech Rep 10
Marazzi ML, Menéndez A (1991) Estudio de las corrientes en los canales de navegación del Río de la Plata. INCYTH LHA-114-003-91
Mazio CA (1987) Modelo hidrodinámico para el Río de la Plata. 5th Simp Científico Com Técn Mixta del Frente Marítimo, Montevideo
Molinari G (1986) Simulación numérica de la circulación en el Río de la Plata. INCYTH Pub S5-017-86
Münchow A, Garvine RW (1993) Buoyancy and wind forcing of a coastal current. J Marine Res 51:293–322
Nagy GJ, López Laborde J, Anastasía LH (1987) Caracterización de ambientes del Río de la Plata Exterior (salinidad y turbiedad óptica). Invest Oceanolog 1:31–56
O'Connor WPO (1991) A numerical model of tides and storm surges in the Río de la Plata. Cont Shelf Res 11:1491–1508
Officer CB (1976) Physical oceanography of estuaries (and associated coastal waters). John Wiley and Sons, New York
Ottmann F, Urien CM (1965) La melange des eaux douces et marines dans le Río de la Plata. Cahiers Oceanographiques 17:213–234
Ottmann F, Urien CM (1966) Sur quelques problemes sedimentologiques dans le Río de la Plata. Rev Geogr Physique et Geol Dinamique 8:209–224
Parker G, Marcolini S, Cavallotto JL, Violante R (1986a) Transporte y dispersión de los sedimentos actuales del Río de la Plata (análisis de texturas). 1ª Reunión de Sedimentología Argentina 38–41
Parker G, Cavallotto JL, Marcolini S, Violante R (1986b) Los registros acústicos en la diferenciación de sedimentos subácueos actuales (Río de la Plata)., 1ª Reunión de Sedimentología Argentina 42–44
Postma H (1967) Sediment transport and sedimentation in estuarine environment. In: Lauff GH (ed) Estuaries. Am Assoc Adv Science, Washington, DC, Publ 83, pp 158–179
Romero SY (1994) Estudio de las ondas de tormenta sobre la costa atlántica argentina. Informe de Beca CONICET
Schubel JR (1968) Turbidity maximum of the Northern Chesapeake Bay. Science 161:1013–1015
Schubel JR, Carter HH (1976) Suspended sediment budget for Chesapeake Bay. In: Wiley ML (ed) Estuarine processes. Academic Press, N.Y., vol II, pp 48–62
Schubel JR, Carter HH (1983) The estuary as a filter for fine-grained suspended sediment. In: Kennedy VS (ed) The estuary as a filter. Academic Press, New York, pp 81–105
Schwiderski EW (1981) Global ocean tides, part IV: The diurnal luni-dolar declination tide (K_1). In: Atlas of Tidal Charts and Maps. Naval Surface Weapons Center NSWC TR 81–142
Seluchi ME (1995) Diagnóstico y pronóstico de situaciones sinópticas conducentes a ciclogénesis sobre el este de Sudamérica. Geofísica Int 34:171–186
Servicio de Hidrografía Naval (1973) Batitermogramas característicos del Mar Argentino. SHN H-702
Servicio de Hidrografía Naval (1996) Tablas de marea. SHN H-610
Servicio Meteorológico Nacional (1982) Estadísticas climatológicas. SMN Buenos Aires
Servicio Meteorológico Nacional (1992) Estadísticas climatológicas. SMN Buenos Aires
Simpson JH (1997) Physical processes in the ROFI regime. J Mar Systems 12:3–15
Taljaard JJ (1967) Development, distribution and movement of cyclones and anticlyclones in the Southern Hemisphere during IGY. J Applied Meteorology 6:973–987
Tossini L (1959) El sistema hidrográfico de la Cuenca del Río de la Plata. Anales Sociedad Científica Argentina 167:41–64
Uncles RJ, Stephens A (1989) Distributions of suspended sediment at high water in a macrotidal estuary. J Geophys Res 94:14395–14405
Uncles RJ, Elliot RCA, Weston SA (1985) Dispersion of salt and suspended sediment in a partly mixed estuary. Estuaries 8:256–269
Urien CM (1966) Distribución de los sedimentos en el Río de la Plata Superior. Bol SHN 3:197–203
Urien CM (1967) Los sedimentos modernos del Río de la Plata Exterior. Bol SHN 4:113–213
Urien CM (1972) Río de la Plata Estuary environments. Geol Soc Am Mem 133:213–234

Valle-Levinson A, Lwiza KMM (1995) The effects of channels and shoals on exchange between the Chesapeake Bay and the adjacent ocean. J Geophys Res 100:18551–18563

Valle-Levinson A, O'Donnell J (1997) Tidal interaction with buoyancy-driven flow in a coastal plain estuary, In: Aubrey DG, Friedrichs CT (eds) Buoyancy effects on coastal and estuarine dynamics. Coastal and Estuarine Studies 53 AGU, Washington, DC, pp 265–281

Van de Kreeke J, Robaczewska K (1989) Effect of wind on the vertical circulation and stratification in the Volkerak Estuary. Netherlands J Sea Res 23:239–253

Wessel P, Smith WHF (1991) Free software helps map and display data. Eos Trans AGU 72:441–446

Wiseman Jr WJ (1986) Estuarine-shelf interactions. In: Mooers CNK (ed) Baroclinic processes on the continental shelves. Coastal and Estuarine Sciences 3 AGU, Washington, DC, pp 63–93

Wong K-C (1994) On the nature of transverse variability in a coastal plain estuary. J Geophys Res 99:14209–14222

Chapter 9

Geomorphological and Physical Characteristics of the Bahía Blanca Estuary, Argentina

Gerardo M.E. Perillo · M. Cintia Piccolo

9.1 Introduction

Located at the southwest of the Buenos Aires Province, Argentina, the Bahía Blanca Estuary (Fig. 9.1) is considered the second largest estuary in the country after the Río de la Plata. However, its geomorphologic and physical characteristics vary significantly from all other estuaries in Argentina and South America. Actually The Wash, located in the United Kingdom, is probably the only estuary in the world that have general characteristics quite similar to Bahía Blanca.

Besides its scientific interest, the estuary is economically very important since the largest deepwater harbour system of Argentina, which manages most of the country's exports of agricultural products, is located along its northern coast. Associated with the Ingeniero White port, a series of subsidiary harbours are related to petrochemical industries, fisheries and the largest naval base of the country.

Although research studies in this estuary date back to 1967, most of them were related to specific biologic and chemical projects. Little information about the geomorphology and physical parameters was available prior to the intense studies that started in 1983 and continues today. About 90% of all the studies made in the estuary were concentrated on its Principal Channel.

Furthermore, during 1990 the navigation route included in the Principal Channel was dredged from 9.5 m (33 ft) to 13.5 m (45 ft) nominal depths. Over 35 million m^3 of sediment were extracted from the channel and harbour areas and disposed on the tidal flats and offshore locations. This dredging has introduced major changes in the circulation pattern of the estuary and the amount of sediment transport that have not been estimated yet.

Even though the now available large data base, only the general characteristics of this very complex environment have been uncovered. The objective of this review is to provide the present status of the knowledge about the geomorphological and physical characteristics of the estuary and to define the main directions that future research should follow to obtain a better picture of its dynamics.

9.2 Geological Setting

The Bahía Blanca Estuary occupies the northern border of the Cretaceous-Tertiary Colorado Sedimentary Basin (Zambrano 1972). Sediments associated to the estuary proper are from Pliocene (Chasicó Fm) and Pleistocene (Pampa Fm) ages plus recent material. Both formations are related to fluvial estuarine sediments (Aliotta et al. 1996).

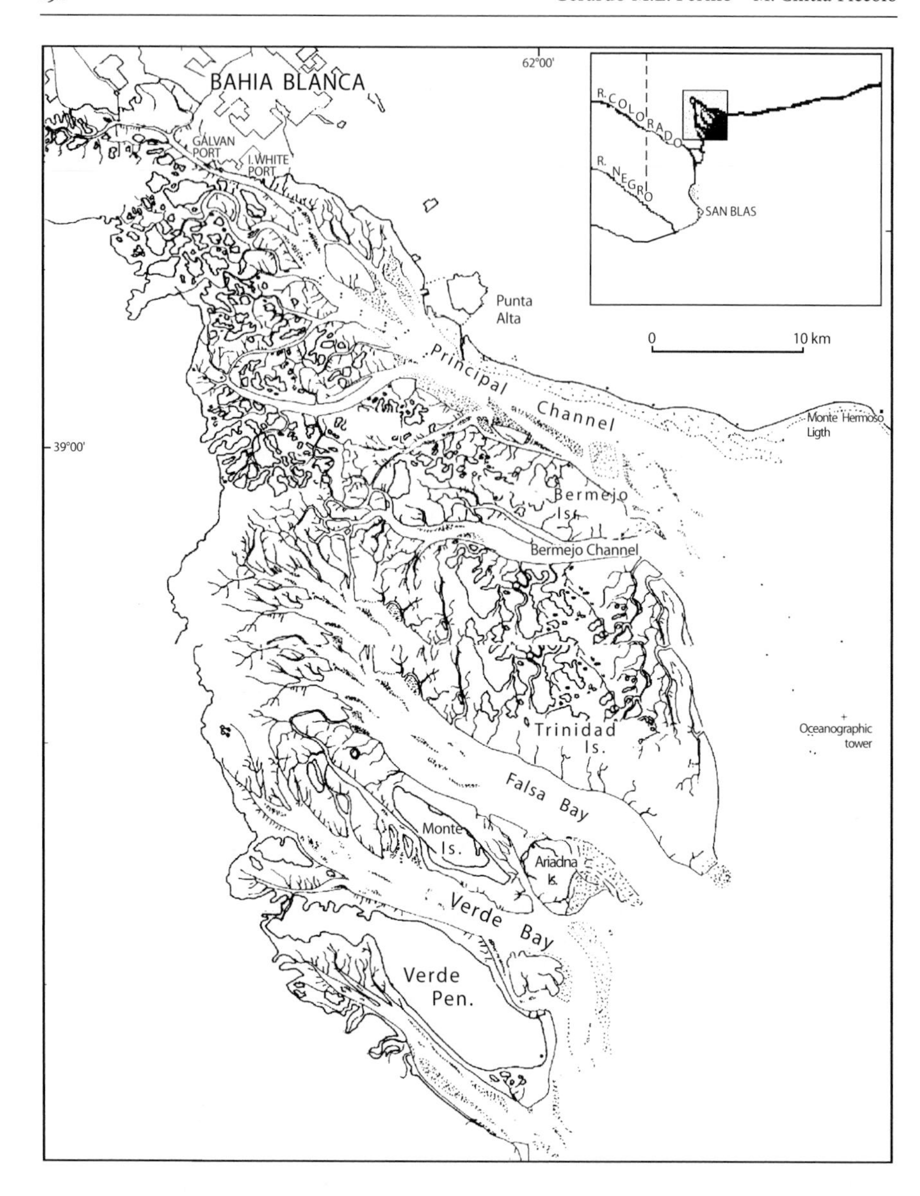

Fig. 9.1. General map of the Bahía Blanca Estuary indicating the main channels (Principal, Bermejo, Verde and Falsa Bays, and Brightman) and the location of the Oceanographic Tower and the various harbours along the Principal Channel. The line along the northern coast of divides the northern sector from the southern sector

The Principal Channel of the estuary is basically the principal valley of a late Pleistocene – early Holocene delta complex generally considered to be the delta of the Colorado River (now flowing about 100 km farther south). Kaasschieter (1965) was the first to assume that the present surficial sediments of the Colorado Basin corresponded to a large delta. However, analysis made by Perillo (1989) and Gonzalez Uriarte (1984) based on LANDSAT imagery, aerial photography and topographic charts suggest that the area was dominated by a complex and wide system formed by the Colorado and Desaguadero Rivers. Perillo (1989) considered that the Principal Channel was formed by the latter, being the Colorado itself or one of its arms, that acted as tributary of the Desaguadero. Clear evidences were found of a later southward migration of this tributary which carved consecutively the other major channels (Falsa, Verde and Brightman Bays). Presently the Desaguadero has an endorheic basin about 300 km west of the estuary; however, its valley is well preserved in today morphology.

The whole original delta span from the northern coast of Bahía Blanca to Bahía San Blas, corresponded to almost 200 km of the southern Buenos Aires Province coast. Both, the northern portion known as Bahía Blanca and the southern portion named Bahía Anegada, have similar characteristics. They are composed by extensive tidal flats with patches of low salt marshes (mostly Spartina grass) and islands. The islands have full continental characteristics both in soil and plant and animal diversity. The only differences between them are that Bahía Anegada islands have a larger proportion of gravel sediments on their surface that is lacking on the northern counterpart, and Bahía Blanca receives some freshwater input whereas Bahía Anegada has no river runoff into its system except for one of the tributaries of Colorado River that flows at the northeast corner of the bay.

Between both bays there is a stretch of the coast of about 60 km almost linear and oriented N-S. It corresponds to the present location of the Colorado River delta. Even though some geomorphologic (Gómez et al. 1994) and seismic (Aliotta 1994) studies were made on the inner shelf in front of this delta, there are no clear indications of where its outer boundary was located. However, all the area have suffered a marine transgression that reached its maximum level about 7 000 years B.P.

In fact, Codignotto and Weiler (1980) have shown for the area between Punta Laberinto and Olga Is. (Fig. 9.1) that the maximum ingression occurred about 6 630 years B.P. in deposits that are between 7.5 and 10 m above present sea level. They also show that the sea has retired reaching the younger ages of 1 300 years B.P. at 2.5 m above present sea level. Similar results were obtained by Gonzalez (1989) and Aliotta and Farinatti (1990) for the inner Bahía Blanca Estuary. They determined that during the maximum transgressive cycle (Flandrian transgression) relative sea level reached also 7–10 m over the present one. For instance, Gonzalez (1989) provided a sequence of five transgressive stages between 5 990 and 3 560 years B.P. Although there are other beach ridges at lower levels that may indicate a regression sequence similar to the one observed by Codignotto and Weiler (1980), no radiocarbon dates were obtained from them.

Studies of submarine outcrops underneath some shoreface-connected sand ridges located on the outer estuary made by Gómez and Perillo (1995) encountered tidal flat deposits at about 15 m below present sea level. Wave-cut terraces were identified by Aliotta and Perillo (1990) and Perillo and Cuadrado (1990) on different parts of the mouth of the Principal Channel also at depths between 13 and 16 m. All these materials and erosional forms were assigned an age of the order of 8 000–9 000 years B.P.

Based on this information, Gómez and Perillo (1995) constructed a minimum sea-level curve for the last 9 000 years and estimated the different rates of variation in that

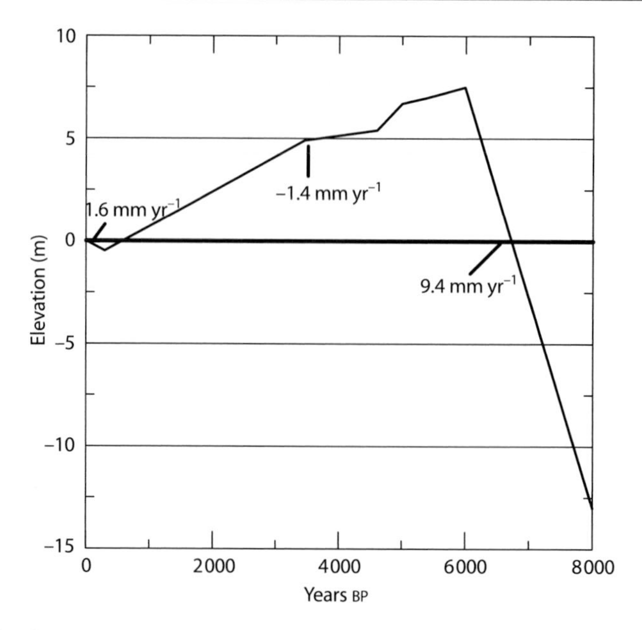

Fig. 9.2. Sea level curve for the last 8000 yr for the Bahía Blanca Estuary. Data for each segment were obtained from published literature only (modified from Gómez and Perillo 1995)

period (Fig. 9.2). In all cases the lowest level of occurrence and their minimum geological age were used. It included the estimation proposed by Lanfredi et al. (1988) that in the last 90 years the sea level in the Argentina coast is rising to an average rate of 1.6 mm yr^{-1}. Using this analysis, the delta sediments were covered by the sea becoming a shallow inner shelf zone for at least 4000 years.

9.3 Geomorphology

The Bahía Blanca Estuary (Fig. 9.1) is a mesotidal system considered by Perillo (1995) as a coastal plain estuary formed by a series of NW-SE major tidal channels separated by extensive tidal flats, patches of low salt marshes and islands. The area has an irregular triangular shape with the apex at the head of the estuary and its base in a NNE-SSW direction. The distance between Monte Hermoso Light and Punta Laberinto, considered the mouth of the area is 56 km.

A general description of the surficial morphology of the area was made by Espósito (1986). He divided the region in three sectors: a) external; b) middle, and c) inner. The external sector corresponds to the outer portion of the line running from about Punta Tejada to Punta Laberinto and it is defined by high tidal flats separated from the sea by sandy barriers. The inner portion is limited by the continent to the west and a imaginary line between Puerto Belgrano and the middle of Verde Peninsula. Extensive tidal flats and some salt marshes are the dominant features. In between both portion is the intermediate one that has large intertidal regions developed mainly parallel to the islands.

However, the analysis of an up to date chart made in 1990 indicates that marked differences exist between two portions located to the north and south, respectively, of a imaginary line running approximately parallel to the northern shore of Bahía Falsa. The northern portion has a funnel shape and, although dominated by the Principal Channel, has a large number of tidal channels and creeks that drain the flats. Numerous small, irregularly shaped islands top the flats. Very few of them have areas in excess of 1.5 km^2. Even the large islands such as Bermejo and Trinidad, are actually a combination of relatively small areas of dry land surrounded by high tidal flats.

On the other hand, the southern portion is dominated by the large channels named Bahía Falsa and Bahía Verde. The head of both channels have the largest lower flat area in the estuary and high flats are only found very close to the islands and in a narrow stretch along the inner shore (Fig. 9.1).

Montesarchio and Lizasoain (1981) estimated the areas of the different environment within the system. The total surface is approximately 2300 km^2, corresponding only 410 km^2 to the islands, the intertidal sector covers 1150 km^2 whereas the subtidal one is 740 km^2. Following the division proposed here, the largest intertidal and island areas are found on the northern sector meanwhile the converse is true for the southern sector. There are no significant differences in areal coverage between neap and spring conditions. The shores of the islands and inner coast are formed, in most cases, by a 1–2 m cliff. Normally the neap tides reach only to the bottom of the cliff, whereas the spring tides rise to the middle or upper level of the cliff.

9.3.1 Geomorphology of the Tidal Channels

Most of the studies made in the Bahía Blanca Estuary have been concentrated on the Principal Channel, the mouth of the major secondary channels flowing into it and on the Bermejo Channel. There is no present bathymetry from the Falsa, Verde and Brightman Bays except for very old surveys of their mouths and middle portions by the Servicio de Hidrografía Naval (Hydrographic Chart H-212).

The first reconnaissance survey of the mouth of the estuary was made by Captain Fitz Roy in 1833 on the HMS Beagle (coincidentally while Darwin was excavating the first *Megatherium* fauna near Monte Hermoso Light, Fig. 9.1). Although maps of the area span from 1854 to the present, last century charts concentrated on the external portion only. The channel was actually surveyed after 1902 when Ingeniero White was established as the main commercial port and Puerto Belgrano became the largest naval base in the country. Several other harbours were constructed within short time, all over the northern margin of the Principal Channel (Puerto Rosales, Puerto Galván and Puerto Cuatreros).

All surveys made of the Principal Channel concentrated on the navigation channel and the entrances to the harbours. In 1983 a program to provide a morphologic characterization of this channel was started. Detailed surveys were made from the mouth and external reaches headward. General descriptions of different sectors were presented by Aliotta and Perillo (1987), Gómez and Perillo (1992), Perillo and Cuadrado (1991), Cuadrado and Perillo (1996), Perillo and Sequeira (1989) and Gómez et al. (1997). All these studies were made with precision echosounder, electronic positioning system and side scan sonar. An integrated analysis of these separate studies will be provided heretofore from the inner portion seaward.

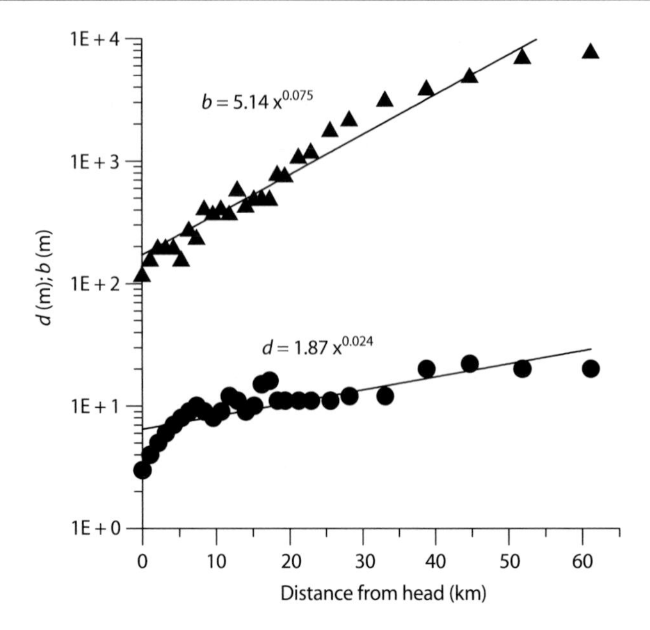

Fig. 9.3. Variation of the depth (*d*) and breadth (*b*) for the Principal Channel of the Bahía Blanca Estuary with distance from the head and their estimated exponential fit. The channel width shows an exponential variation in almost all the channel whereas depth keeps a good fit from 8 km seaward

The Principal Channel has a total length of 60 km and varies in width from about 3–4 km at the mouth to only 200 m at the head. Water depth (all depths are referred to the Datum Plane for the main tidal gage of the area located in Puerto Belgrano) in the channels varies also from 22 m near the mouth to 3 m at the head. However, depth changes are more dramatic within the outer 10–20 km. Taking into consideration only the channel itself, both depth and width show an almost exponential increase from head to mouth (Fig. 9.3). Depth values were estimated before the 1990 dredging and they did not considered the actual dredged depths

In fact, the Principal Channel as well as all other major channels (bays) flowing into the inner shelf, are partly closed by modified ebb deltas which will be described in detail further on. From the shore-connected ebb delta (Cuadrado and Perillo 1997) that closes the Principal Channel headward, depth drops rapidly to 20–22 m for approximately 14 km with minor sills that may reduce depth to 18 m. Within this reach, the largest sand wave field on the estuary was discovered and described by Alliota and Perillo (1987). The general cross section of the channel is steep on the sides with the bottom having a relatively small asymmetry to the right.

The relatively U-shaped bottom is maintained for almost all the channel to some 2 km past Puerto Galván (Fig. 9.1). General depth in this middle section is 6 m. The only change observed there is the presence of the dredged channel from Puerto Belgrano to 2 km before Ingeniero White, with an artificial depth of 14 m. The width of the navigation channel in this section is 400 m, whereas the total width of the Principal Channel is between 2 and 2.5 km. Between this point and Puerto Galván all the complete cross section of the estuary has been dredged to 16 m.

From Puerto Galván headward, the channel has not been ever dredged and maintains the original shape. The channel narrows considerably and keeps an average width of about 500 m almost all the way upstream (Gómez et al. 1997). Also it becomes more V-shaped with the asymmetry following the meandering pattern that is more pronounced in the inner reach. This meandering pattern, together with a steady reduction of depth and width, follows up to its disappearance into the Salitral de la Vidriera. The latter is a backestuary salt flat that becomes partly inundated during very strong storm surges occurring simultaneously with spring high tides.

There are only two freshwater tributaries into the estuary both entering from the northern shore. The most important is the Sauce Chico River that connects with the Principal Channel some 3 km downstream from the head. The Napostá Grande Creek reaches the estuary at about 1 km downstream of Ingeniero White. There are series of small tributaries in between that may input minor quantities of runoff only during local rainfalls. The rest of the time they behave as tidal channels.

The most important tributaries (none of them carry freshwater) occur on the southern margin of the Principal Channel. They range in size from small grooves to major tributaries that can be as large as the Principal Channel itself at their mouth. The geomorphology and time evolution of the largest ones (Cabeza de Buey, La Lista and El Embudo) were described by Ginsberg (1993) and Ginsberg and Perillo (1997).

All these channels have a funnel shape at their mouth becoming strongly meandering headward. The level of meandering has prompted Perillo et al. (1996) to analyse two of them from both fractal and spatial spectral procedures to define the main modes of the meander wavelengths. In all cases, the mouth of the channels at their confluence with the Principal Channel are turned seaward demonstrating the strong ebb dominance that occurs in it. This is also very well seen because of the disposition of the shoals that forms only along the borders of the channels.

A very interesting feature in these channels and that has only been described before by Kjerfve et al. (1979), is the presence of deep scour holes at the confluence of channels. Normally the large channels have maximum depths on the order of 10 m; however, at the holes they reach up to 25 m (Cabeza de Buey hole) (Fig. 9.4). The holes resemble those formed at the confluence of rivers. That is they have elliptical plan view, with the long axis approximately parallel to the main stream, and asymmetrical in cross section. However, deep scour holes on tidal channels differ because they have the steeper side (in the long axis direction) pointing into the tributaries and the gentler away from them; whereas in rivers the sides are reversed. Ginsberg and Perillo (1997) have proposed that even though both types may form first on the imaginary continuation of the tributary channel on the main one, the difference arises because of the effect of the reversing condition of the tidal currents and the headward erosion that occurs at the head going into the tributary.

All tidal channels form a very complex drainage system that in many cases have conflicting channel numbers since the trunk channel may change over the course of a tidal period. The situation arise because several channels go all across the tidal flats from one major channel to the other. The best example is the channel known as Laborde Creek. It connect the Principal Channel with the Bermejo Channel. As the tidal wave moves from south to north in the inner shelf, the higher levels occur first in the Bermejo Channel, then currents in Laborde Creek flow northward. But as the tide rises in the Principal Channel, because of the much larger tidal prism, the cur-

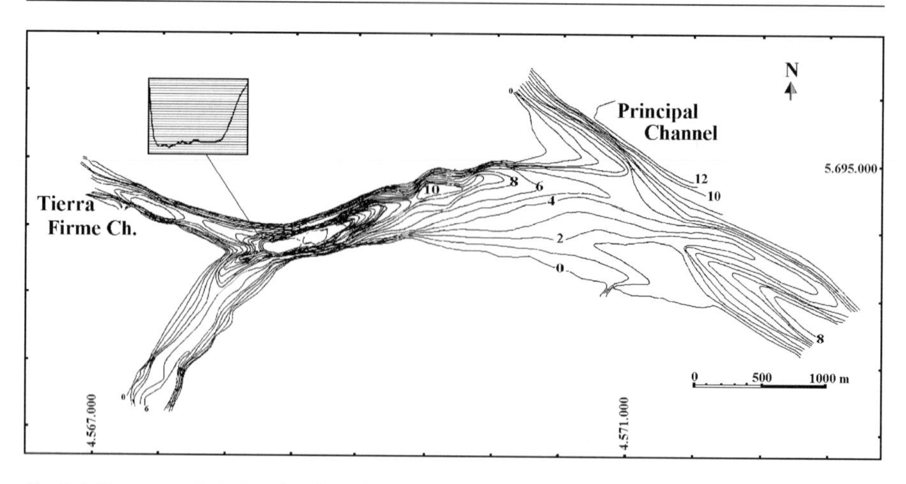

Fig. 9.4. Deep-scour hole found at the Cabeza de Buey Channel. The insert shows a echosounder profile made across the hole. These types of holes are characteristic of the tidal channel confluences at the Bahía Blanca Estuary (modified after Ginsberg and Perillo 1997)

rents in the Laborde Creek reverse. During a single tidal cycle, in this channel currents may change direction up to four times. Unfortunately, only near surface current data is available but it is possible that vertical profiles may be much more complex.

As previously indicated, ebb deltas are observed at the mouth of the Principal and Bermejo Channel and at the southern bays. However, only the one on the Principal Channel still preserves much of its general shape as the others have been largely modified. The latter also present the main orientation of the ebb channel and associated shoals pointing southward due to the ebb dominance that occurs in all the estuary.

Gómez (1989) and Gómez and Perillo (1992) where able to observe the breaking of the Bermejo Channel ebb delta that occurred in 1983. Bathymetric surveys were made just before and after the breaking. It was also possible to identify the consequences over the general geomorphology of the inner shelf. For instance, a shoreface-connected linear shoal, Largo Bank, that has its main axis right across the mouth of the Bermejo Channel, suffered a change in its transversal asymmetry. Before the breaking, the steeper flank was landward; whereas in 1986 the asymmetry has reversed (Gómez and Perillo 1992). This situations has helped to increase the eastward migration rate of the bank as measured from historical charts which reaches up to 37 m yr^{-1}.

The ebb delta partly closing the Principal Channel has very peculiar characteristics. First of all, it is connected to the northern shore which provides a much stronger stability than the ones located at the mouth of the southern channels. On the other hand, Cuadrado and Perillo (1997) have demonstrated that the delta is located on top of a sill of the Chasicó Fm. Using low penetration seismic data it was possible to confirm that the top of this formation moves from a depth of about 24 m on the Principal Channel itself to only 11 m just below of the delta; dipping to 14–16 m offshore. It is evident that this sill was a barrier for sediment transported by the Desaguadero River and also acted as an anchor structure that helped in maintain the delta at the same position.

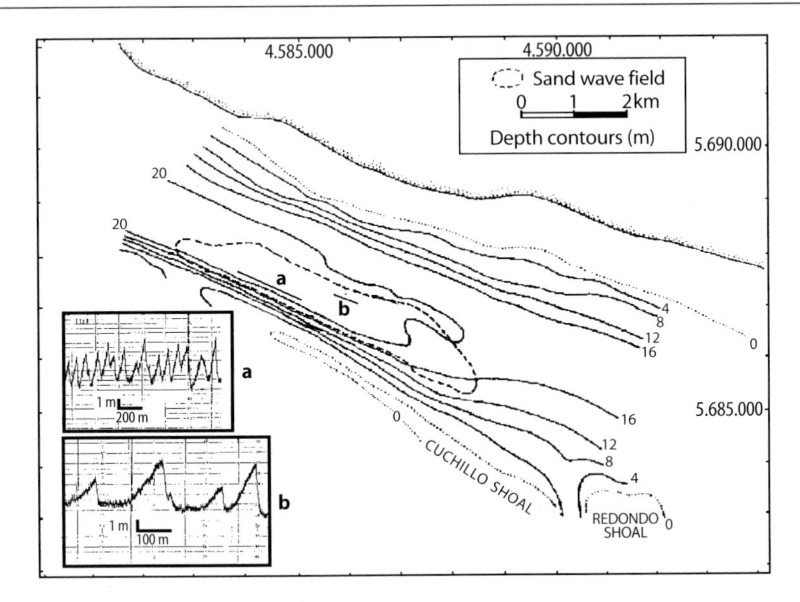

Fig. 9.5. General bathymetry of the Principal Channel near its mouth where the largest sand wave field is located. In the inserts examples of the echosounder records for the regular "saw sooth" type (insert *a*) and the "solitary sand waves" (insert *b*) are shown

Both elements provided an endurance of the delta even though it is subject to strong dynamic marine action. This can be observed because tidal discharge is being produced by a series of channels that were cut by them along the southern lobe of the delta. These channels and the associated shoals are extremely movable whereas the main body of the delta does conserve its position and general shape.

Mobility of the shoals corresponds to the way tidal currents hit the southern lobe of the delta. In fact, tidal currents arrive to the southern lobe with large angles to it. For example, the circulation on the Principal Channel has an angle of about 35° with the channels cutting the lobe, whereas the flood currents have angles of the order of 20° with them. As a result of this situation sedimentation rates in some of the channels, specially El Toro Channel, can reach up to 1 m yr^{-1} (Perillo and Cuadrado 1991).

The complexity of this condition produce that sedimentation within the channels are oblique to the general shape. Although one may expect that this dynamic situation could reach some sort of equilibrium shape, the fact is that in the case of El Toro Channel the wrong orientation is maintained by dredging since it is the only entrance to the Bahía Blanca harbour system. The harbour authorities dredge about 300 000 $m^3 yr^{-1}$ of sediment to allow adequate navigation conditions for 45 ft cargo vessels only in a stretch of the channel less than 4 km in length and 800 m in width.

In several places along the Principal Channel and in the major channels flowing into it, large and medium 3D dunes (sand waves) have been detected and described (Aliotta and Perillo 1987; Aliotta 1988; Gómez and Perillo 1992; Ginsberg and Perillo 1997; Gómez et al. 1997). Because of the sediment distribution, these bedforms are concentrated in specific places within these channels associated with large current velocities and some kind of geomorphologic "trap" that allows for sand deposition. The most common traps are either relatively closed basins and ebb or flood sinus (Ludwick 1974).

Due to both, areal distribution and bedform characteristics, the most important of these fields is the one located on the Principal Channel between Puerto Rosales and the mouth of El Toro Channel (Fig. 9.5). This field covers an elliptical area of about 5.1 km^2 with its main axis (6.5 km) parallel to El Cuchillo shoal and width of over 1.2 km. The basin where the field has developed is about 22 m in depth and it is bordered by the El Cuchillo shoal on the south and areas of lesser depth (18–20 m to the north and west and 14 m to the east).

Even though depth variations are important, there may be other reasons that limit the extent of the field: sediment composition and sand thickness appear as determinant. Both to the north and west of the field as well as we move up El Cuchillo shoal flank, the percentage of silt and clay in the sediment increases from less that 3–5% to 10–15%. On the other hand, from some few (and non conclusive) low penetration seismic data, it appears that the depth of the top of the Chasicó Fm is larger within the field than on the rest of the channel. Aliotta (1988) gives a figure in which bedforms develop atop of a hard rock terrace in a nearby, less important sand wave field. These bedforms become progressively larger with distance from the terrace border. Although with the presently available data no conclusion can be reached, it is possible to speculate that another issue associated with the bedform size increase may be related to the development of a boundary layer on top of the terrace as shown by the classical example of vertical velocity distribution over a flat plate (i.e., Streeter 1948).

Bedforms within this field have heights up to 6 m and wavelengths between 80 and 600 m. The most relevant characteristics are:

1. in all cases these bedforms have their lee side directed to the ebb;
2. lee inclinations have an average of 11°, with maximum values of the order of 31°;
3. on the through and flat areas associated to type of bedforms described for the first time where found: megaripple fans (Aliotta and Perillo 1987, 1994);
4. estimated migration rates for the bedforms vary from 5–90 m yr^{-1} with an average of 33 m yr^{-1}.

The fact that lee sides are steeper than 10° and the formation of the megaripple fans are evidences that flow separation occur on the crest of the 3D dunes. Normally, flow separation in marine environments only has been observed in very shallow depths and intertidal areas normally related to small 2D and 3D dunes (ripples). There is no previous mention in the literature of lee steepness so high in such large number of subtidal deep bedforms. Both elements make this sand wave field unique in the world.

An immediate question is why these forms appear here and nowhere else? First of all, we consider that if they are observed here, that does not prevent that other sand wave field with similar characteristics may occur in other places, they have not been described yet. However, the possible reasons for these characteristics may be associated to the lack of sediment input into the field. One interesting feature of the larger bedforms located closer to the center of the Principal Channel is that they are solitary bedforms in the sense of Perillo and Ludwick (1984). This means that any evolution of the bedform is independent of the neighbour ones.

Since sand is not input from the inner reaches of the channel and sand that may come from the inner shelf cannot bypass the ebb delta or the marginal channels (i.e., El Toro channel), they have to rely only on the material contained within the basin. As these dunes are subject to strong tidal currents, they migrate at a very high velocity in comparison with their size. Therefore, the bedforms use their own material, in a kind of auto feedback, to help them progress developing steeps lee sides in the process to conserve sand and maintain the dynamic equilibrium.

Except for bedforms found in flood-dominated channels, all dunes and shoals observed within the Bahía Blanca Estuary have an ebb dominance. This is a very important factor because when combined with the lack of sediment input from the rivers and the shelf (stopped by the ebb deltas), the relatively large concentration of suspended sediment into the estuary must be maintained from the erosion of the tidal flats and islands' coasts.

The analysis of erosion/accumulation along tidal channels and island borders made by Espósito (1986), Ginsberg (1993) and Ginsberg and Perillo (1997) clearly shows that most places in the estuary are in clear retreating conditions. For instance, Ginsberg and Perillo (1990) have described the presence of erosion cusps along most of the channels in the estuary. A specific study of some of these forms indicated that any one of them may provide up to 18 $m^3 yr^{-1}$ of sediment to circulation. Those erosion cusps have densities of the order of 25 per km in some places. Such figure plus the erosion observed at the cliffy coasts of the islands (up to 1 $m yr^{-1}$) may give a clear picture of the enormous amounts of material that is being extracted from the shores and put into transport.

Perillo and Sequeira (1989) have shown that the southern coast of the Principal Channel across Ingeniero White Harbour has retreated up to 50 m during the period 1980–1986. They have also demonstrated that in the same period about 1.5 million m^3 where net exported from a stretch of about 8 km of the middle reach of the channel.

Because most of the sediments in the estuary are silt and clay results in the fact that it is maintained and transported in suspension. The strong currents and the small intervals in which slack water occurs, impede sediment deposition within the channels. On the tidal flats, wave erosion (as will be described later on) also acts to both erode the old sediment and preclude any accumulation that may occur. The sum of all these factors results in the erosional stage that is found in most of the estuary and the fact that almost all the sediment observed corresponds to the deltaic deposition period.

9.4 Physical Processes

9.4.1 Tides

The larger energy input into the Bahía Blanca system is produced by a standing, semidiurnal tidal wave. The mean, spring and neap amplitudes of the tide at five stations along the channel and one outside the system are plotted in Fig. 9.6. The amplitude and phase angle for the most important semidiurnal and diurnal components of the tide at the main tidal stations are shown in Table 9.1. Although the tide at the Ocea-

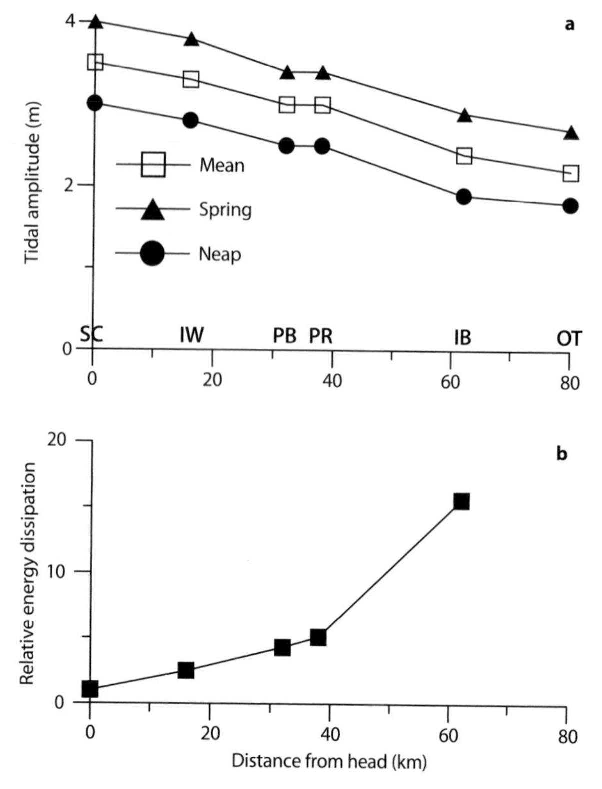

Fig. 9.6. Tidal characteristics of the Principal Channel of the Bahía Blanca Estuary. **a** distribution of the average mean, spring and neap tidal range determined at permanent or temporary tidal gauges along the Principal Channel *SC* Sauce Chico, *IW* Ingeniero White, *PB* Puerto Belgrano, *PR* Puerto Rosales, *IB* Isla Bermejo, *OT* Oceanographic Tower; **b** energy dissipation of the tidal wave due to bottom and border friction (modified after Perillo and Piccolo 1991)

nographic Tower (TO) is mixed semidiurnal, along the Principal Channel it behaves as purely semidiurnal as defined by the Formzahl coefficient (Defant 1961)

The amplitude to depth ratio is large for the inner reach. Although, this ratio may appear small for the outer reach stations, they are caused by a geomorphologic difference. The outer reach from about 2 km headward of Puerto Belgrano to the ebb delta is about twice as deep of all other sectors of the estuary with restricted sedimentation.

Propagation of the tidal wave along a channel is affected by its geometry. The most common effect is the reflection of the tidal wave on the channel flanks and the head, that converts the original progressive form in a standing wave. This occurs when the length of the channel (L_c) is at least one fourth of the wavelength (L) of the tide. For this case $L = P\,(g\,h)^{1/2}$, where P is the tidal period (here equal to the M_2-period = 12.42 h), g is the acceleration of gravity (= 9.8 m s^{-2} and h is the average depth of the channel (= 10 m), resulting in $L = 442$ km; that is over seven times the length of the channel.

Further modifications are produced by the bottom and wall friction that drains energy from the wave, and the geometry of the channel. According to Green's law, tidal amplitude (A) is proportional to $d^{-1/4}\,B^{-1/2}$, in words the amplitude of the tidal wave increases with a decrease in either the depth (d) or the breadth (B) of the channel or both simultaneously. Therefore, the actual amplitude of the tide along an estuary is directly related to the balance between the friction and the convergence.

Table 9.1. Principal tidal components for the three main tidal stations along the Principal Channel

Station	*Zo*	M_2		S_2		O_1		K_1		N_2		*F*
		A	Φ	*A*	Φ	*A*	Φ	*A*	Φ	*A*	Φ	
Ingeniero White	2.32	1.63	271	0.22	48	0.13	43	0.19	110	0.24	172	0.17
Puerto Belgrano	2.43	1.46	266	0.18	44	0.14	51	0.17	81	0.12	149	0.19
Ocean. Tower	2.93	1.09	233	0.17	356	0.15	19	0.2	85	0.17	159	0.28

Based on the relative importance of both processes (friction vs. convergence), Le Floch (1961) classified the estuaries in three categories:

a hyposynchronous (convergence < friction),
b synchronous (convergence = friction), and
c hypersynchronous (convergence > friction).

In the former, the amplitude is reduced headward, while in the synchronous estuaries the amplitude is the same in almost all the channel. From the Fig. 9.6a results that the Bahía Blanca Estuary corresponds to the hypersynchronous type, where the amplitude increases steadily from the mouth to the head. Although not shown in the figure, there is a strong reduction landward of the head of the estuary due to the frictional effect associated to the dampening produced by the river flow for one side and the spreading over the salt flat on the other.

Perillo and Piccolo (1991) estimated the average rate of tidal energy dissipation per unit mass of fluid due to friction using the method presented by Ippen and Harleman (1966) as 0.0017 $m^2 s^{-3}$. Changes in energy dissipation are plotted in Fig. 9.6b. The large negative dissipation rates in the outer reach of the estuary are due to differences in amplitude between the mouth (Bermejo Is.) and Puerto Rosales and Puerto Belgrano. In particular, the mean value of Puerto Rosales is about 20% and 3% larger than Bermejo Is. and Puerto Belgrano, respectively. Even though the latter is only 1 km further inland. Differences in amplitude may be due to the particular setting of the tide gages, since Puerto Rosales (when it operated) was in the open channel, Puerto Belgrano is located in a closed harbour. In the inner reach, which has the less variable storage cross-section, produce an energy dissipation of over 100%.

9.4.2 Wind Influence

The typical weather pattern of the region is dominated by the middle-latitude westerlies and the influence of the Subtropical South Atlantic High. The resulting circulation induces strong NW and N winds for more than 40% of the time. Its influence can be readily assessed when the average wind velocity of 24 km h^{-1} is considered as well as mean hourly maxima of 58 km h^{-1} and gusts of over 100 km h^{-1} (Piccolo 1987).

Wind effect on the estuarine circulation is observed at the level waves, storm surges and subtidal sea level variations. At high frequencies, the wind produces two types of waves a) wind waves and b) interaction waves. The former are normally small waves of about 5–10 cm in height and 1–3 m in wavelength found in the channels and on the tidal flats as they are covered by water. The later are waves formed by the interaction of the incoming tide and the wind blowing from the N and NW direction. These are extremely steep waves, up to 1.5 m high, with wavelengths of the order of 10–20 m.

Although wind waves may be locally important for navigation of small boats, the main effect of the wind is found over the tidal wave at lower frequencies. Comparison of the predicted astronomical tide with the actual tidal records shows large differences both in height and time. In general, winds blowing from the NW sector produce a set down on the water level by pushing the surface water out or preventing the tidal penetration. SE winds generate the opposite effect. However, the predominant winds from NW and N produce the greatest variations by:

a advancing the time of the low water,
b delaying the time of high water, and
c reducing the predicted water levels.

Perillo and Piccolo (1991) have analysed the deviations of two years of hourly simultaneous records of tide both at Ingeniero White and the Oceanographic Tower (Fig. 9.1) from the predicted astronomical tides for both sites. They found 24 cases in which the deviations were larger than 2 m with maximum of −4.01 and 2.39 m at Ingeniero White, whereas at the Oceanographic Tower these maxima were −1.51 and 1.87 m. In both cases the maximum negative values coincide with winds blowing from the NW and the maximum positive with winds from the SW. The latter are very intensive winds that blow normally after a storm and their effect is stronger (as in these cases) when they coincide with high tide. The cross effect on the estuary is to induce a set up on the northern coast.

Piccolo and Perillo (1989) studied the low-frequency sea level response to wind also for the same stations. Spectral analysis indicated that the long term changes (time scales >10 days) prevail. For shorter scales energy peaks are maximum at around 3 day periods. In both cases, corresponding to the passage of fronts over the area. Although both stations showed similar responses to wind forcing, wind effect is greater in the inner estuary.

9.4.3
Salinity and Temperature

The analysis of historical data compiled by Perillo et al. (1987) from several stations located along the axis of the Principal Channel of the Bahía Blanca Estuary presents a mean annual surficial temperature of 13° C varying from 21.6° C in summer to 8.5° C in winter (Piccolo et al. 1987). Average temperature in all seasons is slightly higher at the head. Figure 9.7a shows the along channel distribution of the mean temperature for the four seasons. The distribution of mean surface salinity (with a corresponding correlation with density) shows an exponential growth from the head (Sauce Chico discharge) up to the middle reach of the estuary (Fig. 9.7b). In the middle reach, the

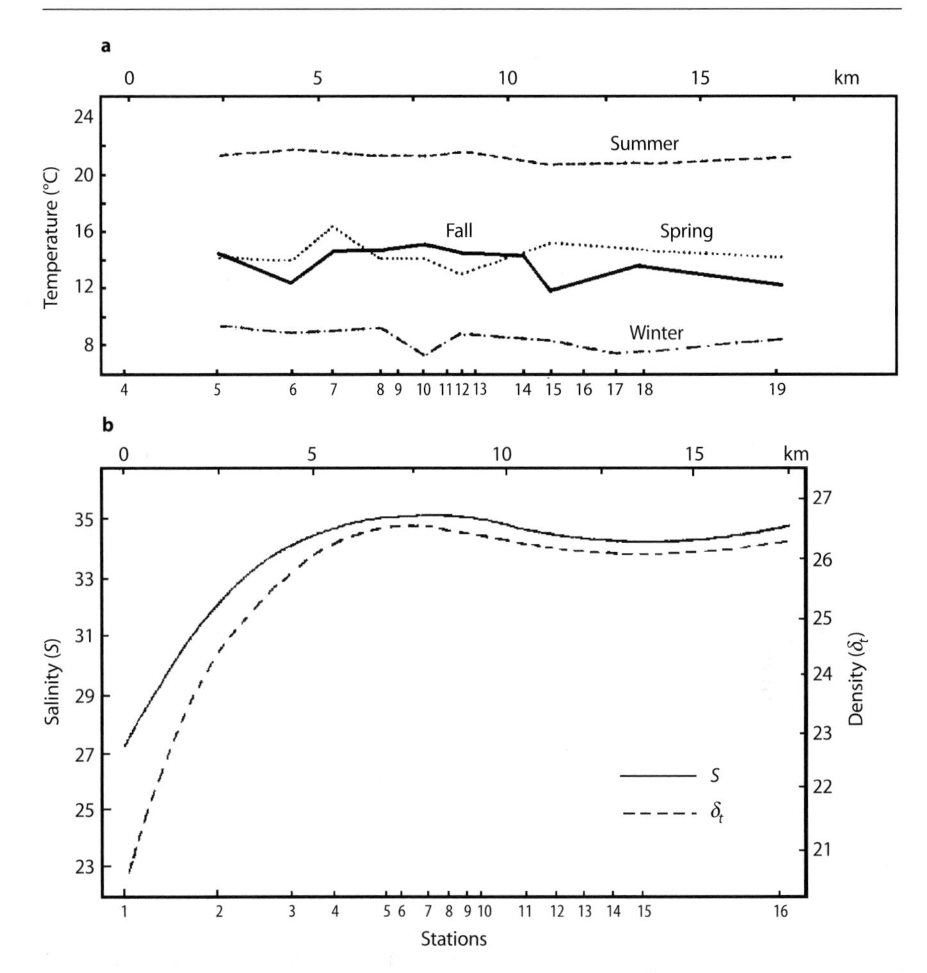

Fig. 9.7. Distribution of mean **a** temperature and **b** salinity and density for the inner and middle reaches of the Bahía Blanca Estuary. Data corresponds averages of all surface values collected since 1967 to 1984 (modified from Piccolo and Perillo 1990)

average salinity has a local minimum produced by the influence of the Napostá Grande Creek and the Bahía Blanca City sewage discharge (estimated in 10 $m^3 s^{-1}$). Mean surface density closely follows the salinity curve, indicating that the influence of temperature in the estimation of this parameter is minimum.

Examples of longitudinal distribution of temperature and salinity are presented in Fig. 9.8.The survey made on October 18, 1984 (Fig. 9.8a,b) represents spring-summer conditions, which are normally associated with rainy periods. A maximum temperature of 17° C is associated to the Napostá and sewage discharges (Fig. 9.8a). The survey made on July 30, 1985 (Fig. 9.8c,d) is typical of winter time with relatively small continental runoff. The thermal vertical structure of the estuary is homogeneous with longitudinal variations smaller than 3° C.

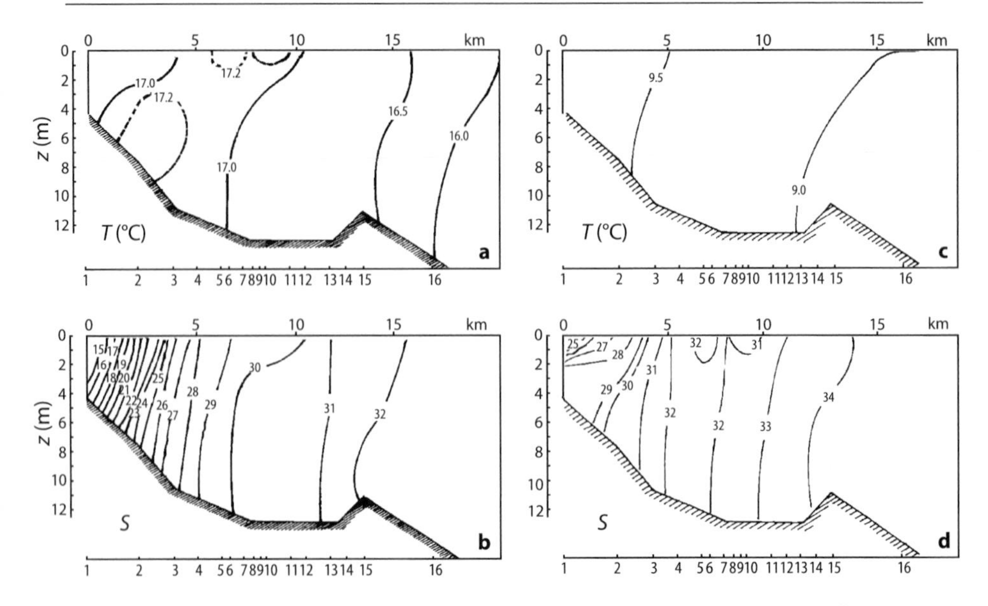

Fig. 9.8. Longitudinal distribution of temperature (*T*) and salinity (*S*) for the inner and middle reaches of the Principal Channel. **a** and **b** correspond to October 18, 1984 representing spring-summer, high freshwater discharge; **c** and **d** corresponds to July 30, 1985, representing winter conditions with low freshwater discharge (modified from Piccolo and Perillo 1990)

The patterns of salinity distribution show that at high water the outer part of the estuary is fairly vertically homogeneous; stratification increases toward the head of the estuary though. Depending on runoff conditions, salinity differences between mouth and head of the estuary may reach 17 and more than 4 between surface and bottom. In the sectors associated to inflow of fresh water, the water column presents a significant stratification marked by an halocline located between 1 and 3 m (Fig. 9.8d). However, in those reaches where little or none fresh water is discharged, the isohalines show small gradients and may result in vertical homogeneity.

The distribution of depth-mean salinity along the inner reach of the estuary is shown in Fig. 9.9a, where a best fit equation of salinity as a function of distance from the estuarine head, for low river discharge, is presented. The time averaged salinity profiles (<S>) (Fig. 9.9b) of the inner stations (1 and 2) present a smooth vertical gradient, whereas in the others stations (3 and 4) they are vertically homogeneous. Also stations 1 and 2 have greater salinities than the typical values observed in the inner continental shelf (33.8, Martos and Piccolo 1988).

The time averaged salinities determined in the innermost station show very high values. Although that station is located in front of the Sauce Chico River mouth, it also receives water coming from the washing of the Salitral de la Vidriera salt flat. The salt flat has a surface of about 30 km^2 further upstream than station 1, it is partially covered during extreme spring tides complemented by the storm surge effect produced by southeast winds. Therefore, the restricted circulation in the inner estuary further increase the salt concentration producing salinities larger than those in the adjacent continental shelf sea.

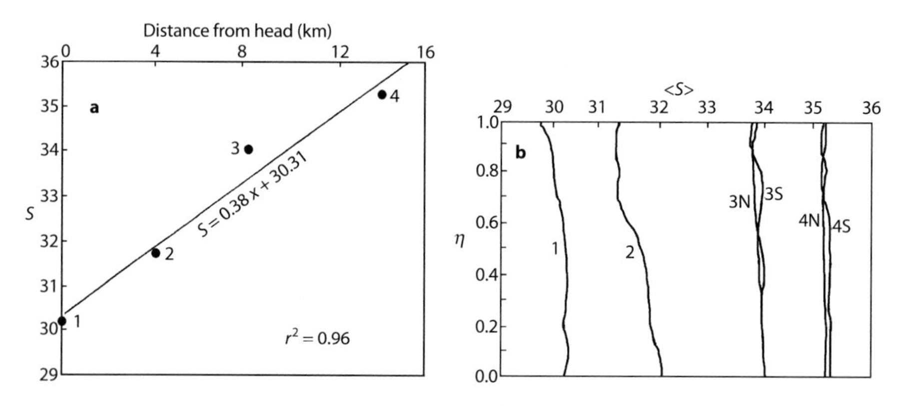

Fig. 9.9. Mean salinity distribution in the inner reach of the Principal Channel. **a** cross-section and tidal averaged salinity; **b** tidal averaged salinity profiles

Based on the vertical salinity distribution, the estuary may be divided in two sectors (Piccolo et al. 1987):

a an inner region where it behaves as a partially mixed estuary with a strong tendency to be sectionally homogeneous (inner and middle reach), and
b an outer one being sectionally homogeneous with mean salinities values similar to the adjacent continental shelf ones (Martos and Piccolo 1988; Piccolo et al. 1994).

9.4.4 Tidal Currents

Depth mean longitudinal velocity (u currents for two stations near Ingeniero White (Fig. 9.10) clearly show a strong asymmetry as indicated by the differences in duration between flood and ebb. The ebb currents ($u > 0$) are stronger than the flood ones. The peak ebb is about twice the maximum flood current in northern station.

The time averaged longitudinal current for the stations located at three sections along the inner reach are plotted in Fig. 9.11 as a function of the normalized depth ($\eta = z / d$). Station 1 (located near the head of the estuary) is the only one that shows a net transport towards the mouth of the estuary in all the water column. On the other hand, stations 3S and 4S (located at 4 and 1 km headward of Ingeniero White over the southern flank of the Principal Channel) exhibit a landward net transport. The stations located near the northern flank of the channel show a downstream flow in the upper third of the water column and upstream in the lower part.

9.4.5 Residual Fluxes

Recent studies of the cross-section distribution of the residual fluxes at the inner and middle reaches of the estuary (based on the methodology deviced by Perillo and Piccolo 1993, 1998) confirm the initial estimations based on the analysis of individual

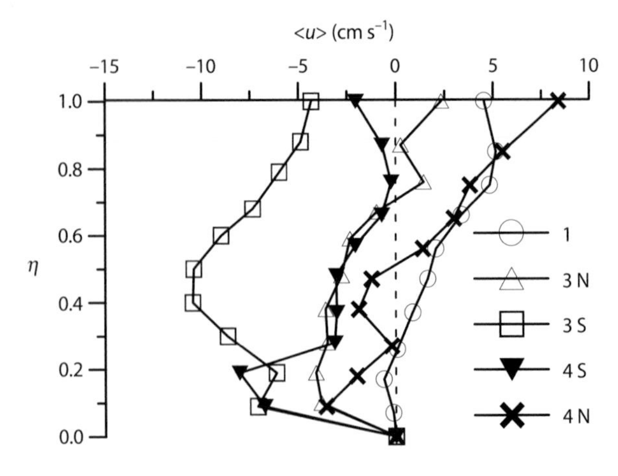

Fig. 9.10. Time averaged velocity profiles obtained on five stations of the inner reach of the Bahía Blanca Estuary. Stations *3S* and *4S* were located on the southern margin of the Principal Channel and show headward residual current in all the water column, whereas stations *3N* and *4N* (located on the northern margin) have an upper seaward flowing layer and a lower headward flowing layer. Station *1* located at the head of the channel is seaward flowing in all the column (modified from Piccolo and Perillo 1990)

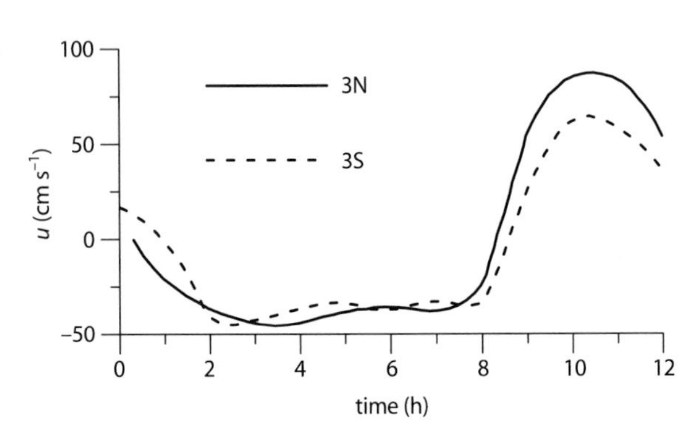

Fig. 9.11. Depth mean longitudinal velocity obtained at stations *3N* and *3S* on the inner reach of the Principal Channel. Note the strong asymmetry between the ebb (positive) currents (larger velocities for a shorter time) and flood (negative) currents (smaller velocities for a longer time)

stations by Piccolo and Perillo (1990) that these reaches of the estuary behave as partly mixed. The new field and data reduction methodology insures an error-free calculation of the fluxes even with complex cross-section morphologies.

Figure 9.12 shows the mass and salt residual flux distributions at a cross-section near Ingeniero White on spring and neap conditions. Depth and cross-section are non-dimensional. The figure shows clearly two layers: an upper one where the residual flux is seaward (positive values) and a lower one with headward flux (negative values). The upper layer becomes deeper from $\chi = 0.6$ northward (Fig. 9.12) indicating the influence of the center of the channel and the higher ebb velocities found at the stations measured there. Meanwhile, the effect of the tidal flats is well demonstrated by the smaller thickness of the upper layer southward of $\chi = 0.4$.

The deeper salt flux (F) on the main part of the section (Fig. 9.12b) is due to the proper salinity distribution. However, the distribution provides a clear evidence that salt is being exported on the upper layer and specially on the center of the channel. The friction produced by the tidal flats and the northern flank of the channel significatively reduce such exportation. On the other hand, salt import is being observed on the lower portion of the channel, less affected by the borders and with the benefit on the low ebb velocities recorded there.

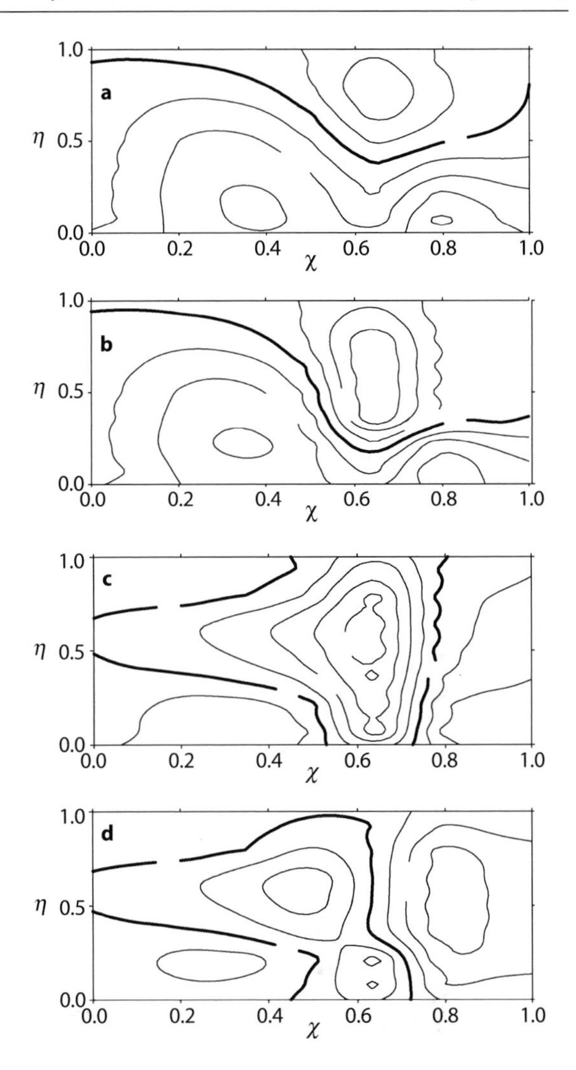

Fig. 9.12. Residual fluxes of mass (**a**, **c**) and salt (**b**, **d**) for the same cross-section near Ingeniero White for spring (**a**, **b**) and neap (**c**, **d**) conditions

9.5 Discussion and Conclusions

Bahía Blanca Estuary displays one of the most complex geomorphological and dynamic conditions in the Argentina coastline. There are major differences with the typical river mouth estuary because of the geological conditions that originated this environment. First of all, the estuary is only part of the remaining of once the largest delta in the country being almost 200 km at the distal point. For the sake of comparison, the Paraná Delta located within the Río de la Plata Estuary is only 45 km wide, the Nile Delta is 250 km and the Niger Delta is 400 km wide.

Along its geological evolution, it went through full fluvial subaerial conditions to a shallow bay/shelf with depths up to 7–10 m over the today flats and more than 30 m in the outer reaches of the channels. As the Flandrian transgression retreated in a much slower pace than the transgression, the coast moved progressively leaving sand bars and spits. Although there are no definitive paleontological features that can provide hard evidence, we consider that the climatic conditions during the retreating period were already as dry as the present ones. Therefore, very little input of sediment was provided by the rivers coming into the receding bay.

This supposition is derived from a comparative geomorphological analysis of Anegada Bay. After its formation, this southern portion of the Colorado Delta had very few fluvial (if any) tributaries coming into it, thus its evolution was dependent only on the input of sediments from the shelf. However, both Anegada and Bahía Blanca have exactly the same geomorphological properties. Normally, sediment from the shelf is not input into Bahía Blanca Estuary (most probably occurs the same with Anegada Bay although there are no studies dealing with this subject there).

Based on the salinity and temperature distribution alone, the Bahía Blanca Estuary is divided in two. The inner one, from the mouth of the Sauce Chico River to Ingeniero White, is classified as a partially mixed estuary during normal runoff conditions, but with a strong tendency to become sectionally homogeneous during low runoff. The outer reach is sectionally homogeneous. The boundary between both reaches is transitional depending on the river discharge. According to the classification of Hansen and Rattray (1966), the estuary may be characterized as Type 1a.

The mixing regime is characterized by $Ri < 2$ and it is associated to strong turbulent processes during maximum ebb and flood conditions, in particular on the northern flank of the estuary. Therefore, full developed mixing is found almost throughout the tidal period. Stratification is established at the beginning of ebb cycle being generated by the combined outflow of river and ebbing tide, producing a seaward tilt of the isohalines. The process is accompanied by low turbulence levels due to the small currents. As the current velocity increases, turbulent mixing also increases. During the flooding stage, the possibilities of developing any stratification are very low since the flood is driving well-mixed water from downestuary. Turbulent mixing at midflood is high and practically constant resulting in low Ri.

The residual circulation shows a significant difference in the direction of the mass transport. On the deeper parts of the sections (northern flank) the flow reverses with depth, being headward near the bottom. The net transport is completely landward on the shallower parts. This behaviour, added to the washing of the backestuary saline, produce a concentration of salt in the inner portion of the estuary, resulting in salinities larger than those observed in the inner continental shelf. Analysis by Piccolo and Perillo (1990) of the current stations located on the southern flank of the Principal Channel may indicate that the asymmetry of the tidal current is mainly originated by the presence of the extensive tidal flats bordering the channel.

Acknowledgements

Funds for this study have been provided by grants from Consejo Nacional de Investigaciones Científicas y Técnicas (CONICET) de la República Argentina (period 1983–1998), and Universidad Nacional del Sur (period 1994–1998). The authors thank Lic Walter D. Melo for drawing some of the figures.

References

Aliotta S (1988) Estudio de la geomorfología y de la dinámica sedimentaria del estuario de Bahía Blanca entre Puerto Rosales y el Fondeadero. PhD thesis Universidad Nacional del Sur

Aliotta S (1994) Geofísica. In: Perillo GME (ed) Estudio oceanográfico de la plataforma continental interior frente al estuario de Bahía Blanca. Instituto Argentino de Oceanografía Contrib Cient 311:23–25

Aliotta S, Perillo GME (1987) A sand wave field in the entrance to Bahía Blanca Estuary, Argentina. Mar Geol 76:1–14

Aliotta S, Farinatti E (1990) Stratigraphy of Holocene sand-shell ridges in the Bahía Blanca Estuary, Argentina. Mar Geol 94:353–360

Aliotta S, Perillo GME (1990) Antigua línea de costa sumergida en el estuario de Bahía Blanca, Provincia de Buenos Aires. Rev Asoc Geol Arg 45:300–305

Aliotta S, Perillo GME (1994) Megaóndulas en abanico submareales de Bahía Blanca: Clasificación y orígen. Workshop on Formas de lecho: Procesos dinámicos y estructuras primarias, Bahía Blanca, Abstract

Aliotta S, Lizasoain GO, Lizazoain WO, Ginsberg SS (1996) Late Quaternary sedimentary sequence in the Bahía Blanca Estuary. J Coastal Res 12:875–882

Codignotto JO, Weiler NE (1980) Evolución morfodinámica del sector costero comprendido entre Punta Laberinto e isla Olga, provincia de Buenos Aires. Proc Simposio Problemas Geológicos del Litoral Atlántico Bonaerenses, Mar del Plata, pp 35–45

Cuadrado DG, Perillo GME (1996) El Toro Channel: A review. Proc Bahía Blanca International Coastal Symposium Bahía Blanca 215–221

Cuadrado DG, Perillo GME (1997) Migration of intertidal sand banks at the entrance of the Bahía Blanca Estuary, Argentina. J Coastal Res 13:139–147

Defant A (1961) Physical Oceanography. Pergamon Press, Oxford

Espósito G (1986) Etude geomorphologique de la zone El Rincón (Argentina). Docteur 3 Cycle thesis Université Paris-Sud

Ginsberg SS (1993) Geomorfología y dinámica de canales de marea del estuario de Bahía Blanca. PhD thesis Universidad Nacional del Sur

Ginsberg SS, Perillo GME (1990) Channel bank recession in the Bahía Blanca Estuary, Argentina. J Coastal Res 6:999–1010

Ginsberg SS, Perillo GME (1997) Deep scour holes at the confluence of tidal channels in the Bahía Blanca Estuary, Argentina. Mar Geol (submitted)

Gómez EA (1989) Geomorfología y sedimentología del sector marítimo externo al Canal Bermejo, Estuario de Bahía Blanca. PhD thesis Universidad Nacional del Sur

Gómez EA, Perillo GME (1992) Largo Bank: A shoreface-connected linear shoal at the Bahía Blanca Estuary entrance, Argentina. Mar Geol 104:193–204

Gómez EA, Perillo GME (1995) Sediment outcrops underneath shoreface-connected sand ridges, outer Bahía Blanca Estuary, Argentina. Quater South Amer Antar Pen 9:27–42

Gómez EA, Ginsberg SS, Perillo GME (1994) Geomorfología del area de El Rincón. In: Perillo GME (ed) Estudio oceanográfico de la plataforma continental interior frente al estuario de Bahía Blanca. Instituto Argentino de Oceanografía Contrib Cient 311:13–15

Gómez EA, Ginsberg SS, Perillo GME (1997) Geomorfología y sedimentología de la zona interior del Canal Principal del Estuario de Bahía Blanca. Rev Asoc Arg Sedimen 3:55–61

Gonzalez MA (1989) Holocene levels in the Bahía Blanca Estuary, Argentine Republic. J Coastal Res 5:65–77

Gonzalez Uriarte M (1984) Características geomorfológicas de la porción continental que rodea la Bahía Blanca, Provincia de Buenos Aires. Actas 9 Congreso Geologico Argentino III:556–576

Hansen DV, Rattray M (1966) New dimensions in estuary classification. Limnol Ocean 11:319–326

Ippen AT, Harleman DRF (1966) Tidal dynamics in estuaries. In: Ippen AT (ed) Estuary and coastlines hydrodynamics. McGraw-Hill, New York pp 493–545

Kaasschieter JPH (1965) Geología de la cuenca del Colorado. Actas II Jornadas Geológicas Argentinas III:251–269

Kjerfve B, Shao C-C, Stapor Jr FW (1979) Formation of deep scour holes at the junction of tidal creeks: an hypothesis. Mar Geol 33:M9–M14

Lanfredi NW, D'Onofrio EE, Mazio CA (1988) Variations of the mean sea level in the southwest Atlantic Ocean. Continental Shelf Res 3:1211–1220

Le Floch J (1961) Propagation de la maree dans l'estuaire de la Seine et en Seine Maritime. DSc thesis Université du Paris

Ludwick JC (1974) Tidal currents and zig-zag sand shoals in a wide estuary entrance. Geol Soc Am Bull 85:717–726
Martos P, Piccolo MC (1988) Hydrography of the argentine continental shelf between 38 and 42°S. Cont Shelf Res 8:1043–1056
Montesarchio L, Lizasoain W (1981) Dinámica sedimentaria en la denominada ría de Bahía Blanca. Instituto Argentino de Oceanografía Contrib Cient 58
Perillo GME (1989) Estuario de Bahía Blanca: Definición y posible origen. Bol Centro Naval 107:333–344
Perillo GME (1995) Definition and geomorphologic classifications of estuaries. In: Perillo GME (ed) Geomorphology and sedimentology of estuaries. Elsevier, Amsterdam, pp 17–49
Perillo GME, Ludwick JC (1984) Geomorphology of a sand wave in lower Chesapeake Bay, Virginia, USA. Geo Marine Lett 4:110–114
Perillo GME, Sequeira ME (1989) Geomorphologic and sediment transport characteristics of the middle reach of the Bahía Blanca Estuary (Argentina). J Geophysical Res Oceans 94:14351–14362
Perillo GME, Cuadrado DG (1990) Nearsurface suspended sediments in Monte Hermoso beach (Argentina): I. Descriptive characteristics. J Coastal Res 6:981–990
Perillo GME, Cuadrado DG (1991) Geomorphologic evolution of El Toro Channel, Bahía Blanca Estuary (Argentina) prior its dredging. Mar Geol 97:405–412
Perillo GME, Piccolo MC (1991) Tidal response in the Bahía Blanca Estuary. J Coastal Res 7:437 449
Perillo GME, Piccolo MC (1993) Methodology to study estuarine cross-section. Rev Geofísica 38:189–206
Perillo GME, Piccolo MC (1998) Importance of grid-cell area in the estimation of residual fluxes. Estuaries 21:14–28
Perillo GME, Piccolo MC, Arango JM, Sequeira ME (1987b) Hidrografía y circulación del estuario de Bahía Blanca (Argentina) en condiciones de baja descarga. Proc 2º Congreso Latinoamericano de Ciencias del Mar, La Molina, II:95–104
Perillo GME, Garcia Martínez MB, Piccolo MC (1996) Geomorfología de canales de marea: análisis de fractales y espectral. Actas VI Reunión Arg Sedimen 155–160
Piccolo MC (1987) Estadística climatológica de Ingeniero White. Período 1980–1985. Instituto Argentino de Oceanografía, Tech Rep
Piccolo MC, Perillo GME (1989) Subtidal sea level response to atmospheric forcing in Bahía. Proc 3 Internat Congress Southern Hemisphere Meteorology and Oceanography 323–324
Piccolo MC, Perillo GME (1990) Physical characteristics of the Bahía Blanca Estuary (Argentina). Estuar Coastal Shelf Sci 31:303–317
Piccolo MC, Perillo GME, Arango JM (1987) Hidrografía del estuario de Bahía Blanca, Argentina. Revista Geofísica 26:75–89
Piccolo MC, Perillo GME, Cuadrado DG, Marcos AO (1994) Oceanografía física. In: Perillo GME (ed) Estudio oceanográfico de la plataforma continental interior frente al estuario de Bahía Blanca. Instituto Argentino de Oceanografía Contrib Cient 311:26–32
Streeter VL (1948) Fluid Dynamics. McGraw-Hill, New York
Zambrano JJ (1972) La cuenca del Colorado. In: Leanza AF (ed) Geología regional argentina. Acad Nacional de Ciencias Córdoba pp 419–437

Subject Index

Geographic Index